**Vortices and Heat Transfer**

Edited by
Martin Fiebig
Nimai K. Mitra

# Notes on Numerical Fluid Mechanics (NNFM)  Volume 63

Series Editors: Ernst Heinrich Hirschel, München (General Editor)
Kozo Fujii, Tokyo
Bram van Leer, Ann Arbor
Michael A. Leschziner, Manchester
Maurizio Pandolfi, Torino
Arthur Rizzi, Stockholm
Bernard Roux, Marseille

Volume 63: Vortices and Heat Transfer (M. Fiebig / Nimai K. Mitra)
Volume 62: Large-Scale Scientific Computations of Engineering and Environmental Problems
(M. Griebel / O. P. Iliev / S. D. Margenov / P. S. Vassilevski)
Volume 61 Optimum Aerodynamic Design & Parallel Navier-Stokes Computations, ECARP-European Computational Aerodynamics Research Project (J. Periaux / G. Bugeda / P. Chaviaropoulos / K. Giannokoglou / S. Lanteri / B. Mantel, Eds.)
Volume 60 New Results in Numerical and Experimental Fluid Mechanics. Contributions to the 10th AG STAB/DGLR Symposium Braunschweig, Germany 1996 (H. Körner / R. Hilbig, Eds.)
Volume 59 Modeling and Computations in Environmental Sciences. Proceedins of the First GAMM-Seminar at ICA Stuttgart, October 12–13, 1995 (R. Helmig / W. Jäger / W. Kinzelbach / P. Knabner / G. Wittum, Eds.)
Volume 58 ECARP – European Computational Aerodynamics Research Project: Validation of CFD Codes and Assessment of Turbulence Models
(W. Haase / E. Chaput / E. Elsholz / M. A. Leschziner / U. R. Müller, Eds.)
Volume 57 Euler and Navier-Stokes Solvers Using Multi-Dimensional Upwind Schemes and Multigrid Acceleration. Results of the BRITE/EURAM Projects AERO-CT89-0003 and AER2-CT92-00040, 1989–1995 (H. Deconinck / B. Koren, Eds.)
Volume 56 EUROSHOCK-Drag Reduktion by Passive Shock Control. Results of the Project EUROSHOCK, AER2-CT92-0049 Supported by the European Union, 1993–1995
(E. Stanewsky / J. Délery / J. Fulker / W. Geißler, Eds.)
Volume 55 EUROPT – A European Initiative on Optimum Design Methods in Aerodynamics. Proceedings of the Brite/Euram Project Workshop „Optimum Design in Aerodynamics", Barcelona, 1992 (J. Periaux / G. Bugeda / P. K. Chaviaropoulos / T. Labrujere / B. Stoufflet, Eds.)
Volume 54 Boundary Elements: Implementation and Analysis of Advanced Algorithms. Proceedings of the Twelfth GAMM-Seminar, Kiel, January 19–21, 1996 (W. Hackbusch / G. Wittum, Eds.)
Volume 53 Computation of Three-Dimensional Complex Flows. Proceedings of the IMACS-COST Conference on Computational Fluid Dynamics, Lausanne, September 13–15, 1995
(M. Deville / S. Gavrilakis / I. L. Ryhming, Eds.)
Volume 52 Flow Simulation with High-Performance Computers II. DFG Priority Research Programme Results 1993–1995 (E. H. Hirschel, Ed.)

Volumes 1 to 51 are out of print.
The addresses of the Editors are listed at the end of the book.

# Vortices and Heat Transfer

Results of a DFG-Supported Research Group

Edited by
Martin Fiebig
Nimai K. Mitra

All rights reserved
© Friedr. Vieweg & Sohn Verlagsgesellschaft mbH, Braunschweig/Wiesbaden, 1998

Vieweg ist a subsidiary company of Bertelsmann Professional Information.

No part of this publication may be reproduced, stored in a retrieval system or transmitted, mechanical, photocopying or otherwise, without prior permission of the copyright holder.

http://www.vieweg.de

Produced by Geronimo GmbH, Rosenheim
Printed on acid-free paper
Printed in Germany

ISSN 0179-9614
ISBN 3-528-06963-5

# Preface

In convective transport phenomena vortices play an important role. Judiciously generated vortices can intensify the transport process. Vortices can be generated by flow separation behind obstacles, by surface curvature or by jet interaction. Vortices are designated as transverse or longitudinal (streamwise), when the axes of the vortices are normal or aligned with the main flow direction respectively. Typically transverse vortices are generated in the shear layers between impinging jets or grooves between fins, whereas longitudinal vortices can be generated behind wing-type vortex generators. Streamwise or longitudinal vortices swirl the flow around the main flow direction and thus cause a continuous mixing of the fluid from far and near the wall. Vortices disturb the boundary layer growth, can destabilize the flow and reduce the Reynolds number for the onset of turbulence and thus become important tools for passive mechanisms of heat transfer enhancement.
Since 1984 the Heat and Mass Transfer group of the Mechanical Engineering Department of the Ruhr-Universitiy, Bochum has been investigating both numerically and experimentally the influence of longitudinal vortices on heat transfer enhancement in channel flows. The scope was much broadened in 1991, when a Research group (Forschergruppe) with the title 'Vortices and Heat Transfer' (Wirbel und Wärmeübertragung) was founded with generous financial support for six years from the Deutsche Forschungsgemeinschaft (DFG). Besides the Heat and Mass transfer group, the Fluid Mechanics group and eventually the chair of Numerical Mathematics of the Ruhr University joined this Reserach Group. The projects encompassed ways to generate both transverse and longitudinal vortices and to study their effects on heat transfer in both laminar and turbulent flows. For the investigations modern computer codes were developed to solve two-dimensional and three-dimensional steady and nonsteady Navier-Stokes and energy equations, and also for large eddy simulations. Experimental investigations consisted of measuring flow fields by laser Doppler anemometry and assessing local heat transfer coefficients by infrared thermography, liquid crystal thermography and ammonia absorption techniques. These techniques have been further developed during the completion of the projects of the research group.
The research group presented its results of the first phase (1991 - 1993) in the Seminar Eurotherm 31, titled also as 'Vortices and Heat Transfer', and held with international participation at Bochum on May 24 - 26, 1993. Besides the research group, scientists from Germany, Great Britain, France, the Netherlands, Russia, USA, Brazil and India presented their results of research on vortex shedding and vortex generation by wavy walls, twisted tape and jets, coherent structures in vortex interaction of multiple jets, interaction of longitudinal vortices with a vortex street behind a tube, vortex breakdown and of course of the effect of vortices on heat transfer. The presented papers were published in proceedings edited by the editors of this volume. The experimental papers were thoroughly revised and meticulously reviewed after the seminar and were published in a special issue of the journal Experimental Thermal and Fluid Science (V. 11, No. 3, 1995).
The research activity of the research group ended formally with a closing symposium titled again 'Vortices and Heat Transfer' held at Bochum on 28 - 29 November, 1996. Besides ten lectures from the research group six invited lectures from Profs. Hanjalic (Delft, the Netherlands), Benocci (VKI, Belgium), Liu (Brown University, USA), Kolovandin (Academy of Sciences, Belarus), Suzuki (Kyoto University, Japan) and Launder (UMIST, GB) were presented.
The research group spokesman Martin Fiebig delivered the first lecture in which he gave an overview of the nine projects. Then the final results of each of the projects were reported in separate lectures.

The present volume contains the overview from Fiebig, the final reports of eight of the nine projects of the research group and three papers based on the invited lectures. The final report of the project of D. Braess has been published elsewhere and is not included here. The manuscripts of the invited lectures from Hanjalic, Benocci and Launder were not available for publication.

The first paper of this volume is the overview from Fiebig, which is a revised version of his lecture in the closing seminar. It gives a comprehensive view of all the projects and their interconnectedness, discusses the results and summarizes their main findings. Next follow the project reports A1, A2, A3, A4, A6 and B1, B2 and B3. The projects numbered with A and B concentrate on transverse and longitudinal vortices respectively. The project A5 was discontinued after the first phase. The project reports are also revised versions of the lectures presented in the closing seminar. In the seminar primarily the research results of the final phase of the project were presented. In the project reports the reseach results of the first phase were also summarized.

After the project reports the papers of Liu, Kolovandin et al. and Suzuki et al. are presented. Liu uses nonlinearily developing longitudinal vortices that originate from upstream weak Görtler vortices on semi-infinite concave walls as the prototype vortices. He reviews the experimental aspects of these longitudinal vortices and the analogy between heat and mass transfer with that of momentum transfer. Kolovandin et al. discuss the fundamentals of turbulent mixing theory based on the supposition that the main mechanism of turbulent mixing at high turbulence Reynolds numbers is attributed to persistent vortex structures. Suzuki et al. discuss how heat transfer enhancement can be achieved by a large eddy break-up manipulator attached with a triangular winglet type vortex generator, which is inserted into a flat plate turbulent boundary layer. They also discuss how longitudinal vortices, targeting heat transfer enhancement on a flat plate, can be generated by an impinging jet in a cross flow. In the final part of the paper they treat Taylor-Görtler vortices in a curved channel. They show how by mounting fine wires at a spanwise interval to channel height, the position of the Taylor-Görtler vortices can be fixed in order to facilitate an experimental study of the cross sectional structure of the vortices. The volume ends with a list of publications of the research group.

The members of the research group would like to acknowledge their gratefulness to DFG for its generous financial support. The science ministry of the State of Nordrhein-Westfalen helped financially in acquiring research equipment (workstations). Finally we thank the Vieweg Verlag and Prof. E.H. Hirschel for making it possible to publish our research results in form of this monograph.

Nimai Mitra                                                                                   Martin Fiebig

Bochum, December 1997

# Contents

Page

**M. Fiebig**
Vortices & Heat Transfer:
DFG Research Group - Topics & Insights..................................................1

**H. Herwig, J. Severin, P. Schäfer**
Heat Transfer in Laminar Flow with Finite
Regions of Pressure Induced Separation.....................................................25

**D. Vieth, R. Kiel, K. Gersten**
Two-Dimensional Turbulent Boundary Layers
with Separation and Reattachment
Including Heat Transfer..............................................................................63

**W. Leiner, S. Lorenz, M. Dietrich, J. Torkar**
Turbulent Flow Structure and Local Heat Transfer
in Asymmetrically Ribbed Channels..........................................................104

**A. Grosse-Gorgemann, H.-W. Hahne, H. Neumann,
D. Weber, M. Fiebig**
Flows and Heat Transfer in Ribbed Channels with
Self-Sustained Oscillating Transverse Vortices.........................................131

**N. K. Mitra, H. Laschefski, T. Cziesla**
Flow Structure and Heat Transfer of Impinging Jets.................................169

**H. Neumann, H.-W. Hahne, U. Müller, M. Fiebig**
Heat Transfer and Flow Losses
in Steady and Self-Oscillating Channel Flows
with Rectangular Vortex Generators (RVG)..............................................214

**S. Lau, V. Vasanta Ram**
Measurement and Analysis of the Turbulent
Flow Quantities in a Channel with Embedded
Longitudinal Vortices.................................................................................252

**H. Neumann, H. Braun, M. Fiebig**
Vortex Structure, Heat Transfer and Flow
Losses in Turbulent Channel Flow with
Periodic Longitudinal Vortex Generators..................................................288

## Contents (contiuned)

**J. T. C. Liu**
Longitudinal Vortices in Boundary Layer
Heat Transfer Augmentation ..................................................................328

**B. A. Kolovandin, I. A. Vatutin**
A New Look into the Turbulent Mixing
in Viscous Fluids:
Implication to Heat and Mass Transfer ....................................................338

**K. Suzuki, K. Inaoka**
**M. Kobayashi, H. Maekawa, K. Matsubara**
Flow Control and Heat Transfer Enhancement
with Vortices ...........................................................................................356

List of Publications and Dissertations
from the Forschergruppe .........................................................................374

# VORTICES & HEAT TRANSFER

# DFG RESEARCH GROUP - TOPICS & INSIGHTS

Martin Fiebig
Institut für Thermo- & Fluiddynamik, Ruhr-Universität Bochum
D-44780 Bochum, Germany

## SUMMARY

The background, methods of investigation, and major results of the nine projects of the research group VORTICES & HEAT TRANSFER are outlined. Locally separated turbulent boundary layers are described by universal wall functions. For embedded transverse and longitudinal vortices extensive data bases are established. Transverse vortex generators (TVGs) generate local but essentially no global heat transfer enhancement as long as the flow remains steady. When the flow becomes self-oscillating or turbulent large increases in heat transfer occur. Longitudinal vortex generators (LVGs) swirl the flow. This corkscrew motion may increase heat transfer already locally and globally for steady flow conditions. Vortex generators (VGs) cause transition at much smaller Reynolds numbers and thereby allow the same heat transfer at only a small fraction of the flow losses of the corresponding flow without VGs.

## INTRODUCTION

The research group 'VORTICES & HEAT TRANSFER' was established in 1990 and has been funded by the Deutsche Forschungsgemeinschaft (DFG) since 1991. It grew out of a nucleus studying the influence of vortices generated by wing-type vortex generators (WVGs) and cylindrical inserts on heat transfer and flow losses in channel flow [1 to 9].

At the close of the funding period we want to give an overview of the background, concept and objectives, topics and tools, and major insights and conclusions of our joint research efforts. Here the general and common aspects will be stressed. The specifics of the nine research projects and the corresponding literature will be given in the special contributions [10 to 18]. Projects A1 to A4 investigate theoretically and experimentally the flow structure and heat transfer generated by different transverse vortex generators (TVGs) with the laminar and turbulent flat plate boundary layer and channel flows as the base flows, i.e. the flows with vorticity but without vortices, see Table 1, A. Projects B1 to B3 study experimentally and numerically the effects of embedded longitudinal vortices (LVs) generated by variations of one common longitudinal vortex generator (LVG) on velocity and temperature fields, see Table 1, B. Here laminar or turbulent channel flow is the base flow. Project A6 explores numerically the flow and temperature fields when the vortices are generated by impinging jets. For better insight into fast Navier-Stokes solvers project A7 studies different multigrid algorithms.

## BACKGROUND

Even though no agreed upon mathematical definition exists for a vortex [19,20], common understanding is that a vortex is swirling motion around an axis. Principally transverse vortices (TVs) and longitudinal vortices (LVs) can be distinguished. TVs have their axes transverse to the flow and are consistent with 2D flow, while LVs have their axes along the flow mainly in the streamwise direction and imply 3D flow. Vortices have to be generated and I shall call their generators correspondingly transverse vortex generators (TVGs) and longitudinal vortex generators (LVGs). TVGs have a two dimensional geometry and LVGs a three dimensional geometry.

Fig. 1: Karman vortex street behind a circular cylinder in cross flow at Re=140. Photograph by Sadatoshi Taneda from van Dyke [21].

The Karman vortex street in the wake of a circular cylinder in cross flow is perhaps the most thoroughly investigated TV system, Fig.1. For Reynolds numbers based on the cylinder diameter and the free stream velocity between 5 and 47 the two dimensional flow separates close to plus and minus 90° from the stagnation line and forms two TVs with a closed recirculation region [22]. For higher Re the interaction of the two shear layers starts. The separation lines oscillate and shed alternating shear layers which form into the TVs of the Karman vortex street seen in Fig.1. Up to Re equal 170 the vortex axes and the vorticity vector are aligned pointing out of the figure. When Re is increased further three dimensional instabilities are amplified and the wake becomes three dimensional. Secondary vortices which are not any more pure TVs develop [22]. More and more frequencies are amplified and the wake is well on its way to turbulence. In internal flows Karman vortex streets with TVs are generated for example by ribs in channels.

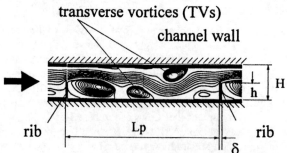

Fig. 2: Karman vortex street behind periodic ribs in a channel, $Re_h$=87.5, from [31].

For the ribbed channel shown in Fig.2 self-sustained large amplitude oscillations are generated for Reynolds numbers based on the mean velocity and rib height greater than about 45 [31]. Figure 2 shows instantaneous streamlines and associated TVs. These unsteady TVs enhance heat transfer by 'Reynolds' averaged transport. Whether stationary TVs will increase heat transfer remains an open question. The steady TVs transport energy from the wall to the core of the fluid and back to the wall. No additional convective mechanism is generated to transport energy downstream by stationary TVs. Self-oscillating periodic channel flows and heat transfer have been investigated during the last 10 years by Patera and Mikic and co-workers [23 to 26], Mayinger and co-workers [27,28] and our group [29 to 31].

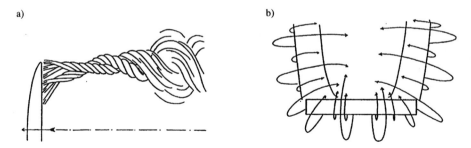

Fig. 3: a) Trailing vortices and their tendency to intercoil. From Lanchester [32].
b) Equivalent of circulation of bound and trailing vortices. From Lanchester [33].

In contrast to TVs, LVs spiral the flow around their axes in the streamwise direction. The wingtip trailing vortices and the leading edge vortices of delta wings at angle of attack are classical longitudinal vortices and have been studied intensively in aerodynamics for nearly 100 years. For wingtip vortices it was shown by Lanchester [32,33] that they have the tendency to intercoil and that their circulation equals the bound circulation of the wing, see Fig.3. The leading edge vortices of delta wings at moderate angles of attack are generated by the separation along the edges which result in detached shear layers rolling up to longitudinal vortices, see Fig.4. The induced circumferential and streamwise velocities are in these cases of the same magnitude as the freestream velocity [34]. No self-oscillating LVs have so far been reported. The intense corkscrew motions induced by wing tip and leading edge vortices prompted the use of LVGs to enhance heat transfer, first in exploratory studies [35,36] and then in more detailed investigations [37 to 39,1 to 9].

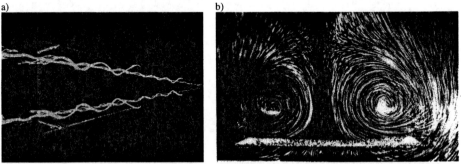

Fig. 4: Longitudinal vortices above a delta wing with 30° vertex angle at 20° angle of attack and Re=20000. (a) Lines of coloured fluid in water show the symmetrical pair of vortices. (b) Tiny air bubbles in water show the cross flow of the vortex pair. Photograph by Werle, from van Dyke [21].

Longitudinal vortices result also from centrifugal instabilities, they are called Taylor, Dean, and Goertler vortices depending on the type of concave surface by which they are generated. They will not be considered here.

## CONCEPT AND OBJECTIVES

The research concept of the group has been to investigate how vortices are generated and how they influence the original velocity and temperature fields, heat transfer and flow losses. For local investigations of two-dimensional, steady and unsteady flow and temperature fields theoretical (numerical) and experimantal tools had to be further developed. To establish confidence into the methods it was envisioned to compare experimental and theoretical (numerical) results for typical flow situations. To allow in-depth studies with these tools the number of base configurations, i.e. VGs and base flows was kept to a minimum. The three TV base configurations and the one LV base configuration are shown in table 1. Apart from the fundamental studies of the vortical flow and temperature fields the following questions were of major interest: What are the wall friction and heat flux distributions for the base configurations? Under what kind of steady flow conditions lead TVG and LVG configurations to global heat transfer enhancement and what are the associated flow losses? What are the conditions that TV and LV flow fields become unsteady? What heat transfer enhancement mechanisms are associated with TVs and LVs and what are the induced flow losses?

Table 1: Characteristics of the base configurations. A) Transverse vortices and heat transfer, 2D geometry; B) Longitudinal vortices and heat transfer, 3D geometry

| Project | Lead Scientist | Scheme | l. | t. | st. | un. | th. | ex. |
|---|---|---|---|---|---|---|---|---|
| A1 | Herwig / Gersten |  | x |  | x | • | x |  |
| A2 | Gersten / Herwig |  |  | x | x |  | x | x |
| A3 | Leiner |  |  | x | x |  |  | x |
| A4 | Fiebig |  | x |  | • | x | x | • |
| A7 | Braess |  | x |  | x |  | x |  |
| A6 | Mitra / Page |  | • | x |  | x | x | x |

Table 1 continued

| Project | Lead Scientist | Scheme | l. | t. | st. | un. | th. | ex. |
|---|---|---|---|---|---|---|---|---|
| B1 | Fiebig | | x | | x | x | x | |
| B2 | Vasanta Ram | | | x | x | | | x |
| B3 | Fiebig | | x | x | • | | x | • |

l. = laminar,    st. = steady,    th. = theory,    x = emphasis on
t. = turbulent,    un. = unsteady,    ex. = experiment    • = also

Because of its technical importance air was chosen as the heat transfer medium. Mach number effects were neglected because most internal flows are low Mach number flows. The effects of variable fluid properties were considered of secondary importance and only exemplary investigations were contemplated. As thermal boundary conditions constant wall temperature and constant heat flux were chosen.

## TOPICS AND SPECIFIC TOOLS

The nine specific research projects A1 to A4, A6, A7 and B1 to B3 are depicted in table 1. For the boundary layer and channel flow TV projects A1 through A4 two dimensional time averaged velocity and temperature fields are expected. The flow and temperature fields generated by impinging jets in A6 are two dimensional or axisymmetric for single jets and three dimensional for jet arrays. A7 investigates different smoothers for NS solvers to enhance their convergence [10]. The LV projects B1 to B3 are all based on the same array of LVGs in the form of rectangular winglets at angle of attack.

In A1 and A2 steady two dimensional boundary layers with local pressure induced separation regions are studied. The location of the separation point is unknown. In the laminar regime asymptotic theories and NS solvers are used to determine the flow and temperature fields for asymptotically small and shallow contour changes [11]. Flat plate boundary layer flow and slender channel flow are the base flows without TVs. A2 studies the equivalent turbulent boundary layer experimentally and theoretically. Here, the aim is to determine universal wall functions for the velocity and temperature profiles which are also valid at the separation point and in the separation region. The asymptotic theory determines the functional dependence of the temperature and velocity profiles, experiments and measurements are needed to determine the functions which appear in the universal wall functions [12].

A3 and A4 consider fully developed periodic channel flow with massive separation caused by grooves and ribs on one channel wall. A3 studies turbulent flow and heat transfer with relatively large rectangular and semicircular ribs of different height and pitch [13]. The aim is to measure detailed velocity profiles, local heat transfer, and wall pressure distributions as a benchmark and to assess the influences of form, pitch, and height of the ribs. As measurement technique Laser Dopppler Anenometry (LDA) is used for velocity and turbulence measurements and a highly sensitive infrared technology (IRT) is further developed for local heat transfer measurements. A4 investigates steady (laminar) and self-oscillating (transitional) periodic channel flow and heat transfer, where the periodicity and the self-sustained oscillations are generated by periodic rectangular ribs, their height and pitch are varied [14]. Of special interest are the following questions: What is the lower critical Reynolds number for the onset of oscillations? What is their frequency? How does the number of frequencies increase with Reynolds number? What is the influence of these oscillations on heat transfer and flow losses? Does the 'Reynolds' averaged transport lead to the same heat transfer for much lower flow losses than a plane channel? Here FIVO, a 3D unsteady finite volume code developed by our group is the main tool for the investigations. A specially designed wind tunnel which allowed in-situ-calibration of the hot-wire probe was developed to validate the code with respect to its prediction of the Reynolds number at transition, i.e. the onset of self-exited oscillations.

A6 studies numerically flow field, vortex structure, transition and heat transfer of a single two-dimensional circular inline and radial impinging jet and of a three-dimensional slot jet of a jet array. For the laminar single round jet the influence of initial swirl at the nozzle exit and of free convection on flow and heat transfer have also been investigated [15]. Direct numerical simulations (DNS), large eddy simulations (LES), and low-Reynolds number-k-$\varepsilon$ model computations are carried out.

Projects B1 to B3 (Table 1B) study channel flow where LVs dominate the flow structure and heat transfer. As LVGs an array of rectangular low aspect ratio winglets at angles of attack are chosen. These LVGs are termed wing-type vortex generators (WVGs). Seven geometrical parameters determine a periodic channel element. As base configuration they were fixed as follows: longitudinal and lateral pitch to channel height equal 5 and 4 respectively, winglet angle of attack equals 45°, winglet length to height and thickness to height ratio equal 4 and 0.1 respectively, winglet height to channel height equals 0.5, and winglet tip seperation s to channel height H equals $s/H=(1-\sin\beta)l/2H$ where $\beta$ is the angle of attack and l the winglet lenght. This LVG geometry of table 1B was chosen so that it transforms into the TVG geometry of projects A3 and A4 (Table 1A) for identical longitudinal pitch, height, and thickness ratios. This can be seen from the winglet tip seperation. It becomes 0 for $\beta=\pi/2$. The trailing wingtip vortices are the strongest vortices. For small angles of attack their axes are oriented along the flow direction. But a whole system of different vortices is present consisting besides the wing tip vortex of the leading and trailing edge vortices and the juncture vortex (horse shoe vortex). These vortices are initially TVs and are bent later into the flow direction.

B1 studies the full range of angles of attack from zero to 90° with laminar and transitional three dimensional flow and heat transfer [16]. The methods are the same as in project A4. In addition a high resolution technique to determine local heat transfer was needed. The emphasis here is however more on the effects of LVs on the flow structure and local heat transfer. Of special interest are the relative contributions of swirl and self-sustained oscillations to heat transfer and flow losses. Will the LVG or the TVG geometry generate higher heat transfer for the same flow losses? And which of the two geometries will lead to higher heat transfer and flow losses at the same Re? The answer to these questions should give hints in which direction the development of heat transfer surfaces for compact heat exchangers should go.

B2 starts with a fully developed turbulent channel flow and uses the base configuration described earlier at an angle of attack of 30° as LVG [17]. The main aim was to establish an experimental data base for further investigations into the realm of embedded LVs. For this it was necessary to develop a technique for measurement of the instantaneous velocity vector and temperature through a multi wire hot-wire probe with 4 sensors. This task constituted, in fabrication of the probe, in developing computer assisted methods for its calibration and use, and in the development of algorithms to deduce from the data the different parts contributing to the flow structure.

B3 also uses the same basic WVG geometry. In its theoretical part the LES code FRACAS was developed and then used to determine numerically the unsteady turbulent flow and temperature field [18]. In the experimental part optical techniques - IRT and ammonia absorption method (AAM), were further refined with respect to high accuracy and resolution [13,16,18]. The aims were mainly to extend the Re domain of B1, to provide local heat transfer data for the flow of B2, to investigate the existence of dominant self-excited frequencies at higher Re, and to reduce the vortex generator height into the roughness domain. The latter is of special technical interest.

## RESULTS

Here only some specific results of the projects can be reported, detailed descriptions of the findings will be given in the special contributions.

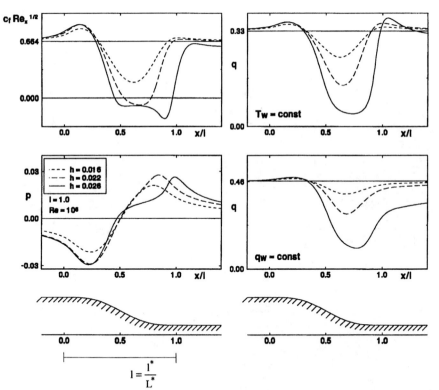

Fig. 5: Friction coefficient, local induced pressure and heat transfer for 'soft' reclining step for step height of 1.6, 2.2 and 2.6% of the step lenght for a Reynolds number $Re=ul/\nu=10^6$.

## Transverse Vortices and Heat Transfer

For two dimensional, steady, laminar boundary layers with pressure induced TVs asymptotic theory reveals that no global heat transfer enhancement and no additional drag result. Locally strong deviations in heat transfer and friction however occur. For the 'soft' reclining step with step heights between 1.6 and 2.6% of the step length considered in A1 the friction coefficient, pressure and heat transfer along the wall are shown in Fig.5. The flat plate values are shown as straight lines. That the 'soft' step induces upstream influences on friction, pressure and heat transfer can well be seen. The 1.6% step height does not yet generate a vortex because no reversed flow region results. The calculations were based on the theory of interacting boundary layers and a Reynolds number of $10^6$. More details can be found in [11,40,41].

Fully developed laminar channel flow with periodic ribs on one channel wall was calculated with the Navier-Stokes code FIVO. The geometry is the baseline configuration described earlier with rib height 0.5 of the channel height and longitudinal pitch and rib thickness 10 and 0.1 of the rib height respectively. For a Reynolds number based on the mean velocity and twice the channel height of 100 and constant wall temperature the results are shown in Fig.6. The flow is still steady and the large TV behind the rib with the reattachment point $R_1$ and the small TV in front of the rib with the separation point $S_1$ can clearly be seen in the streamline picture.

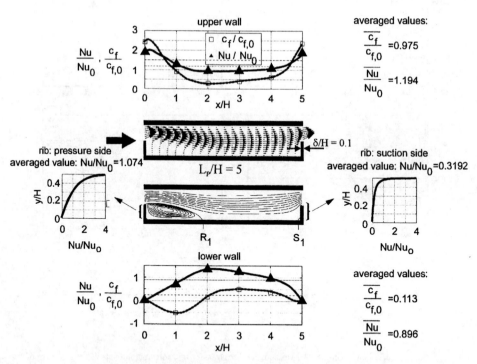

Fig. 6: Steady streamlines, friction coefficients and Nusselt number for the ribbed channel at $Re_{2H}=100$, $Nu_0=7.54$, $f_0=24/Re$, $f/f_0=2.88$, $Nu/Nu_0=1.015$, $Re_{2H}=u2H/\nu$.

On the unribbed wall no separation occurs and the Nusselt number and friction distribution look similar. On the ribbed wall with separation the corresponding distributions are very dissimilar. At the reattachment point the friction is zero and the heat transfer has a maximum. On the pressure and suction side of the rib the Nusselt numbers are different by more than a factor of three. Compared to the plane channel the overall Nusselt number changes by less than 2%, while the pressure loss is increased by a factor of 2.88 due to the form drag of the ribs despite the decrease of the friction coefficient. For laminar steady flow the conclusion for TVGs can be drawn: (1) Globally TVs do essentially cause no change in heat transfer but may cause considerably increased flow losses. (2) Locally TVs cause large changes in heat transfer and friction. (3) TVGs embedded in boundary layers may be used to change local heat transfer, pressure, and wall shear distributions.

If the Reynolds number is increased the flow becomes unsteady. For the rib geometry shown in Fig.6 hotwire measurements and calculations gave already large amplitude oscillations at $Re_{2H}=200$. The early start of transition is caused by the velocity profiles with a large reversed flow region and an inflection point in the core of the channel. These velocity profiles amplify small disturbances at much lower Re than a fully developed plane channel flow. This destabilisation process has been discussed in detail in [42]. Instantaneous streamlines and isotherms are shown at 5 instances of one time period in Fig.7 at $Re_{2H}=350$ for the geometry of Fig.6. TVs are formed at the rib and on the plane wall. During their transport through the channel they are dissipated. The isotherms show the unsteady transport of thermal energy and the spatial and temporal variation of the wall temperature gradient. The unsteady velocity and temperature distributions give rise to 'Reynolds' averaged transport which in turn leads to higher wall gradients. This can be seen in Fig.8 where the heat transfer and pressure loss enhancement is shown for a variation in rib height at the $Re_{2H}=350$ as in Fig.7. As long as the flow is steady only the flow losses increase and the heat transfer is essentially unchanged. But when the flow becomes unsteady the heat transfer also increases dramatically. For the situation of Fig.7 the heat transfer is increased by 80%. Comparison between numerics and experiments can be found in [30], parameter variations in [14] and [31].

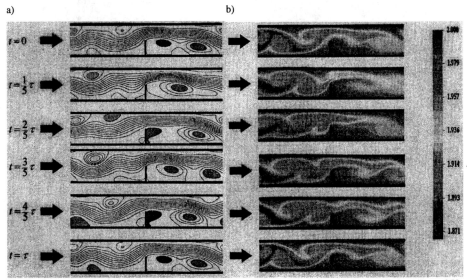

Fig. 7: Instantaneous streamlines (a) and isothermes (b) at different times of a period of oscillations, $Re_{2H}=350$, for ribbed channel of Fig. 6.

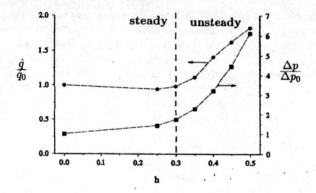

Fig. 8: Heat transfer and pressure enhancement for variable rib height, LP/H=5, δ/H=0.01.

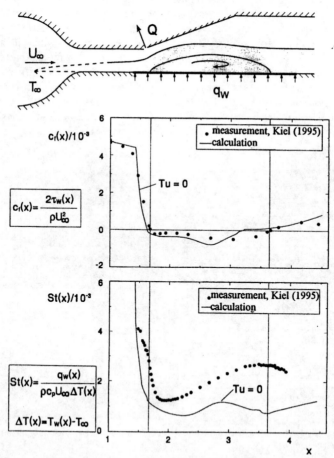

Fig. 9: Schematic of wind tunnel and comparison between calculated and measured distribution of dimensionless wall friction and heat flux.

When the base flow is already turbulent we arrive at the situation of projects A2 and A3, studied in [12,13]. In the limit of very large Re universal wall functions exist for the pressure induced finite separation region of the two dimensional turbulent boundary layer [12]. These universal wall functions for velocity, temperature, turbulent kinetic energy, and dissipation of turbulent kinetic energy are made dimensionless with the unknown wall shear and wall pressure gradient. So an iterative determination is necessary. The universal wall functions contain universal functions which have to be determined experimentally [43]. From theory it can only be deduced that they depend on the ratio of wall pressure gradient to wall shear stress [44]. A special windtunnel was built to determine these universal functions. Fig. 9 shows a schematic of the tunnel and a comparison of measured and calculated skin friction and Stanton number [45,46].

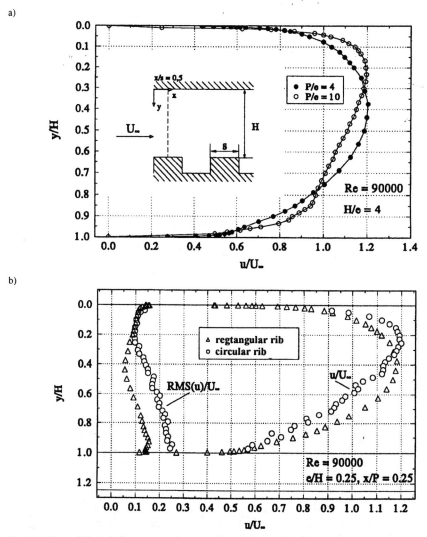

Fig. 10: Effect of rib pitch form on the velocity profile, (a) pitch to rib of 4 and 10 for rectangular rib and (b) circular and rectangular rib for pitch to rib hight of 4.

The turbulent channel flow with ribs on one side is characterised by geometry induced separation. The pitch has a large influence on the velocity profile. The asymmetry of the profile is much more pronounced for the circular than the rectangular ribs, and for a pitch to rib height of 10 than 4, Fig.10. The TVs generated by the ribs penetrate more into the channel and cause higher turbulence levels for the larger pitch. As a consequence the pressure coefficient and Stanton number are higher for the ribbed channel side with the larger pitch, Fig.11 [13,47]. The IRT technique used for local heat transfer measurements is detailed in [48].

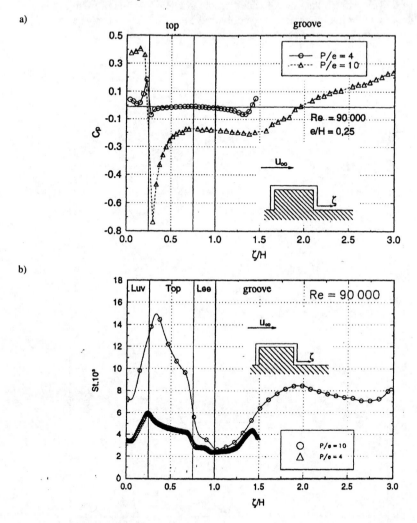

Fig. 11: Influence of pitch on local pressure coefficient and Stanton number.

Figure 12 shows a schematic of a radial slit nozzle considered in [15] which might be of special interest for drying delicate products which do not stand large forces. The DNS code was used for low Reynolds numbers i.e. laminar and transitional flow. For $Re_d=100$ Fig. 13 shows the Nusselt number distribution as a function of the distance along the impingement plate for three nozzles, the axial nozzle (90°), the radial nozzle (60°), and the horizontal nozzle (0°). The axial nozzle results in a heat transfer maximum at the center of the plate, while the radial nozzles result in heat transfer maxima at their impingement lines away from the centre. It is interesting to note that the 60° nozzle gives a higher maximum and average heat transfer than the axial nozzle [49,50,51]. For turbulent flow cases a special dynamic subgrid model was incorporated into the LES code which allows for embedded laminar flow regions as they may occur in the stagnation region of an impinging jet.

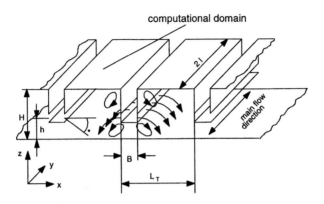

Fig. 12: Schematic of an array of radial rectangular impinging jets.

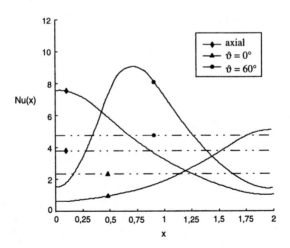

Fig. 13: Comparison of Nu(x)-distributions in x-direction for axial and radial laminar jets ($\vartheta=0°$ and $60°$), $Re_d=100$, $h/2B=1$.

For the solution of the steady Navier-Stokes equation a large non-linear system of equations has to be solved. To enhance the convergence multigrid schemes are advantageous. In

A7 the question was in the forefront whether well known iterative solvers from engineering, as the SimpleR, SimpleC, and Vanka solver, can be used as efficient smoothers in multigrid schemes. In the solvers of Simple-type the iteration consists of two steps. In the first step a new velocity field is determined with the old pressure distribution, and in the second step the pressure is corrected with a Poisson equation. With a suitable adoption of the second step an efficient smoother could be generated, while the first step always resulted in a roughening [54,55,56]. The situation is more complex with the Vanka-iteration and is discussed in [10].

## Longitudinal Vortices and Heat Transfer

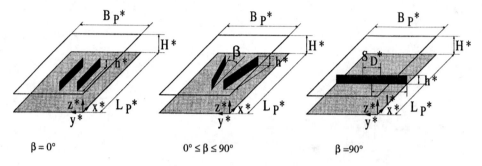

### DIMENSIONLESS PARAMETERS

| | | | | | |
|---|---|---|---|---|---|
| form: | rectangular | angle of attack: | $0 \leq \beta \leq \pi/2$ | Reynolds number: | |
| lateral pitch: | $B_P = B_P^*/h^* = 8$ | length ratio: | $l^*/h^* = 4$ | $100 \leq Re = \dfrac{\overline{u}^* d_h^*}{\nu^*} \leq 40000$ | |
| longitudinal pitch: | $7.5 \leq L_P = L_P^*/h^* \leq 15$ | height ratio: | $0.05 \leq h = h^*/H^* \leq 0.5$ | Prandtl number: $Pr = \dfrac{\nu^*}{a^*} = 0.71$ | |
| thickness ratio: | $0.1 \leq \delta = \delta^*/h^* \leq 0.14$ | point of rotation: | $S_D = S_D^*/l^* = 1/4; 1/2$ | hydraulic diameter: $d_h^* = \dfrac{4V}{A} \approx 2H^*$ | |

Fig. 14: Periodic element of the wing-type vortex generator (WVG) array and dimensionless parameters.

The three LV projects all have the same base configuration shown in Fig.14 where the range of variation of the 7 dimensionless geometric parameters and the Reynolds number is also indicated. In B1 the Reynolds number has been changed from the laminar to the transitional flow regime, while in B2 and B3 turbulent flow situations with Re up to 40000 were studied.

The influence of angle of attack as the control parameter for the vortices is of primary concern in B1 [16,42,55,56,57]. Figure 15 shows five cross sections of one periodic element with the WVGs at an angle of attack of 45°. For a Re of 350 the dimensionless temperature distribution and the Nusselt number enhancement on the finned and smooth wall are presented together with the cross flow velocity vector plot and the friction coefficients. Because the WVGs are attached to the wall they may be considered as fins. All values are referred to the plane channel values and constant wall temperature. The calculations were done with symmetry side boundary conditions for the periodic element and resulted for this angle of attack and Re in a steady solution. The two counter-rotating vortices extend nearly over the whole cross section. The induced circumferential velocities are of the same order as the axial mean velocity. The temperature boundary layer is thin and local heat transfer is high in the stagnation or downwash regions and lower in the upwash regions, Fig.15a,b,c. This does not imply heat transfer enhancement by itself. Only the corkscrew motion of the fluid guarantees that sensitive heat taken up by a fluid particle at the wall is transported away from the wall and downstream. By the time it comes again close to the wall it has been cooled by the bulk of the fluid and can again take up considerable heat from the wall. The local Nusselt number enhancement is given in Fig.15b. On the finned wall it reaches values of three and higher in the downwash areas immediately behind and between the WVGs. It is lowest at the juncture between wall and WVG and low in the upwash regions outside the WVGs. On the smooth wall Nusselt number enhancement reaches also values above three in the strong downwash regions outside the WVGs and low values in regions between the WVGs. The local streamwise friction coefficient enhancement in Fig.15d does not correspond to the local Nusselt number enhancement. This shows that the swirl is primarily responsible for the increased heat transfer. Notice that the reversed flow regions are here much smaller than they were for the ribbed channel.

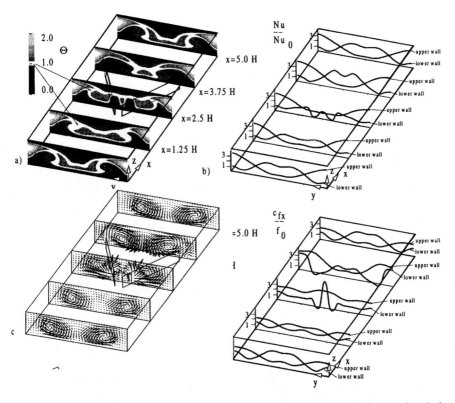

Fig. 15: Steady thermal and flow characteristics of a periodic channel element with a pair of rectangular winglet VGs with $L_P=10h$, $B_P=8h$, $h=0.5H$, $l=0.1h$, $s02h(1-\sin\beta)$, a) dimensionless temperature distribution $\Theta = (T-T_W)/(T_B-T_W)$ in five cross sections, b) corresponding local Nusselt number enhancement $Nu/Nu_0$ on the finned and unfinned wall, c) vector plot showing the counter-rotating vortices in the same five cross sections, d) corresponding local streamwise friction coefficient enhancement $c_f/f_0$ on the finned (lower) and unfinned (upper) wall at $Re_{2H} = 350$. Steady flow is realized by using symmetric side boundary conditions.

The calculations were also performed for periodic side boundary conditions. All results for Nusselt number, flow losses, and friction coefficient enhancement as a function of angle of attack are displayed in Fig.16. For the flow losses and friction coefficient the differences for different side boundary conditions were always small. For the Nusselt number the differences were more than 15% for angle of attack 30° and 45°. The reason for this difference is that for these angles of attack the flow was steady for symmetric side boundary conditions and oscillatory for the periodic side boundary conditions. These large self-sustained oscillations in the lateral and longitudinal directions caused the increase in heat transfer while they influenced the flow losses only little. The specific distribution of the Nusselt number and its global value will depend mainly on the vortex generator area to heat transfer area and the angle of attack. As long as the VGs may be considered as slender and thin, their thickness and height are of secondary importance.

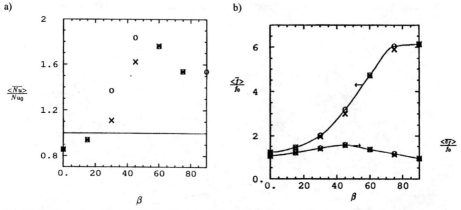

Fig. 16: Heat transfer enhancement, flow losses and friction coefficient at $Re_{2H} = 350$ for one periodic element with $h/H = 0.5$, $\delta/H = 0.05$, $L_p/H = 5$ and $Pr = 0.71$ for symmetric (x) and periodic (o) side boundary conditions.

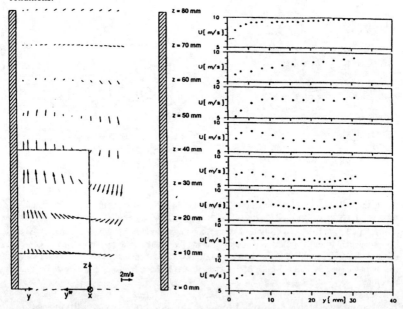

Fig. 17: Velocities in a cross section between two winglet rows (a) streamvice velocity profiles in different lateral positions and (b) circumferential velocities. Configuration as in fig. 15 with angle of attack 30°.

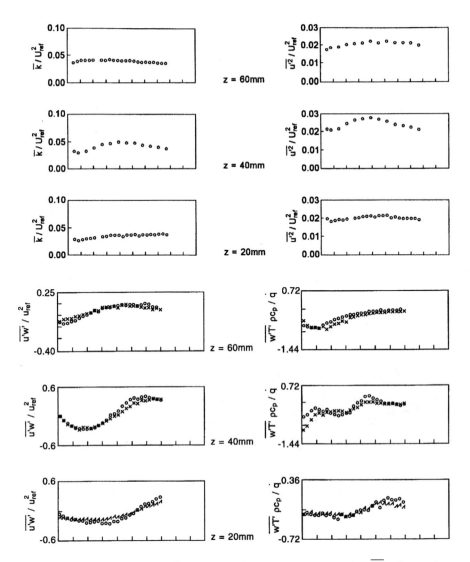

Fig. 18: Turbulent kinetic energy $k/u_{ref}^2$, root mean square axial velocity fluctuation $\overline{u'^2}/u_{ref}^2$, velocity and temperature correlations $\overline{u'w'}$ and $\overline{w'T'}$ for the cross sections and the lateral positions of fig. 17. $u_{ref}$ equals mean velocity

In B2 an angle of attack of 30° was considered for a single row and for an array of WVGs as shown in Fig.14. The four wire HDA technique allowed the determination of the three velocity components, and the six velocity and three temperature correlations. The refinement of the measuring technique, the measurements, and their interpretation were the major points of the investigation. Parts of these results have already been published in [58] and [59]. The longitudinal and cross velocity vector plot of Fig 17 shows for the cross section halfway between two winglet rows the velocity defect of the longitudinal velocity in the vortex core and the strong cross flow velocities. It is then tacitly assumed that the criteria for defining the vortex core is the non-vanishing of cross flow velocities. As in laminar flow the vortex extends over nearly the whole cross section and the highest circumferential velocities are of the same magnitude as the mean axial velocity. The streamwise velocity correlation, turbulent kinetic energy, and $\overline{u'w'}$ and $\overline{w'T'}$ correlations are represented in Fig.18 for the same cross section and in addition for a cross section half a winglet height behind a winglet row. The profiles in the direction of the winglet height are shown for four lateral positions starting from the centerline between the winglets. The vortex core has the highest turbulent kinetic energy and streamwise velocity fluctuations. For longitudinal vortices in external flow this is known for vortices after vortex breakup. The $\overline{u'w'}$ correlation changes sign in the vortex core region. If the analogy between heat and momentum transfer would hold the $\overline{w'T'}$ correlation should look similar to the $\overline{u'w'}$ correlation. That this is not the case is another indication that the induced circumferential velocities and their fluctuations are of major importance for the heat transfer enhancement by LVs.

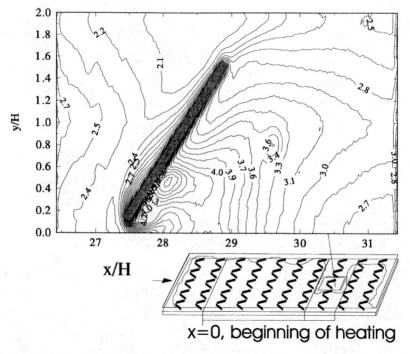

Fig. 19: Local Nusselt number enhancement $Nu/Nu_0$ for a periodically fully developed flow condition at $Re_{2H}=3\cdot10^4$; $e/H=0.5$, $L_P/e=10$, $B_P/e=8$, $l/e=4$, $s/e=1.172$ for $\beta=45°$; constant heat flux on the lower ribbed wall and an adiabatic upper wall.

In B3 Re, angle of attack, and height of the WVGs have been varied. For the same geometry as in B2 but at an angle of attack of 45° the local heat transfer enhancement has been determined by IRT as shown in Fig.19 for Re=30 000. The local heat transfer enhancement varies between a factor of 2 and 4. The distribution looks similar to the corresponding laminar flow but the enhancement is higher. The influence of angle of attack and Reynolds number on heat transfer enhancement and flow losses are displayed in Fig.20 for thermally developing but hydrodynamically fully developed conditions. The results are normalized with the conditions at zero angle of attack. Already at Re=1000 a continuous spectrum of frequencies was measured with a hotwire. No dominant frequencies exist any more at Re=1000, while for the cylinder in cross flow dominant frequencies can be detected up to Re of several million. As in laminar flow the flow losses increase monotonously with angle of attack while the heat transfer peaks at angles of attack around 60°. The high flow losses for Re=40,000 indicate, that WVGs with a large height ratio are not suitable for practical applications [60].

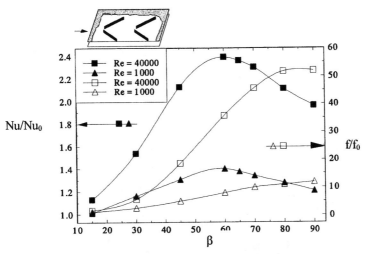

Fig. 20: Time and area averaged standardized Nusselt number and apparent friction factor as a function of the angle of attack; e/H=0.5, $L_P/e$=10, $B_P/e$=8, l/e=4, s/e=(1-sin β)·l/2e, constant heat flux on the lower ribbed wall and an adiabatic upper wall, the flow is periodically fully developed and has a thermal entrance condition, $Nu_0$ and $f_0$ are the measured $Nu_0$ and $f_0$ of the channel with winglets at 0° angle of attack at the same Reynolds number.

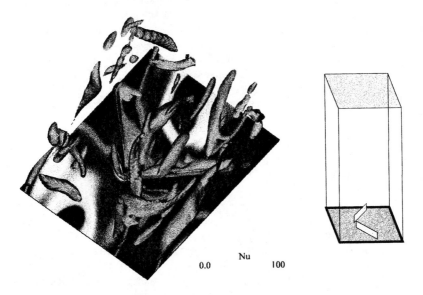

Fig. 21: Instantaneous vortex structure and Nusselt number distribution in turbulent flow; β=45°, h/H=0.05, $Re_{2H}$=12000

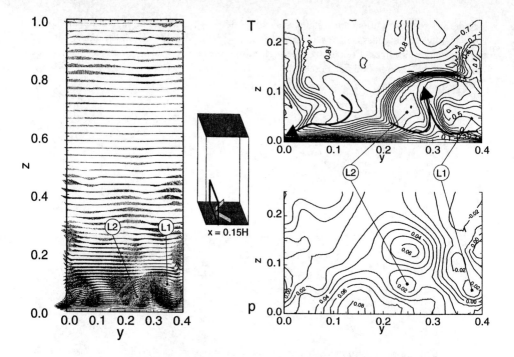

Fig. 22: Instantaneous vortex structure, temperature and pressure field in turbulent flow; $\beta=45°$, h/H=0.05, $Re_{2H}=12000$.

The height ratio was reduced from 0.5 to 0.05 for the LES calculations. The Reynolds number was chosen to be 12,000 because extensive DNS and LES results exist for that Reynolds number for plane channel flow and could be used for program validation. Figure 21 shows the geometry, instantaneous vortex structure, and Nusselt number distribution on the finned wall. The variety of vortices with respect to orientation and size is remarkable. The instantaneous cross velocity and vortex structure as well as the temperature and pressure distribution for a cross section in front of a winglet row are shown in Fig.22. Two longitudinal vortices L1 and L2 are pointed out, their cores are associated with local pressure minima.

## CONCLUDING REMARKS

The following experimental and numerical / theoretical tools were developed or further improved in connection with the different projects:
(1) A multi-wire probe with 4 sensors, its computer assisted calibration and evaluation algorithms for the determination of the instantaneous velocity vector and fluid temperature [17];
(2) a measurement technique for the temperature velocity correlations which employs the simulataneous measurements of a Laser Doppler anemometer and hot-wire probe [12];

(3) a software package for the evaluation of heat transfer measurements with infrared thermography (IRT) which takes into account the radiation properties of different materials and view factors of different geometries and the radiation part of the transferred heat flux [13];
(4) a software package of the ammonia absorption method (AAM) using a flatbed scanner with a local resolution of 400 dpi and a 8 bit digitization depth [14,16];
(5) a special wind tunnel for insitu-calibration of hot-wire probes at air velocities between 0.05 and 1 m/s [14];
(6) smoothers for multigrid Navier-Stokes solvers which improve their rate of convergence [10];
(7) a fast finite volume Navier-Stokes solver for three dimensional unsteady flow [14];
(8) a second order accurate Navier-Stokes solver with a dynamic subgrid model for large eddy simulations and the associated software for the evaluation of the flow structure, Reynolds fluxes, wall shear and heat transfer [15,18].

These tools were essential to arrive at the results which are summarized in the following paragraphs.

For two dimensional steady boundary layers with pressure induced separation regions - TVs embedded in the boundary layer - it can be stated:
(1) In laminar flow large local deviations in friction, pressure and heat transfer distribution occur, but no global changes in drag and heat transfer result, according to asymptotic theories.
(2) In turbulent flow universal wall functions for velocity and temperature exist which depend on the wall shear and pressure distribution. Large variations in wall friction and heat transfer result.

For channel flows with TVGs in the form of ribs on one channel wall - geometry induced separation - it was found:
(1) Steady TVs imply practically no global heat transfer enhancement but additional flow losses. The additional flow losses stem from the form drag of the TVGs. The practically unchanged global heat transfer is the result of the complex interaction of the TVs with the channel walls and ribs.
(2) Self-sustained oscillations, i.e. transition, are induced at much lower Reynolds numbers than in plane channel flow.
(3) Unsteady TVs imply 'Reynolds' averaged transport and hence increased heat transfer and flow losses. For the same heat transfer as in plane turbulent channel flow, flow losses are however reduced by two orders of magnitude.
(4) In turbulent flow an experimental data base for circular and rectangualr ribs of identical pitch ratio has been established which is suitable to test advanced turbulence models.

For channel flows with WVGs in the form of rectangular winglets which generate mainly LVs it could be established:
(1) Heat transfer enhancement is generated by developing boundary layers, swirl and unsteadiness.
(2) WVG to wall area and angle of attack of the WVGs are the major parameters which control heat transfer and flow loss enhancement.
(3) For steady flow the developing boundary layers on the WVGs and the corkscrew motion induced by LVs can increase the global heat transfer by more than 50% at Re=350.
(4) For self-oscillating flow the heat transfer may be further increased by about 20% at Re=500.
(5) For turbulent flow WVGs in the roughness domain generate for the same heat transfer enhancement considerably lower flow losses than WVGs which are much larger than the viscous sublayer.

(6) For the same flow loss WVGs which generate mainly LVs produce higher heat transfer than VGs which generate mainly TVs.

(7) With the present results and wing theory in mind dense packaging WVGs at relatively small angles of attack should result in the highest heat transfer enhancement for a given flow loss.

(8) For large LVs embedded in turbulent flow a data base has been established which may be used to test turbulence models.

For impinging radial jets TVs are generated, for impinging radial jet arrays the TVs are transformed into LVs. The major results are:

(1) For inclination angles higher than 60° the radial jets give higher heat transfer than the axial jets.

(2) The radial jets become turbulent at smaller Re than the axial jets.

(3) Swirl increases heat transfer for radial jets but not for axial jets.

(4) Free convection increases heat transfer more for axial than for radial jets.

## REFERENCES

1. Fiebig, M., Kallweit, P., and Mitra, N.K., Wing Type Vortex Generators for Heat Transfer Enhancement, *Heat Transfer* 1986, Proc. Eight Int. Heat Transfer Conf., Vol. 6, pp. 2909-2913, Hemisphere, New York, 1986.
2. Biswas, G., Mitra, N.K., and Fiebig, M., Computation of Laminar Mixed Convection Flow in a Channel with Wing Type Built-in Obstacles, *J. Thermophys.*, **3**, 447-453, 1989.
3. Fiebig, M., Brockmeier, U., Mitra, N.K., and Güntermann, T., Structure of Velocity and Temperature Fields in Laminar Channel Flows with Longitudinal Vortex Generators, *Num. Heat Transfer, Part A*, **15**, 281-302, 1989.
4. Brockmeier, U., Fiebig, M., Güntermann, T., and Mitra, N.K., Heat Transfer Enhancement in Fin-Plate Heat Exchangers by Wing Type Vortex Generators, *Chem. Eng. Technol.*, **12**, 288-294, 1989.
5. Fiebig, M., Kallweit, P., Mitra, N.K., and Tiggelbeck, S., Heat Transfer Enhancement and Drag by Longitudinal Vortex Generators in Channel Flow, *Exp. Thermal Fluid Sci.*, **4**, 103-114, 1991.
6. Tiggelbeck, S., Mitra, N.K., and Fiebig, M., Flow Structure and Heat Transfer in a Channel with Multiple Longitudinal Vortex Generators, *Exp. Thermal Fluid Sci.*, **5**, 425-436, 1992.
7. Tiggelbeck, S., Mitra, N.K., and Fiebig, M., Experimental Investigations of Heat Transfer Enhancement and Flow Losses in a Channel with Double Rows of Longitudinal Vortex Generators, *Int. J. Heat Mass Transfer*, **36**, 2327-2337, 1993.
8. Fiebig, M., Mitra, N.K., and Dong, Y., Simultaneous Heat Transfer Enhancement and Flow Loss Reduction of Fin-Tubes, *Heat Transfer* 1990, Proc. Ninth Int. Heat Transfer Conf., Vol. 4, pp. 51-56, Hemisphere, New York, 1990.
9. Fiebig, M., Mitra, N.K., and Dong, Y., Einfluß ausgestanzter Deltaflügel-Wirbelerzeuger auf Wärmeübergang und Strömungswiderstand von Rippenrohren, *Wärme Stoffübertrag.*, **25**, 33-43, 1990.
10. Braess, D., How can Algorithms of SIMPLE-Type be Used as Smoothers in Multigrid Iterations, (Project A7 in this volume).
11. Herwig, H., Severin, J., and Schäfer, P., Heat Transfer in Laminar Flow with Finite Regions of Pressure Induced Seperation, (Project A1 in this volume).
12. Gersten, K., Herwig, H., Kiel, R., and Vieth, D., Two Dimensional Turbulent Boundary Layers with Seperation and Reattachment Including Heat Transfer, (Project A2 in this volume).
13. Lorenz, S., and Leiner, W., Flow Structure and Local Heat Transfer in an Asymmetric Grooved Channel in the Turbulent Regime, (Project A3 in this volume).
14. Hahne, H.-W., Weber, D., and Fiebig, M., Flow and Heat Transfer in Ribbed Channels with Self-Sustained Oscillating Transverse Vortices, (Project A4 in this volume).
15. Cziesla, T., Laschefski, H., and Mitra, N.K., Direct and Large Eddy Simulation of Impinging Jets, (Project A6 in this volume).

16. Neumann, H., Grosse-Gorgemann, A., Weber, D., Hahne, H.-W., Müller, U., and Fiebig, M., Heat Transfer and Flow Losses in Transition from Longitudinal to Transverse Vortices in Steady and OscillatingChannel Flow, (Project B1 in this volume).
17. Vasanta Ram, V., Lau, S., Experimental Investigation of Momentum and Heat Transport in the Turbulent Channel with Embedded Longitudinal Vortices, (Project B2 in this volume).
18. Braun, H., Neumann, H., and Fiebig, M., Vortex Structure, Heat Transfer and Flow Losses in Turbulent Channel Flow with Periodic Longitudinal Vortex Generators, (Project B3 in this volume).
19. Treffethen, L.M., and Panton, R.B., Some Unanswered Questions in Fluid Mechanics, *Appl. Mech. Rev.*, Vol. 110, pp. 175-183, 1990.
20. Jeong, J., and Hussain, F., On the Identification of a Vortex, *J. Fluid Mech.*, Vol. 285, pp. 69-94, 1995.
21. van Dyke, M., An Album of Fluid Motion, *The Parabolic Press*, Stanford, Fourth Printing, 1988.
22. Noack, B.R., and Eckelmann, H., A Low-Dimensional Galerkin Method for the Three-Dimensional Flow around a Circular Cylinder, *Phys. Fluids*, Vol. 6, No. 1, pp. 124-143, 1994.
23. Ghaddar, N.K., Korzak, K.Z., Mikic, B.B., and Patera, A.T., Numerical Investigation of Incompressible Flow in Grooved Channels, Part 1, Stability and Self-Sustained Oscillations, *J. Fluid Mech.*, Vol. 163, pp. 99-127, 1986.
24. Ghaddar, N.K., Mikic, B.B., and Patera, A.T., Numerical Investigation of Incompressible Flow in Grooved Channels, Part 2, Resonance and Oscillatory Heat Transfer Enhancement, *J. Fluid Mech.*, Vol. 168, pp. 541-567, 1986.
25. Greiner, M., Ghaddar, N.K., Mikic, B.B., and Patera, A.T., Resonant Convective Heat Transfer in Grooved Channels, *Proc. 8th Int. Heat Transfer Conf.*, San Francisco, USA, Vol. 6, pp. 2867-2872, 1986.
26. Amon, C.H., and Mikic, B.B., Numerical Prediction of Convective Heat Transfer in Self-Sustained Oscillatory Flows, *J. Thermophysics*, Vol. 4, No. 2, pp. 239-246, 1990.
27. Hermann, C.V., and Mayinger, F., Experimental Investigation of the Heat Transfer in Laminar Forced Convection Flow in a Grooved Channel, *2nd World Conf. on Exp. Heat Transfer, Fluid Mechanics and Thermodynamics*, Dubovnik, pp. 387-392, 1991.
28. Hermann, C.V., Mayinger, F., Mikic, B.B., and Deculic, D.B., Numerical and Experimental Studies of Self-Sustained Oscillatory Flows in Communicating Channels, *Int. J. Heat Mass Transfer*, Vol. 35, pp. 3115-3129, 1992.
29. Weber, D., Grosse-Gorgemann, A., Mitra, N.K., and Fiebig, M., Selbsterregte Schwingungen in Kanalströmungen mit Wirbelerzeugern, GAMM Tagung, Braunschweig, 1994.
30. Fiebig, M., Grosse-Gorgemann, A., Hahne, W., Leiner, W., Mitra, N.K., and Weber, D., Local Heat Transfer and Flow Structure in Grooved Channels, Measurements and Computations, *Heat Transfer*, Vol. 4, ed. G.F. Hewitt, Francis and Taylor, pp. 237-242, 1994.
31. Grosse-Gorgemann, A., Weber, D., and Fiebig, M., Experimental and Numerical Investigation of Self-Sustained Oscillations in Channels with Periodic Structures, *Experimental Thermal and Fluid Science*, Vol. 11, No. 3, pp. 226-233, 1995.
32. Lanchester, F.W., *Aerodynamics*, Constable & Co, London, 1907.
33. Lanchester, F.W., *Proc. Inst Auto Eng*, **9**, 171-259, 1915.
34. Delery, J., Horowitz, E., Leuchter, O., and Solignac, J.-L., Fundamental Studies on Vortex Flow, *La Recerche Aerospatiale*, No. 2, pp. 1-24, 1984.
35. Edwards, F.J., and Alker, C.J.R., The Improvement of Forced Convection Surface Heat Transfer Using Surface Protrusions in the Form of (A) Cubes and (B) Vortex Generators, *Heat Transfer 1974*, Proc. Fifth Int. Heat Transfer Conf., Vol. 2, pp. 244-248, JSME, Tokyo, 1974.
36. Russel, C.M.B., Jones, T.V., and Lee, G.H., Heat Transfer Enhancement Using Vortex Generators, *Heat Transfer 1982*, Proc. Seventh Heat Transfer Conf., Vol. 3, pp. 283-288, Hemisphere, New York, 1982.
37. Turk, A.Y., and Junkhan, G.H., Heat Transfer Enhancement Downstream of Vortex Generators on a Flat Plate, *Heat Transfer 1986*, Proc. Eighth Int. Heat Transfer Conf., Vol. 6, pp. 2903-2908, Hemisphere, New York, 1986.
38. Torii, K., Yanagihara, J.I., and Nagai, Y., Heat Transfer Enhancement by Vortex Generators, *Proc. of the ASME / JSME Thermal Engineering Conference*, J.R. Lloyd and Y. Kurosaki, Eds., Book No. 10309C, ASME, New York, 1992.
39. Subramanian, C.S., Ligrani, P.M., and Tuzzolo, M.F., Surface Heat Transfer and Flow Properties of Vortex Arrays Induced Artificially and from Centrifugal Instabilities, *Int. J. Heat Fluid Flow*, **13**(3), 210-223, 1992.
40. Schäfer, P., Untersuchung von Mehrfachlösungen bei laminaren Strömungen, Dissertation, Ruhr-Universität Bochum, 1995.
41. Schäfer, P., and Herwig, H., Drag and Heat Transfer in Laminar Boundary Layers with Finite Regions of Seperation, *Proc. of Eurotherm 31*, Bochum, pp. 17-22, 1993.

42. Fiebig, M., Embedded Vortices in Internal Flow: Heat Transfer and Pressure Loss Enhancement, *Int. J. Heat and Fluid Flow*, Vol. 16, No. 5, pp. 376-388, 1995.
43. Gersten. K., Klauer, J., and Vieth, D., Asymptotic Analysis of Two-Dimensional Turbulent Seperating Flows, *NNFM*, Vol. 40, Vieweg Verlag, Braunschweig, pp. 125-132, 1993.
44. Kiel, R., and Vieth, D., Experimental and Theoretical Investigations of the Near-Wall Region in a Turbulent Seperated and Reattached Flow, *Experimental Thermal and Fluid Science*, Vol. 11, pp. 243-254, 1995.
45. Vieth, D., Berechnung der Impuls- und Wärmeübertragung in ebenen turbulenten Strömungen mit Ablösung bei hohen Reynolds-Zahlen, Dissertation, Ruhr-Universität Bochum, 1996.
46. Kiel, R., Experimentelle Untersuchung einer Strömung mit beheiztem lokalen Ablösewirbel an einer geraden Wand, Dissertation, Ruhr-Universität Bochum, 1995.
47. Lorenz, S., Lokaler Wärmeübergang und Strömungsstruktur bei turbulenter Strömung in einseitig querberippten Kanälen, Dissertation, Ruhr-Universität Bochum, 1995.
48. Neumann, H., Lorenz, S., and Leiner, W., Infrarot-Thermographie durch Fenster bei niedrigen Temperaturen, *Wärme- und Stoffübertragung*, Vol. 29, pp. 219-225, 1994.
49. Laschefski, H., Numerische Untersuchung der dreidimensionalen Strömungsstruktur und des Wärmeüberganges bei ungeführten und geführten, laminaren Freistrahlen mit Prallplatte, Dissertation, Ruhr-Universität Bochum, 1994.
50. Laschefski, H., Braess, D., Haneke, H., and Mitra, N.K., Numerical Investigations of Radial Jet Reattachment Flows, *Int. J. for Numerical Methods in Fluids*, Vol. 18, pp. 629-646, 1994.
51. Owsenek, B.L., Cziesla, T., Biswas, G., and Mitra, N.K., Numerical Investigation of Heat Transfer in Impinging Axial and Radial Jets with Superimposed Swirl, to be published in *Int. J. Heat and Mass Transfer*, 1996.
52. Braess, D., and Sarazin, R., An Efficient Smoother for the Stokes Problem, to be published in *Numerical Applied Mathematics*, 1997.
53. Hemforth, F., Die Behandlung der instationären Navier-Stokes-Gleichungen mit P1/P2-Elementen: Diskretisierung und Löser, Dissertation, Ruhr-Universität Bochum, 1996.
54. Sarazin, R., Eine Klasse von effizienten Glättern vom Jacobi-Typ für das Stokes Problem, Dissertation, Ruhr-Universität Bochum, 1996.
55. Grosse-Gorgemann, A., Numerische Untersuchung der laminaren oszillierenden Strömung und des Wärmeüberganges in Kanälen mit rippenförmigen Einbauten, Dissertation, Ruhr-Universität Bochum, 1995.
56. Weber, D., Experimente zu selbsterregt instationären Spaltströmungen mit Wirbelerzeugern und Wärmeübertragung, Dissertation, Ruhr-Universität Bochum, 1995.
57. Fiebig, M., Vortices: Tools to Influence Heat Transfer - Recent Developments, *Proc. of 2nd European Thermal Sciences and 14th UIT National Heat Transfer Conference 1996*, ed G.P. Celata, P. Di Marco and A. Mariani, 1996.
58. Lau, S., Experimental Study of the Turbulent Flow in a Channel with Periodically Arranged Longitudinal Vortex Generators, *Experimental Thermal and Fluid Science*, Vol. 11, No. 3, pp. 225-261, 1995.
59. Lau, S., Meiritz, K., and Vasanta Ram, V.I., Measurement of Momentum and Heat Transport in the Channel Flow with Embedded Longitudinal Vortices, to be published *Int. J. Heat and Mass Transfer*, 1996.
60. Braun, H., Grobstruktursimulation turbulenter Geschwindigkeits- und Temperaturfelder in Spaltströmungen mit Wirbelerzeugern, Dissertation, Ruhr-Universität Bochum, 1996.

# HEAT TRANSFER IN LAMINAR FLOW WITH FINITE REGIONS OF PRESSURE INDUCED SEPARATION

H. Herwig, J. Severin, P. Schäfer
TU Chemnitz-Zwickau, Reichenhainer Str. 70
D-09126 Chemnitz, Germany

## SUMMARY

The influence of pressure induced lateral vortices on heat transfer over contoured walls is investigated. The flow is laminar and steady for most cases with an extension to unsteady flows for some geometries. Asymptotic theories like interacting boundary layer theory, triple deck theory and slender channel theory are preferred. Their results are compared to corresponding Navier–Stokes solutions. Special attention is given to heat transfer over smooth backward facing steps, through diffusers and through channels with wavy wall contours.

## INTRODUCTION

In most engineering applications convective heat transfer is caused by turbulent flow. However, there are examples of practical interest where convective heat transfer occurs in the laminar flow regime. Typical examples are heat exchangers with small flow rates, see for example and heat transfer in small devices like cooling of electronic equipment and the wide field of microsystem technology, see for example [1], [2]. Aside from the aspect of applicability the laminar case is especially suitable for the elucidation of fundamental problems since it is not loaded with the turbulence problem which untill today is not satisfactorily solved.

Futhermore, various asymptotic theories like boundary layer theory, triple deck theory and slender channel theory can be applied which all analytically account for the Reynolds number dependence. Thus, the influence of the flow in convective heat transfer situations is clearly revealed.

In the part of the projekt reported here, we restrict ourselves to two–dimensional flows so that vortices that appear in conection with finite regions of separation are transverse vortices. Since we are interested in pressure induced separation the geometries considered are always smooth without any sharp corners. Typical geometries are smooth backward facing "steps", smooth diffusers and wavy walls.

After the basic equations and theories are provided, results for a variety of parameters are presented. The fundamental aspects of the vortex affected momentum and heat transfer are addressed. The model geometry for this purpose is the backward facing smooth "step". Internal flows are addressed afterwards with special emphasis on heat transfer results.

# BASIC EQUATIONS AND CONCEPTS

## Basic equations

The basic equations for two-dimensional laminar convective heat transfer of a Newtonian fluid are:

Continuity equation:
$$\frac{\partial \varrho}{\partial t} + \frac{\partial \varrho u}{\partial x} + \frac{\partial \varrho v}{\partial y} = 0. \tag{1}$$

Momentum equations (Navier-Stokes):
$$\varrho \left[ \frac{\partial u}{\partial t} + u\frac{\partial u}{\partial x} + v\frac{\partial u}{\partial y} \right] =$$
$$-\frac{\partial p}{\partial x} + \frac{1}{Re} \left[ \frac{\partial}{\partial x} \left( \mu \left( 2\frac{\partial u}{\partial x} - \frac{2}{3} \left( \frac{\partial u}{\partial x} + \frac{\partial v}{\partial y} \right) \right) \right) + \frac{\partial}{\partial y} \left( \mu \left( \frac{\partial u}{\partial y} + \frac{\partial v}{\partial x} \right) \right) \right] \tag{2}$$

$$\varrho \left[ \frac{\partial v}{\partial t} + u\frac{\partial v}{\partial x} + v\frac{\partial v}{\partial y} \right] =$$
$$-\frac{\partial p}{\partial y} + \frac{1}{Re} \left[ \frac{\partial}{\partial x} \left( \mu \left( \frac{\partial u}{\partial y} + \frac{\partial v}{\partial x} \right) \right) + \frac{\partial}{\partial y} \left( \mu \left( 2\frac{\partial v}{\partial y} - \frac{2}{3} \left( \frac{\partial u}{\partial x} + \frac{\partial v}{\partial y} \right) \right) \right) \right]. \tag{3}$$

Thermal energy equation:
$$\varrho c_p \left[ \frac{\partial \Theta}{\partial t} + u\frac{\partial \Theta}{\partial x} + v\frac{\partial \Theta}{\partial y} \right] = \frac{1}{RePr} \left[ \frac{\partial}{\partial x} \left( k\frac{\partial \Theta}{\partial x} \right) + \frac{\partial}{\partial y} \left( k\frac{\partial \Theta}{\partial y} \right) \right]. \tag{4}$$

Equations (1)–(4) are nondimensionalized according to table 1. The reference state $R$ will be fixed later together with $\Delta T_R^*$ in $\Theta$. The dimensionless groups are

$$Re = \frac{\varrho_R^* U_R^* L_R^*}{\mu_R^*} = \frac{U_R^* L_R^*}{\nu_R^*} \quad \text{(Reynolds number)} \tag{5}$$

$$Pr = \frac{\mu_R^* c_{p_R}^*}{k_R^*} \quad \text{(Prandtl number)}. \tag{6}$$

In the Navier–Stokes equations buoyancy effects have been neglected, equation (4) does not account for pressure work and dissipation effects. For details see for example [3].

Boundary conditions will be provided for specific geometries and flow conditions only.

Table 1: Nondimensional quantities in the basic equations
\* $\hat{=}$ dimensional quantity

| $x, y$ | $t$ | $u, v$ | $p$ | $\Theta$ | $\varrho, \mu, k, c_p$ |
|---|---|---|---|---|---|
| $\dfrac{x^*, y^*}{L_R^*}$ | $\dfrac{t^*}{L_R^*/U_R^*}$ | $\dfrac{u^*, v^*}{U_R^*}$ | $\dfrac{p^* - p_R^*}{\varrho_R^* U_R^{*2}}$ | $\dfrac{T^* - T_R^*}{\Delta T_R^*}$ | $\dfrac{\varrho^*}{\varrho_R^*}, \dfrac{\mu^*}{\mu_R^*}, \dfrac{k^*}{k_R^*}, \dfrac{c_p^*}{c_{pR}^*}$ |

## Interacting boundary layer theory (IBT)

In the high Reynolds number limit, $Re \to \infty$, the Navier–Stokes equations can be approximated by the boundary layer theory. In a systematic approach by asymptotic expansions this singular pertubation problem is solved exactly in the limit $Re = \infty$ and approximately for $Re \to \infty$. For details see for example [4] or [3].

A crucial feature of the classical boundary layer theory is the hierachical matching order for inner and outer expansions. In the course of matching it turns out that displacement and curvature are the most important higher order effects. However, asymptotically they are not always on the same level of approximation. In [5] it was shown that curvature effects are negligible compared to displacement effects for the geometries investigated in this study.

Instead of accounting for the displacement by the second order boundary layer equations it can be taken into account by an iteration procedure between the first order inner and outer solution. This way of treating the displacement effect was proposed by Prandtl already. But, its justification can only be provided by triple deck theory which will be introduced in the next chapter.

The equations for what from now on will be called the "interacting boundary layer theory", abbreviated as IBT, for the steady two–dimensional case are ($y_B = y\, Re^{1/2}$, $v_B = v\, Re^{1/2}$) :

$$\frac{\partial u}{\partial x} + \frac{\partial v_B}{\partial y_B} = 0, \tag{7}$$

$$u\frac{\partial u}{\partial x} + v_B \frac{\partial u}{\partial y_B} = -\frac{dp}{dx} + \frac{\partial^2 u}{\partial y_B^2}, \tag{8}$$

$$u\frac{\partial \Theta}{\partial x} + v_B \frac{\partial \Theta}{\partial y_B} = \frac{1}{Pr}\frac{\partial^2 \Theta}{\partial y_B^2}. \tag{9}$$

$$U(x) = 1 + \frac{1}{\pi} \int_{-\infty}^{\infty} \frac{dy_C/dz}{x-z} dz + \frac{1}{\pi} \int_{-\infty}^{\infty} \frac{d\delta_1/dz}{x-z} dz. \quad (10)$$

Equation (10) basically is the outer inviscid solution in terms of the so-called Hilbert integral by which the pressure distribution (here in terms of $U(x)$, due to linearisation) is determined over the wall contour $y_C$ enlarged by the displacement thickness $\delta_1$, see [6] for details. The associated boundary conditions are:

$$y_B = 0 : \quad u = v_B = 0, \quad (11)$$

$$y_B \to \infty : \quad u - U(x) = \Theta = 0. \quad (12)$$

The wall boundary condition for $\Theta$ is fixed with a specific thermal boundary condition. Since in our study wall contours are investigated which behave like flat plates away from the contour deformations, a transformation with $\sqrt{x}$ is performed. Introducing

$$\eta = \frac{y_B}{\sqrt{x}}, \quad f(x,\eta) = \frac{\Psi}{\sqrt{x}} \quad \text{with} \quad \frac{\partial \Psi}{\partial y_B} = u, \quad \frac{\partial \Psi}{\partial x} = -v_B, \quad (13)$$

the boundary momentum equation reads:

$$f_{\eta\eta\eta} + \frac{1}{2} f f_{\eta\eta} = x \left( p_x + f_\eta f_{\eta x} - f_{\eta\eta} f_x \right). \quad (14)$$

Choosing standard thermal boundary conditions, we get for the energy equation:

$$T_w = const : \quad \vartheta(x,\eta) = \Theta(x,y_B) \quad \to \quad \vartheta_{\eta\eta} + \frac{Pr}{2} f \vartheta_\eta = x Pr \left( f_\eta \vartheta_{\eta x} - \vartheta_\eta f_x \right). \quad (15)$$

$q_w = const$:

$$\vartheta(x,\eta) = \Theta(x,y_B)/\sqrt{x} \quad \to \quad \vartheta_{\eta\eta} + \frac{Pr}{2} (f \vartheta_\eta - \vartheta f_\eta) = x Pr \left( f_\eta \vartheta_{\eta x} - \vartheta_\eta f_x \right). \quad (16)$$

The associated boundary conditions are:

$$\eta = 0 : \quad f = f_\eta = 0, \quad (17)$$

$$\eta \to \infty : \quad f_\eta - U(x) = \vartheta = 0 \quad (18)$$

and

$$\vartheta(x,0) = 1 \quad \text{for} \quad T_w = const, \quad (19)$$

$$\vartheta_\eta(x,0) = 1 \quad \text{for} \quad q_w = const. \quad (20)$$

The solution procedure is straightforward as long as no separation occurs. With $\tau_W < 0$ a so-called inverse procedure is needed by which the displacement thickness

$$\delta_1(x) \sqrt{Re} = \sqrt{x} \lim_{\eta \to \infty} [\eta - \frac{f(x,\eta)}{U(x)}]. \quad (21)$$

is prescribed for the boundary layer calculation, see [7] for details.

# Triple deck theory (TDT)

The asymptotic theory for viscous–inviscid interaction between a boundary layer and the outer inviscid flow is called triple deck theory, abbreviated TDT. It is named after the three–layered structure of the flow field in the vicinity of the interaction process. It was first developed to scope with the problem of trailing edge flow, see [8], but soon turned out to be an asymptotic theory for all kinds of viscous–inviscid interactions. Its main features are:

- TDT for $Re \to \infty$ has no explicit $Re$-dependence. The Reynolds number is scaled out of the problem and only appears in terms of gauge functions in a higher order triple deck theory.

- All three layers are asymptotically small, i.e. the whole structure vanishes identically in the limit $Re = \infty$. Finite layers only exist for finite Reynolds numbers.

- All three layers have the same streamwise extension $\sim Re^{-3/8}$. The *main deck* as a continuation of the oncoming boundary layer has a lateral extension $\sim Re^{-1/2} = Re^{-4/8}$. The *lower deck*, compared to the main deck is asymptotically small ($\sim Re^{-5/8}$) whereas the *upper deck* is asymptotically large ($\sim Re^{-3/8}$), see figure 1.

- The interaction process is described by an asymptotic version of the Hilbert integral, see equation (10), adjusting the pressure determined by the outer flow in the upper deck to the displacement determined by viscous effects in the lower deck.

Details of this theory can be found in [8], [3] and [6]. For a physically motivated derivation of the tripple deck scaling see [9].

Here, only the final equations are provided. They are the lower deck equations together with the interaction law (Hilbert integral).

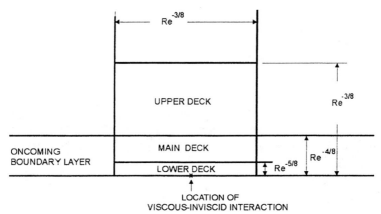

Fig. 1    Triple deck structure of a viscous–inviscid interaction process

The fundamental expansion parameter of the triple deck theory is

$$\varepsilon = Re^{-1/8}.$$

In table 2 the lower deck scaling is shown. For conveniance the parameters $\lambda$ and $\lambda_\vartheta$ of the oncoming flow ($\lambda = 0.332$, $\lambda_\vartheta = \lambda_\vartheta(Pr)$ in our case) is scaled out.

Table 2: Triple deck variables; $x, y, u, v, p, \vartheta = \Theta$ according to table 1

| | |
|---|---|
| coordinates | $X = x \dfrac{\lambda^{5/4}}{\varepsilon^3}$ , $Y = (y - y_C) \dfrac{\lambda^{3/4}}{\varepsilon^5}$ |
| velocity components | $U = u \dfrac{\lambda^{-1/4}}{\varepsilon}$ , $V = v \dfrac{\lambda^{-3/4}}{\varepsilon^3}$ |
| pressure | $P = p \dfrac{\lambda^{-1/2}}{\varepsilon^2}$ |
| temperature | $\Theta = (\vartheta - \vartheta_0) \dfrac{\lambda^{3/4}}{\varepsilon \lambda_\vartheta}$ |
| auxiliary function | $A = A_1 \lambda^{3/4}$ |

Introducing the stream function $\Psi$ by $\Psi_Y = U$ and $\Psi_X = -V$ the lower deck equations read:

$$\Psi_Y \Psi_{YX} - \Psi_{YY} \Psi_X = -P_X + \Psi_{YYY}, \qquad (22)$$

$$Pr(\Psi_Y \Theta_X - \Theta_Y \Psi_X) = \Theta_{YY}, \qquad (23)$$

with the associated boundary conditions:

$$X \to -\infty : \quad \Psi_Y = \Theta = Y, \qquad (24)$$

$$Y \to \infty : \quad \Psi_Y = \Theta_Y = Y - D, \qquad (25)$$

$$Y = 0 : \quad \Psi = \Psi_Y = 0, \qquad (26)$$

$$\Theta = 0 \quad (T_w = const).$$

$$\Theta_Y = 1 \quad (q_w = const).$$

The upstream boundary condition (24) is the undisturbed momentum and thermal boundary layer asymptotically close to the wall. The outer boundary condition (25) reflects the displacement effect

$$D(X) = -A(X) - y_C(X) \qquad (27)$$

through the auxiliary function $A(X)$ determined iteratively from the simultaneous solution of equations (22)–(27) and the Hilbert integral ($x_T = X/\varepsilon^3$)

$$P(x_T) = -\frac{1}{\pi}\int_{-\infty}^{\infty}\frac{dy_C/dz}{x_T - z}dz - \frac{1}{\pi}\int_{-\infty}^{\infty}\frac{dD/dz}{x_T - z}dz. \qquad (28)$$

It is important to note that in the framework of TDT the wall contour that gives rise to the interaction process is scaled according to the variables $X$ and $Y$ in table 2.

With a typical length $l = l^*/L_R^*$ and a height $h = h^*/L_R^*$ of a wall deformation the corresponding parameters in TDT are:

$$l_T = \frac{\lambda^{5/4}}{\varepsilon^3}l, \qquad (29)$$

$$h_T = \frac{\lambda^{3/4}}{\varepsilon^5}h. \qquad (30)$$

Table 3 shows how TDT results are transformed back to the physical quantities like wall shear stress ($c_f$), pressure and Nußelt number, respectively.

Table 3: Physical quantities from TDT

| shear stress $\tau_T$ | $\tau_T = \dfrac{c_f}{c_{f_0}} = \Psi_{YY}$ |
|---|---|
| pressure $P_T$ | $p = P_T\, Re^{-1/4}\, \lambda^{1/2}$ |
| Nußelt number $q_{T,T}$ ($T_w = const$) | $q_{T,T} = \dfrac{Nu}{Nu_0} = \Theta_Y(X,0)$ ($Nu = \dfrac{q_w^*(x^*)x^*}{k^*(T_w^* - T_\infty^*)}$) |
| Nußelt number $q_{T,q}$ ($q_w = const$) | $q_{T,q} = \left(\dfrac{1}{Nu} - \dfrac{1}{Nu_0}\right)\lambda^{3/4}\, Re^{5/8} = -\Theta(0.X)$ ($Nu = \dfrac{q_w^* x^*}{k^*(T_w^*(x^*) - T_\infty^*)}$) |

## Slender channel theory (SCT)

Especially for internal flows through slowly changing geometries the so-called "slender channel equations" are appropriate (For the discussion of slow versus slight variations see [10]). Asymptotically, they are a distinguished limit of a two parameter pertubation problem. One parameter is the inverse of the Reynolds number, i.e. $\varepsilon_1 = Re^{-1} \to 0$, the other one is the ratio of the two different characteristic lengths in streamwise direction and normal to it, respectively, i.e. $\varepsilon_2 = L_y^*/L_x^* \to 0$. The distinguished limit is

$$\lambda = \frac{\varepsilon_2}{\varepsilon_1} = O(1) \quad \text{for } \varepsilon_1 \to 0, \; \varepsilon_2 \to 0, \tag{31}$$

from which the slender channel scaling

$$X = \varepsilon_2 x, \quad \bar{v} = \frac{v}{\varepsilon_2}, \tag{32}$$

follows immediately.

The leading order equations (with respect to an asymptotic expansion in $\varepsilon_2$) are:

$$\frac{\partial u}{\partial X} + \frac{\partial \bar{v}}{\partial y} = 0, \tag{33}$$

$$u\frac{\partial u}{\partial X} + \bar{v}\frac{\partial u}{\partial y} = -\frac{\partial p}{\partial X} + \frac{1}{\lambda}\frac{\partial^2 u}{\partial y^2}, \quad \frac{\partial p}{\partial y} = 0, \tag{34}$$

$$u\frac{\partial \Theta}{\partial X} + \bar{v}\frac{\partial \Theta}{\partial y} = \frac{1}{\lambda Pr}\frac{\partial^2 \Theta}{\partial y^2}. \tag{35}$$

Here, we kept the parameter $\lambda = O(1)$ as a constant by which different degrees of slenderness can be fixed for a certain shape of the wall geometry $H(X)$. The associated boundary conditions are ($T_w = const$):

$$y = 0: \quad \frac{\partial u}{\partial y} = \bar{v} = \Theta = 0, \tag{36}$$

$$y = H(X): \quad u = \bar{v} = \Theta - 1 = 0. \tag{37}$$

Conditions (36) hold for symmetrical profiles for which either the lower or the upper half of the flowfield is calculated. For the unsymmetrical case the whole profile must be determined and (36) is replaced by equation (37) for $y = -H(X)$.

Equations (33) to (35) have the same form as the (leading order) boundary layer equations. But, the physical meaning as well as the underlying scaling is different. Whereas the pressure $p$ is prescribed in the boundary layer theory it is part of the solution for the slender channel case. The extra equation needed to determine $p$ is the $y$–momentum equation $\partial p/\partial y = 0$, see equation (34).

Introducing the transformed coordinate $\eta$ and a stream function $F(X, \eta)$ according to

$$\eta = \frac{y}{H(X)}; \; u = F_\eta/H(X), \; \bar{v} = \eta\frac{dH/dX}{H}F_\eta - F_X, \; F_X = \partial F/\partial X, \; F_\eta = \partial F/\partial \eta, \tag{38}$$

transforms the slender channel equations into the following form which is more convenient for numerical solutions:

$$F_{\eta\eta\eta} + \lambda[H(F_{\eta\eta}F_X - F_\eta F_{\eta X}) + \frac{dH}{dX}F_\eta^2] - \lambda H^3 \frac{dp}{dX} = 0, \tag{39}$$

$$\Theta_{\eta\eta} + \lambda H Pr(\Theta_\eta F_X - F_\eta \Theta_X) = 0. \tag{40}$$

with the associated boundary conditions:

$$\eta = 0: \quad F = F_{\eta\eta} = \Theta = 0, \qquad \eta = 1: \quad F - 1 = F_\eta = \Theta - 1 = 0. \tag{41}$$

## Linear stability theory (LST)

All three theories introduced so far (IBT, TDT, SCT) are asymptotic theories for steady laminar flows which are exact in the limit $Re = \infty$ and can provide approximate solutions for large but finite Reynolds numbers.

Real flows, however, become turbulent at finite Reynolds numbers $Re_{tr}$ and often unsteady (oscillating) at Reynolds numbers $Re_{OS} < Re_{tr}$. In the range $Re_{OS} < Re < Re_{tr}$ a laminar oscillating flow can be observed, a phenomenon which is called *self-sustained oscillation*. The transition process at Reynolds numbers close to $Re_{tr}$ is intimately linked to the stability of laminar flow. Its first stage has been sucsessfully described by the linear stability theory (LST).

Our hypothesis is that the phenomenon of self–sustained oscillations is a first step in the transition process and thus is somehow related to the findings of linear stability theory. For example, we expect the frequencies of self–sustained oscillations to be related to those of the most unstable modes in linear stability theory.

The basic idea behind the LST is a decomposition of flow quantities into mean and superimposed fluctuating parts, i.e. $a = \bar{a} + a'$; $a = u, v, p$. Fluctuations are assumed to be small so that terms with nonlinear combinations of fluctuating quantities can be neglected. Based on this decomposition an abitrary disturbance can be expanded in a Fourier series with a single oscillation of the disturbance being of the form

$$a'(x, y, t) = \hat{a}(y) exp[i\alpha(x - \hat{c}t)] + cc. \tag{42}$$

In (42) a two–dimensional disturbance is assumed that develops in time (temporal stability). The shape of the disturbance is given by the complex shape function $\hat{a}$ which is assumed to be independent of $x$. Its damping or amplification is controlled by the imaginary part of the complex quantity $\hat{c}$, with $c_i > 0$ for amplification and $c_i < 0$ for damping. For further details see for example [11]. With a stream function $\Psi(x, y, t) = \hat{\varphi}(y) \exp[i\alpha(x - \hat{c}t)] + cc.$, the shape function $\hat{\varphi}$ of a single oscillation of wave number $\alpha$ is determined from the so-called Orr Sommerfeld equation

$$(U - \hat{c})(\hat{\varphi}'' - \alpha^2 \hat{\varphi}) - U''\hat{\varphi} = -\frac{i}{\alpha Re}(\hat{\varphi}'''' - 2\alpha^2 \hat{\varphi}'' + \alpha^4 \hat{\varphi}), \tag{43}$$

which for a boundary layer, for example, is subject to the boundary conditions

$$y = 0 \;\;:\;\; u' = v' = 0 \;\;\rightarrow\;\; \hat{\varphi} = \hat{\varphi}' = 0, \tag{44}$$

$$y \rightarrow \infty \;\;:\;\; u' = v' = 0 \;\;\rightarrow\;\; \hat{\varphi} = \hat{\varphi}' = 0. \tag{45}$$

In equation (43) the mean flow velocity profile $U$ is assumed to be independent of $x$. This so–called parallel flow assumption neglects the change of $U$ in streamwise direction.

This change is accounted for approximately if flow stability is analysed on the basis of equation (43) with local velocity profiles $U(y)$ which change with $x$. As far as the mean velocity profile is concerned $x$ then is a parameter rather than an independent variable. This procedure will be called quasi–nonparallel flow stability analysis.

With this kind of stability analysis applied to boundary layer flow it turns out that the shape of $U(y)$ alone determines the stability of the flow. All other mechanisms like pressure gradient or blowing/suction rate at the wall affect the stability only indirectly through changes in $U(y)$ at a certain $x$–position. Thus, if $U(y)$ can at least approximately be characterized by just one parameter, this will be the only controlling parameter of the boundary layer stability. Such a parameter is

$$\beta_3 = \int_0^\infty (1 - U^2)\, U\, dy. \tag{46}$$

which measures the energy thickness of a boundary layer (see [3] for details).

Figure 2 shows a "universal" stability criterion for laminar boundary layers. The critical Reynolds number as function of $\beta_3$ is the lowest Reynolds number of a flow for which at least one elementary wave is not damped (i.e. for Reynolds numbers above the critical Reynolds number elementary waves in a certain range of wave numbers are amplified).

The index 2 at $Re_{2,cr}$ indicates that $Re$ is defined with the momentum thickness $\delta_2^*$ of the boundary layer, i.e. $Re_2 = U_R^* \delta_2^* / \nu_R^*$. In figure 2 the critical Reynolds number of various flows with different pressure gradients and different rates of wall mass flux are shown. They very closely gather around one single curve which we will call the "neutral line" because all flows with $(Re_2, \beta_3)$ combinations below this line are stable, all with those above it are unstable.

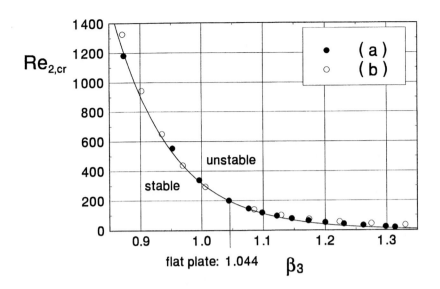

Fig. 2  "Universal" stability criterion for laminar boundary layers
(a) critical Reynolds numbers for various pressure gradients
(b) critical Reynolds numbers for various rates of blowing and suction
— neutral line

**Expected flow regimes**

The various flow regimes we expect by varying the parameters of the actual problem under consideration are:

- unique steady solutions
- multiple steady solutions
- unsteady periodic solutions
- unsteady nonperiodic solutions
- turbulent solutions.

Turbulent solutions are not in the scope of this study. We also exclude 3D effects. The basis of our investigations are the 2D unsteady Navier–Stokes equations (1)–(3) together with the thermal energy equation (4). Emphasis is on asymptotic considerations ($Re \to \infty$) as an approximation for Reynolds numbers small enough to be in the parameter regime "prior" to turbulent flow.

# MOMENTUM AND HEAT TRANSFER IN BOUNDARY LAYERS OVER CONTOURED WALLS: APPLYING THE IBT AND TDT CONCEPTS

## Model geometry

As mentioned in the introduction already we have chosen the backward facing smooth "step" as model geometry for boundary layer flows. In figure 3 the wall contour is shown together with the geometric parameters $L^*$ (distance from the leading edge), $l^*$ (step length) and $h^*$ (step height). The shape of contour is given by

$$\frac{y_c(x)}{h} = 20\left(\frac{x}{l}\right)^7 - 70\left(\frac{x}{l}\right)^6 + 84\left(\frac{x}{l}\right)^5 - 35\left(\frac{x}{l}\right)^4. \tag{47}$$

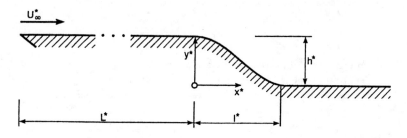

Fig. 3    Wall contour; note that the step always is in the range $0 \leq x/l \leq 1$; for $h = 0$ the wall geometry is that of the flat plate

## Parameters of the problem

In a nondimensional representation parameters of the problem are (note: the overall reference length is $L_R^* = L^*$):

(1) Step length

$$l = \frac{l^*}{L^*}. \tag{48}$$

(2) Step height

$$h = \frac{h^*}{L^*}. \tag{49}$$

(3) Reynolds number

$$Re = \frac{U_\infty^* L^*}{\nu^*}. \tag{50}$$

(4) Prandtl number
$$Pr = \frac{\mu^* c_p^*}{k^*} . \tag{51}$$

In most cases the Reynolds number will be scaled out of the problem (asymptotic theories for $Re \to \infty$). However, it is a parameter of the interaction process in the IBL procedure, see equation (10) together with equation (21), and of course it is an explicit parameter of the Navier–Stokes equations (1)–(3).

## Limiting solutions for $Re \to \infty$

Asymptotic theories for $Re \to \infty$ are all based on perturbations of the (often singular) solutions at $Re = \infty$. Therefore it is important to know how the flow looks like in this limit. In figure 4 two possible solutions for $Re \to \infty$ are shown together with the pressure distribution for a particular set of parameters $l$, $h$.

Which of the two solutions is relevant in the limit $Re \to \infty$ depends on which limit of the parameter
$$\kappa = \frac{\delta_1^*(x^* = 0)}{h^*} \tag{52}$$
is assumed. For $\kappa \to \infty$ the boundary layer thickness (displacement thickness) is large compared to the step height. As a consequence the recirculation region (if it exists) is part of the boundary layer.

For $\kappa \to 0$ the oncoming boundary layer is small compared to the step height. It separates as a whole and surrounds a basically non–viscous recirculation region.

Both asymptotic limits are quite different in character. For details of the two limits and their asymptotic structure see [12] or [13].

In this study $\kappa \to \infty$ is assumed which corresponds to case (a) in figure 4. Since $\delta_1^* \to 0$ for $Re \to \infty$, $h^*$ must be linked to the Reynolds number in a rational theory for $Re \to \infty$ under the assumed limit behaviour $\kappa \to \infty$. This explicitely is done in the TDT concept, see equations (29) and (30).

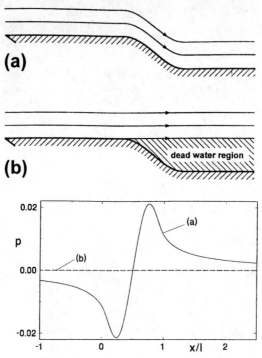

Fig. 4  Limiting solutions for $Re \to \infty$
(a) $\kappa \to \infty$ (this study)
(b) $\kappa \to 0$ (one of several possible solutions, see [12])

## Results of the IBL approach (unique steady solutions)

In figures 5a–c flow and heat transfer results in terms of $c_f Re_x^{1/2}$, $p$ and $q = Nu Re_x^{-1/2}$ with $Re_x = Re(1 + x^*/L^*)$ are shown for systematic parameter variations around the standard case

$$\text{S1:} \quad l = 1.0\,; \quad h = 0.016\,; \quad Re = 10^6\,; \quad Pr = 1.0. \tag{53}$$

This standard case S1 is considerably influenced by the wall contour but still fully attached. A qualitatively similar behaviour (further decrease in minimum wall shear, onset of recirculation) occurs for

(a) increasing step height
(b) decreasing step length
(c) increasing Reynolds number.

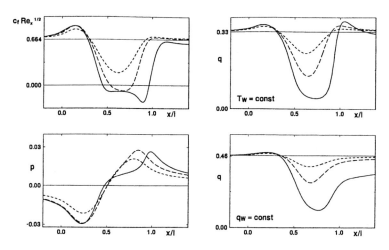

Fig. 5a   Variation of $h$ with respect to case S1
- - - - - $h = 0.016$ (standard case S1)
- - - - $h = 0.022$
——— $h = 0.025$

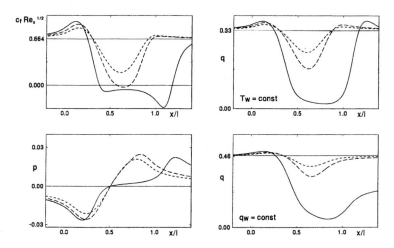

Fig. 5b   Variation of $l$ with respect to case S1
- - - - - $l = 1.0$ (standard case S1)
- - - - $l = 0.8$
——— $l = 0.6$

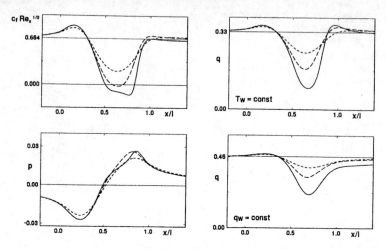

Fig. 5c  Variation of $Re$ with respect to case S1
- - - - - $Re = 1.0 \cdot 10^6$ (standard case S1)
- - - - $Re = 1.5 \cdot 10^6$
——— $Re = 3.0 \cdot 10^6$

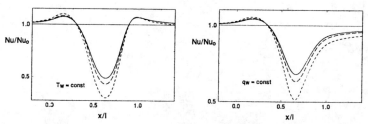

Fig. 5d  Variation of $Pr$ with respect to case S2
——— $Pr = 0.5$
- - - - $Pr = 1.0$ (standard case S2)
- - - - - $Pr = 5.0$
$Nu_0$:  flat plate heat transfer

The physical explanation for this behaviour is that in all three cases the steps become steeper (increase of $h/l$). In cases a and b this is obvious, in case c the increase of Reynolds number reduces the "smoothing" effect of finite Reynolds numbers on the so-called effective wall contour, i.e. the physical wall contour $y_c$ enlarged by displacement thickness $\delta_1$. Thus, effectively the contour becomes steeper.

Further details of the results are discussed later, especially details of heat transfer in figures 5a–c or in figure 5d, where the influence of Prandtl number is shown with respect to the standard case

$$\underline{S2}: \quad l = 1.0\, ;\quad h = 0.022\, ;\quad Re = 10^6\, ;\quad Pr = 1.0. \tag{54}$$

This standard case is characterized by a recirculation region within the step area.

## Results of the TDT approach (unique steady solutions)

In the TDT approach only three parameters are left, since the Reynolds number is completely scaled out of the problem (as long as we restrict ourselves to the leading order of triple deck equations). They are the step length $l_T$ and the step height $h_T$ according to equation (29) and (30), respectively, and the Prandtl number.

In figures 6a,b flow and heat transfer results for a parameter variation around the standard case

$$\underline{S3}: \quad l_T = 1.0\, ;\quad h_T = 1.0\, ;\quad Pr = 1.0 \tag{55}$$

are shown in terms of $\tau_T$, $p_T$, $q_{T,T}$ and $q_{q,T}$ according to table 3. Since $X$ and $l_T$ are scaled likewise results can be shown as a function of $x/l = X/l_T$. Like with the IBL approach the onset of recirculation is provoked by increasing step heights as well as by decreasing step lengths.

Results from the IBL and TDT approach can be compared for particular finite Reynolds numbers taking into account the different scalings of the two theories. For example, the triple deck case $h_T = 3.0$, $l_T = 1.0$ corresponds to the interacting boundary layer solutions

(1) $\quad H = 0.005:\quad l = 0.053\ \text{for}\ Re = 10^5$,

(2) $\quad H = 0.022:\quad l = 0.125\ \text{for}\ Re = 10^4$.

In figure 7 the corresponding results from both theories are shown. As a general trend coincidence increases for increasing Reynolds numbers. This is to be expected since both theories are exact in the limit $Re = \infty$.

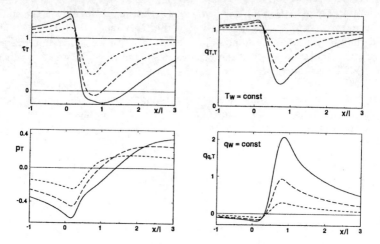

Fig. 6a  Variation of $h_T$ with respect to case S3
- - - - - $h_T = 1.0$ (standard case S3)
- - - - $h_T = 2.0$
——— $h_T = 3.0$

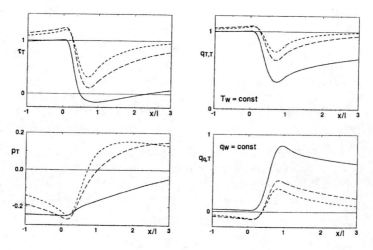

Fig. 6b  Variation of $l_T$ with respect to case S3
- - - - - $l_T = 1.0$ (standard case S3)
- - - - $l_T = 0.5$
——— $l_T = 0.2$

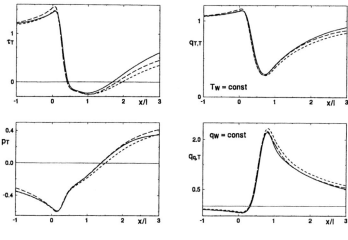

Fig. 7  Comparison of asymptotic theories ; $h_T = 3.0$; $l_T = 1.0$; $Pr = 1.0$
——— TDT ($Re \to \infty$)
- - - - IBL ($Re = 10^5$)
· · · · · IBL ($Re = 10^4$)

## Results of TDT approach (multiple steady solutions)

If for a certain parameter combination more than one solution exists this indicates a folding of the surface of solutions in a three–dimensional representation of results. In figure 8 a sketch of $p_{max}(h_T, l_T)$ is shown which exhibits such a behaviour. Usually only solutions A and C are stable so that two different solutions are expected in parameter ranges in which more than one solution is possible. In [5] multiple solutions over contoured walls were found by the IBL approach. From these results we draw the conclusion that multiple solutions in the TDT approach should exist for rather large values of the parameters $h_T$ and $l_T$.

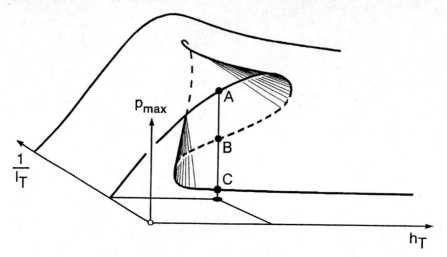

Fig. 8    Sketch of multiple solutions with a folded surface of solutions

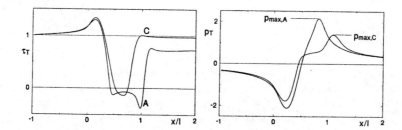

Fig. 9    Multiple triple deck solutions for the parameter combination $h_T = 79;\ l_T = 50$.

Indeed they exist, for example, for the parameter combination $h_T = 79;\ l_T = 50$, see figure 9. In figure 10 this solution is imbedded in the parameter plane $(h_T, l_T)$. The shaded area is the projection of the folded part, c.f. figure 9. Above the broken line flow seperation occurs, below the flow is completely attached. Again, these results are provided for a later interpretation.

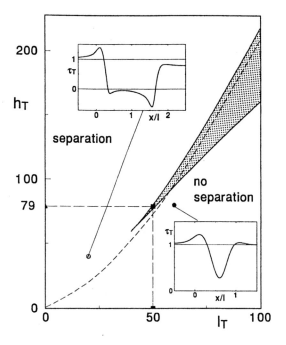

Fig. 10  Characteristic of solutions in the $(h_T, l_T)$-plane
shaded area: multiple solutions
- - - - boundary between separated and non-separated flows
■ multiple solution, see figure 9

# MOMENTUM AND HEAT TRANSFER IN BOUNDARY LAYERS OVER CONTOURED WALLS: APPLYING THE STEADY AND UNSTEADY NAVIER–STOKES EQUATIONS

### Steady solutions compared to asymptotic results

In a steady version the Navier–Stokes equations (1)–(3) together with the thermal energy equation can be used to find steady solutions for finite Reynolds numbers. They don't neglect any terms like the asymptotic theories do. But, they are still approximations with respect to their numerical solutions on finite grids.

In figure 11 two triple deck solutions are compared to the corresponding Navier–Stokes solutions. In figure 11a the coincidence is good with the expected trend: an improved coincidence for increasing Reynolds numbers. But, this trend is not observed in figure 11b which shows results for a steeper step ($h_T = 3.0$ instead $h_T = 1.0$ in figure 11a). Obviously the expected trend with increasing Reynolds numbers is concealed by the rapidly decreasing coincedence with increasing step height as figure 12 shows (increasing step height at constant Reynolds number). This happens because finite step heights are an approximation with respect to the second parameter involved in the asymptotic process, namely $\kappa$ according to equation (52). With $\delta_1 \sim Re^{-1/2}$ and $h \sim Re^{-5/8}$ the asymptotic order of $\kappa$ is $\kappa = O(Re^{1/8})$.

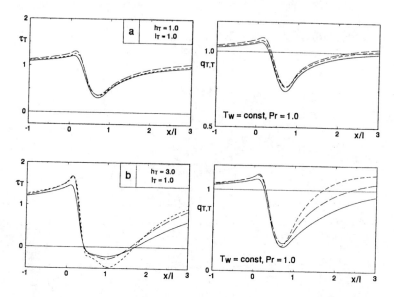

Fig. 11  Comparison of TDT and Navier–Stokes solutions
———— TDT ($Re \to \infty$)
- - - - - NS ($Re = 5 \cdot 10^5$)
– – – – NS ($Re = 10^4$)

That means, $\kappa \to \infty$ for $Re \to \infty$. but with a small exponent so that for finite Reynolds numbers $\kappa$ is still small if $h_T$ (c.f. equation (30)) is not a small number. Descrepancies due to finite values of $\kappa$ especially occur in the downstream region. This is important for the interpretation of global momentum and heat transfer effects later on.

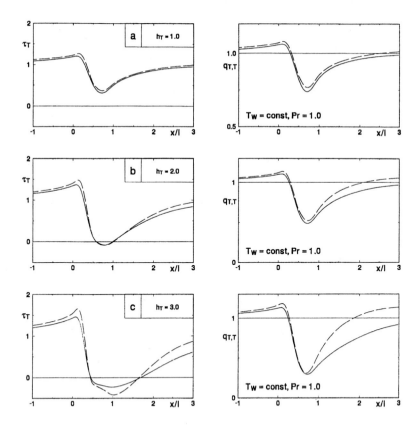

Fig. 12   Comparison of TDT and Navier–Stokes solutions for increasing step height; $l_T = 1.0$
- - - - NS ($Re = 10^5$)
———— TDT ($Re \to \infty$)

**Unsteady periodic solutions**

So far only steady solutions have been calculated. They are solutions of the basic equations for abitrarily high Reynolds numbers. However, they are not observed in real flows if the Reynolds number exeeds certain values. Then, obviously unsteady solutions also exist and prevail against the steady ones.

As mentioned before already, our hypothesis is that linear stability behaviour of a flow is intimately related to the existence and probably onset of unsteady, oscillating solutions. In order to examine this hypothesis we choose two steady solutions which are close to the onset of instabilities, one being stable everywhere and one with unstable regions.

Stability considerations were based on the "universal" stability criterion provided in figure 2. From this figure with the local value of $\beta_3(x)$ the critical value $Re_{2,cr}(x)$ can be deduced. From this, together with the global Reynolds number of the flow, the local critical Reynolds number follows according to

$$Re_{cr}(x) = \frac{(Re_{2,cr}(x)/\beta_2(x))^2}{1+x}. \tag{56}$$

If $Re_{cr}(x) > Re$ the flow is locally unstable. The basic assumption of this analysis is the local validity of the parallel flow assumption. It is known that non–parallel effects modify the stability behaviour only moderately, see for example [14].

In figure 13 the flow with $h = 0.02$ is stable everywhere, whereas the flow with $h = 0.05$ is unstable within the region $0.61 \leq x/l \leq 1.29$.

Solutions of the unsteady Navier–Stokes equations (1)–(3) for these two cases show the expected behaviour. With initial conditions compatible with the steady version of equations (1)–(3) for $h = 0.02$ only the steady solution could be found for large times. For $h = 0.05$, however, under the the same initial conditions an unsteady solution evolves with increasing amplitudes for small and moderate times and an amplitude saturation for large times t. In figure 14 two parts of time dependent oscillating wall shear stress results are shown. one in the developing range ($100 \leq t \leq 110$) and one for saturation ($400 \leq t \leq 410$).

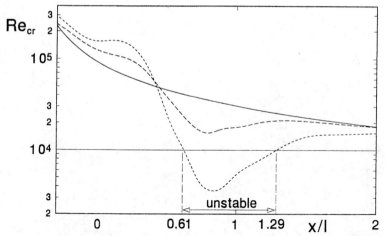

Fig. 13   Stability behaviour locally based on the universal stability criterion according to figure 2 ( $Re = 10^4$; $l = 2.0$)
——— flat plate ($h = 0$)
- - - - $h = 0.02$
· · · · · $h = 0.05$

Since no time dependence is introduced by means of the initial or boundary conditions we call this behaviour "self–sustained oscillations".

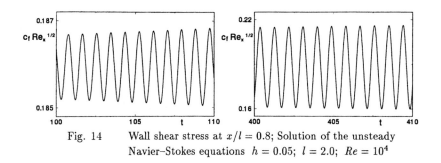

Fig. 14   Wall shear stress at $x/l = 0.8$; Solution of the unsteady Navier–Stokes equations $h = 0.05$; $l = 2.0$; $Re = 10^4$

# MOMENTUM AND HEAT TRANSFER IN BOUNDARY LAYERS OVER CONTOURED WALLS: DISCUSSION OF FIVE PARTICULAR ASPECTS

In the previous chapters results from different theories and for a variety of parameters are provided. Several conclusions of general importance with respect to boundary layer flows over contoured walls can be drawn from these examples. Five aspects will be discussed in the following.

### (1) Asymptotic versus Navier–Stokes solutions

In the Reynolds number range $0 < Re < \infty$ numerical solutions of the Navier–Stokes equations and asymptotic solutions are best suited at different ends. Asymptotic solutions are exact at $Re = \infty$ with increasing degrees of approximation for smaller Reynolds numbers. Navier–Stokes solutions on the other hand get into trouble for increasing Reynolds numbers mainly due to the occurance of steep gradients close to walls and shear layers.

In figures 11 and 12 we nevertheless could demonstrate the existence of a Reynolds number range in which results of both theories satisfactorily coincide. Deviations must be judged in the light of numerical as well as of asymptotic approximations with respect to the two parameters $Re \to \infty$, $\kappa \to \infty$.

Conclusion (1): Triple deck results are appropriate asymptotes for Navier–Stokes solutions in the high Reynolds number limit, the better, the smaller the contour parameter (here: $h_T$) is.

### (2) Global effects of transverse vortices

An important question with respect to practical applications is how transverse vortices affect the overall momentum and heat transfer in the vicinity of a contoured wall. The answer given by the asymptotic triple deck theory is as clear as unexpected: not at all! Though there are $O(1)$ changes in pressure, wall shear stress and temperature the overall effect is zero, at least to the leading triple deck order (finite effects can only be higher order effects in terms of the triple deck theory). As far as the flow field is concerned this result was deduced in [15] and [16]. For a numerical example with a dented wall see [17].

In [18] this triple deck result could be confirmed on the basis of the IBL approach and extended to the heat transfer problem. The crucial part of the analysis is that downstream of the contour changes $lim_{x\to\infty}(I_1 - I_2) = lim_{x\to\infty}(I_3) = 0$ must hold with the definitions

$$I_1(x) = \sqrt{Re} \int_0^x p \frac{dy_c}{dx} dx \, , \quad I_2(x) = 2\int_0^x \sqrt{x} \frac{df_w''}{dx} dx \, , \quad I_3(x) = 2\int_0^x \sqrt{x} \frac{d\Theta_w'}{dx} dx. \quad (57)$$

For a certain set of parameters $I_1$, $I_2$ and $I_3$ can be evaluated numerically, see figure 15. Though an extremely high numerical resolution is necessary the asymptotic findings are well confirmed.

From these results we can conclude that finite, nonzero global effects can only occur due to finite Reynolds number influence (asymptotically: higher order effects) or due to a restricted range of integration. In figure 15 integration must reach very far downstream to meet the asymptotic requirements.

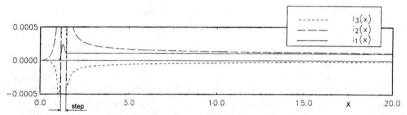

Fig. 15    Numerical integration of auxiliary integrals $h = 0.005$; $l = 0.5$; $Re = 10^6$; $Pr = 1.0$; $T_W = const$

If, now, integration ends a few characteristic lengths $l$ downstream of the step because the trailing edge of a finite plate is reached, for example, non-zero global effects are obvious. (Note that the trailing edge effect of a finite plate on total drag and heat transfer is asymptotically small, see for example [8]).

Based on the Navier-Stokes solutions of this study global effects over contoured walls of finite length are determined in terms of

$$c_{total} = \int_0^{L_t} c_f(x)dx + \int_0^{L_t} c_p(x)\frac{dy_c}{dx}dx \, , \quad Nu_{total} = \int_0^{L_t} Nu(x)dx \quad (58)$$

with

$$c_f = \frac{\tau_w^*}{\varrho_\infty^* U_\infty^{*\,2}} \, , \quad c_p = \frac{p^*(x) - p_\infty^*}{\varrho_\infty^* U_\infty^{*\,2}} \, , \quad Nu(x) = \frac{q_w^*(x) L^*}{\lambda^*(T_w^* - T_\infty^*)}. \quad (59)$$

Here, $L_t$ is the trailing edge distance from the leading edge of the contoured wall.

As an example figure 16 shows deviations from the flat plate results for $c_{total}$ and $Nu_{total}$ due to the contour effect on a plate of finite length.

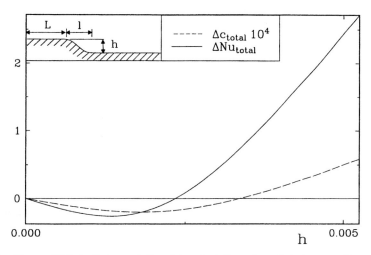

Fig. 16  Global effects on finite plates; deviations from flat plate results;
$l = 0.0529$; $Re = 10^5$; $Pr = 1.0$; $L_t = 1.5$
- - - - $\Delta c_{total} = c_{total} - c_{total,flatplate}$
——— $\Delta Nu_{total} = Nu_{total} - Nu_{total,flatplate}$

For small values of $h$ both global values are smaller for the contoured wall. Especially for the heat transfer result this means a decrease in heat transfer! This was to be expected from the local heat transfer distribution, see figure 5a for the corresponding IBL results, since heat transfer is considerably decreased in the recirculation zone with only minor overshoot effects upstream and a slow regain of the overall heat transfer rate in a downstream region far beyond the integration range of figure 16. for example. From figure 5d we conclude that this result is only weakly affected by variation of the Prandtl number.

For larger values of $h$, however. from the Navier–Stokes results a different trend follows: Heat transfer is increased by a contoured wall. Obviously effects that are of higher order in the triple deck theory for larger values of $h$ become important. They manifest themselves in an increasing deviation of downstream heat transfer for increasing values of $h$, c.f. figure 12.

Conclusion (2): The global effect of wall contour changes (that can lead to steady laminar transverse vortices) on heat transfer over a plate of finite length is:

- a decrease for small values of $h$
- an increase for larger values of $h$.

An increase is not predicted by (first order) triple deck theory, but by the Navier–Stokes equations.

For practical applications this means: Accidental small contour changes must be avoided in situations where high heat transfer rates are important since they may lead to a decrease in heat transfer. On the other hand, wall contouring which leads to strong recirculation can be used for heat transfer enhancement.

## (3) Influence of variable properties

Assuming constant properties as we did so far is an approximation when temperature and/or pressure vary within the flow field. Asymptotically this assumption corresponds to the leading order term of a regular pertubation expansion with respect to two small parameters, one for temperature and one for pressure effects.

According to the regular nature of the expansion moderate temperature changes, for example, will cause only moderate changes in flow and heat transfer quantities. As an example we calculated the influence of temperature dependent viscosity assuming all other properties to be constant. Water will basically behave according to this assumption. For small or moderate temperature changes there are two ways of treating the problem asymptotically, see [21]:

a) by the expansion method: based on the Taylor series expansion of the properties all quantities are expanded likewise. A hierarchy of equations is deduced with the constant property case corresponding to the leading order.

b) by the combined method: based on the known asymptotic structure of the final result the unknown coefficients are determined from a finite number of accurate numerical results of the full problem (including variable properties).

Applying method b) in a first step we determine the asymptotic form, for example, of the skin friction:

$$\frac{c_f}{c_{f_0}} = 1 + \varepsilon K_\mu A_\mu + \varepsilon^2 (K_{\mu 2} A_{\mu 2} + K_\mu^2 A_{\mu\mu}) + O(\varepsilon^3), \tag{60}$$

with

$$\varepsilon = \frac{T_w^* - T_\infty^*}{T_\infty^*}, \quad K_\mu = [\frac{\partial \mu^*}{\partial T^*} \frac{T^*}{\mu^*}]_\infty, \quad K_{\mu 2} = [\frac{\partial^2 \mu^*}{\partial T^{*2}} \frac{T^{*2}}{\mu^*}]_\infty. \tag{61}$$

In a second step we determine $A_\mu(x)$ (and $A_{\mu 2}(x)$, $A_{\mu\mu}$ for second order results) from numerical results shown in figure 17 for $\tau_T(x)$ by evaluating

$$A_\mu = \left[\frac{\partial(c_f/c_{f0})}{\partial(\varepsilon K_\mu)}\right]_0, \quad A_{\mu 2} = \left[\frac{\partial(c_f/c_{f0})}{\partial(\varepsilon^2 K_{\mu 2})}\right]_0, \quad A_{\mu\mu} = \frac{1}{2}\left[\frac{\partial^2(c_f/c_{f0})}{\partial(\varepsilon K_\mu)^2}\right]_0. \tag{62}$$

For the minimum skin friction, for example, we thus get

$$\frac{c_{fmin}}{c_{fmin_0}} = 1 - 2.36 K_\mu \varepsilon + O(\varepsilon^2). \tag{63}$$

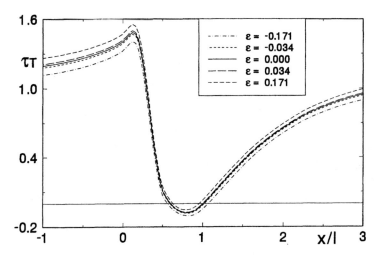

Fig. 17    Influence of temperature dependent viscosity on skin friction results
$h = 0.0034$; $l = 0.0529$; $Re = 10^5$; $Pr = 0.7$; $K_\mu = 0.775$

Conclusion (3): The effect of variable properties is asymptotically small. It can be accounted for in a systematic way by applying regular pertubation procedures.

## (4) Multiple solutions and their physical interpretation

In figures 8, 9 and 10 multiple steady solutions of the basic equations were shown. In the theory of nonlinear equations multiple solutions are classified in different categories, see for example [20]. The category which is relevant for our case is the so-called cusp singularity (or: cusp catastrophe). In figure 18 it is shown how the cusp appears as a projection of the folded surface of the solutions. Comparing this to figure 10 elucidates the character of the multiple solutions. However, one should keep in mind that by this classification no deeper understanding of the physical mechanisms involved is achieved.

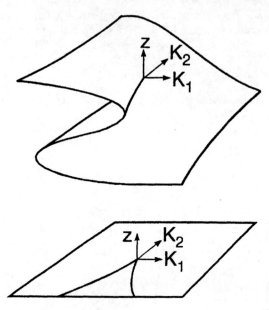

Fig. 18   Cusp singularity; the cusp appears after a projection into the $K_1$, $K_2$-plane

The basically topological information about the solutions is nevertheless important to understand why in certain parameter ranges more than one solution is possible.
Conclusion (4): Multiple solutions of the kind found here can be interpreted by means of the cusp singularity.

### (5) Unsteady solution of 2D equations

In figure 14 an unsteady two-dimensional solution was shown for a particular parameter combination. If these parameters are kept fixed except the Reynolds number, three distinct parts of the Reynolds number range can be identified when steady boundary and initial conditions are imposed:

(a) $Re < 9 \cdot 10^3$: Only steady solutions are found

(b) $9 \cdot 10^3 < Re < 2.5 \cdot 10^4$: Due to a locally unstable flow field (c.f. figure 13) in the downstream part of the step geometry unsteady, oscillating solutions are found in this part of the flow field (self-sustained oscillations). Further downstream they are damped by a mean flow which is stable in character.

(c) $2.5 \cdot 10^4 < Re$: No solutions of the unsteady two–dimensional equations are found. Obviously the equations are no longer adequate for the underlying physics which probably is three–dimensional in character. Corresponding steady solutions are also not found with the unsteady equations. They only emerge from the steady equations, at least in this parameter range.

The overall behaviour in the Reynolds number ranges (a)–(c) is compatible with the physics of the transition process on the flat plate, for example: steady laminar flow for small Reynolds numbers $Re_x$, two–dimensional (Tollmien–Schlichting) waves for moderate $Re_x$ and three–dimensional turbulent flow for large enough $Re_x$.

Conclusion (5): By means of the two–dimensional unsteady Navier–Stokes equations solutions with self-sustained oscillations are found in a rather narrow Reynolds-number range only. Their existance is compatible with predictions made by the linear stability theory.

# MOMENTUM AND HEAT TRANSFER OF INTERNAL FLOWS THROUGH DIFFERENT GEOMETRIES

Two different types of geometry will be considered for internal flows: local contour changes of finite extend and periodically changing contours of infinite extend.

A diffuser geometry is chosen as a prototype for the first category a wavy wall for the second. In both cases separation is pressure induced. Slender channel theory (SCT), see the chapter "Basic equations and concepts", is used as a basis for asymptotic considerations in the double (distinguished) limit of infinite Reynolds numbers and asymptotically slender channels.

**Contour changes of finite extend: diffuser flow**

Taking into account the slender channel scaling, c.f. equations (31) and (32), the model diffuser is:

$$H(x) = \frac{1}{2}\left[3 + \tanh\left(\frac{\lambda x}{Re}\right)\right] \tag{64}$$

with

$$H = \frac{H^*}{L_R^*}; \qquad x = \frac{x^*}{L_R^*}; \qquad Re = \frac{Q^*/B^*}{2\nu^*}; \qquad Q^* = 2\bar{u}^* B^* H^*. \tag{65}$$

The reference length $L_R^*$ is the minimum half channel width, $B^*$ is the lateral extend (being infinite in the 2D case), the oncoming profile far upstream is assumed to be of Hagen–Poiseuille type, i.e. $u = 1.5(1 - y^2)$.

With the additional auxiliary parameter $\lambda$ ($\lambda = O(1)$ asymptotically) arbitrary degrees of slenderness can be chosen for each particular Reynolds number under consideration. In figure 19 three different diffuser flows are shown in terms of $\tau = H^2 c_f Re$ ($c_f = 2\tau_w^*/(\varrho^* \bar{U}^{*2})$, $\bar{U}^*$: mean velocity over the channel) and $Nu_H = -q_W^* H^*(x^*)/k^* \Delta T_W^*$; $\Delta T_W^* = T_{Wu}^* - T_{Wl}^* = const$, for the thermal boundary condition of constant, but of course different temperatures of the upper and the lower wall, respectively. As far as heat transfer is concerned, again, like with the step geometry, a considerable decrease occurs in the recirculation zone which is compensated by slight to moderate overshoots downstream. Though there is no analytical verification for a complete compensation (c.f. the smooth step case) physical considerations lead to the conclusion that asymptotically small single contour changes for internal flows should behave like those in external flows.

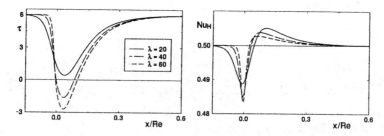

Fig. 19    Shear stress and heat transfer in symmetrical diffuser flows; slender channel theory ($Re \to \infty$); $Pr = 1.0$

In table 4 results of a numerical integration over the range shown in figure 19 are listed. Deviations from the straight channel case ($Nu_H = 0.5$) are calculated in two parts: $\Delta Nu_{loss}$ for the x-range in which $Nu_H < 0.5$ and $\Delta Nu_{gain}$ for the rest of it.

Table 4:    Global heat transfer effect of the diffuser flows in figure 18: $\Delta Nu = \int (Nu_H - \frac{1}{2}) dx$

| $\lambda$ | $\Delta Nu_{loss}/Re$ | $\Delta Nu_{gain}/Re$ | $\Delta Nu_{total}/Re$ |
|---|---|---|---|
| 20 | $9.006 \cdot 10^{-4}$ | $8.912 \cdot 10^{-4}$ | $-9.425 \cdot 10^{-6}$ |
| 40 | $6.520 \cdot 10^{-4}$ | $6.445 \cdot 10^{-4}$ | $-7.525 \cdot 10^{-6}$ |
| 60 | $4.840 \cdot 10^{-4}$ | $4.780 \cdot 10^{-4}$ | $-6.040 \cdot 10^{-6}$ |

**Contour changes of infinite extend: wavy wall flow**

In figure 20 a channel with periodically wavy walls is shown with contour changes according to

$$H(x) = 0.75 + 0.25 \cos(2\pi \frac{\lambda x}{Re}) \quad \text{for} \quad x > -L_E. \tag{66}$$

For $x < -L_E$ the channel has straight walls ($H = 1$). The oncoming velocity profile again is $u = 1.5(1 - y^2)$. In figure 21 the periodicity of the (skin friction and heat transfer) results after only a few "waves" can be seen ($\tau$ and $Nu_H$ have the same definition as for the diffuser geometry).

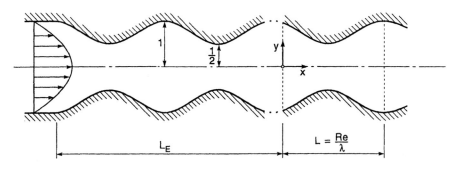

Fig. 20    Channel with wavy walls

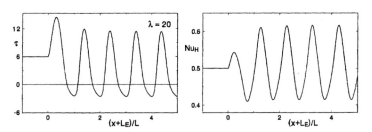

Fig. 21    Shear stress and heat transfer in wavy wall flows; slender channel theory ($Re \to \infty$); $Pr = 1.0$

As far as heat transfer is concerned the "gain part" of the integrated effect obviously is larger than the "loss part". This is evaluated numerically within one period shown in figure 22. For four different degrees of slenderness the overall heat transfer is numerically analyzed in table 5 in the same way as for the diffuser case (see table 4). Obviously, for contour changes of infinite extend global heat transfer effects are non-zero.

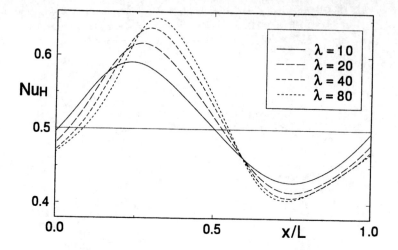

Fig. 22  Heat transfer in one period of a wavy wall; slender channel theory ($Re \to \infty$); $Pr = 1.0$

This is not realy surprising if one takes into account that compensation of heat transfer effects from contour changes of finite extend only occurs over ranges that reach very far downstream. If, due to continuous changes in wall geometry this compensation is suppressed obviously finite effects are possible.

Table 5:  Global heat transfer effect of the wavy wall flow
$\Delta Nu = \int (Nu_H - \frac{1}{2}) dx$

| $\lambda$ | $\Delta Nu_{loss}/Re$ | $\Delta Nu_{gain}/Re$ | $\Delta Nu_{total}/Re$ |
|---|---|---|---|
| 10 | 0.027 | 0.023 | $-0.004$ |
| 20 | 0.034 | 0.028 | $-0.006$ |
| 40 | 0.038 | 0.030 | $-0.008$ |
| 80 | 0.039 | 0.031 | $-0.008$ |

Our results for internal flows all were gained by the slender channel theory, i.e. they hold for asymptotically small contour changes. Whether Navier–Stokes solutions for finite contour changes will show a different trend is still an open question.

# CONCLUSIONS

Two main conclusions governing laminar heat transfer over smooth wall contours with separation can be drawn from the result of this study:

- Asymptotic theories for $Re \to \infty$ taking into account the viscous–inviscid interaction are adequate tools for analyzing this kind of flow situations. For large but finite Reynolds numbers the asymptotic results are approximations to the problem, still good enough to give reasonable coincidence with the solutions of the full Navier–Stokes equations which due to their inherent numerical approximation become less accurate for increasing Reynolds numbers. Thus no "Reynolds number gap" appears between Navier–Stokes and asymptotic solutions. This also means, that the asymptotic results may serve as asymptots to numerical solutions for increasing Reynolds numbers. Since, however, asymptotic theories for viscous–inviscid interaction also require $\kappa \to \infty$, i.e. contour changes occur in the near wall region of the boundary layer, there is a rather strong restriction on admissible heights of the contour changes. For larger heights especially heat transfer results should be based on the full Navier–Stokes equations.

- As far as heat transfer is concerned transverse vortices lead to a considerable decrease in heat transfer rate in the vicinity of the recirculation region. This is counteracted by a moderate but far reaching increase of heat transfer downstream of the recirculation region. Within the (first order) asymptotic theory integration to $x \to \infty$ shows a complete compensation of these two effects with the consequence of zero change of global heat transfer. Integration to finite $x$ only, as a consequence, results in a decrease of global heat transfer.

For larger wall contour changes solutions of the Navier–Stokes equations show that obviously higher order effects become important. Global heat transfer then can be increased above the level of heat transfer over a plane wall.

The situation, even for asymptotically small contour changes, obviously is different for periodic wall contour changes of infinite extend downstream. This was demonstrated for a wavy channel geometry. This contours should be studied in future investigations with respect to their potential for heat transfer enhancement.

# APPENDIX: SOME DETAILS OF THE NUMERICAL SOLUTIONS

- **Navier–Stokes equations**

    (1) program:     FIVO 2D (Lehrstuhl für Wärme– und Stoffübertragung)
        grid:        body fitted grid
        algorithm:   SIP (SIMPLEC)

    (2) program:     FLOW 3D (Computational Fluid Dynamics Services (CFDS))
        grid:        body fitted grid (300*150)
        algorithm:   SIP (momentum, continuity and energy equation)
                     ICCG (pressure)
                     SIMPLEC (velocity–pressure coupling algorithm)

- **Triple deck theory (TDT)/Interacting boundary layer theory (IBT)**

    program:        —
    discretization: Keller–Box scheme
    grid:           250*150
    algorithm:      quasi simultaneous method

- **Slender channel theory (SCT)**

    program:        —
    discretization: Keller–Box scheme
    grid:           200*100
    algorithm:      Newton method

- **Linear stability theory (LST)**

    program:   —
    method:    Gram-Schmidt Orthonormalization
    algorithm: Runge–Kutta shooting method

# References

[1] Kakac, S., Yüncü, H., Hijikata, K.: "Cooling of Electronic Systems", Proc. Nato Adv. Study Institute on Cooling of Electronic Systems, Cesme, Izmir, Turkey, June 21– July 2, (1993).

[2] KIM, S.J., LEE, S.W.: "Air Cooling Technology", CRC Press, Boca Raton, FL, (1996).

[3] GERSTEN, K., HERWIG, H. (1984):"Strömungsmechanik/Grundlagen der Impuls–, Wärme– und Stoffübertragung aus asymptotischer Sicht", Vieweg Verlag, Braunschweig (1992).

[4] VAN DYKE, M.:"Pertubation Methods in Fluid Mechanics", The Parabolic Pess, Stanford (1975).

[5] SOMMER, F.: "Mehrfachlösungen bei laminaren Strömungen mit druckinduzierter Ablösung: Eine Kuspen-Katastrophe. Fortschr.-Ber. VDI Reihe 7, Nr. 206, VDI-Verlag, Düsseldorf (1992).

[6] SCHÄFER, P.:"Untersuchung von Mehrfachlösungen bei laminaren Strömungen", Dissertation, Ruhr–Universität Bochum (1995).

[7] VELDMAN, A.E.P.:"A numerical method for the calculation of laminar, incompressible boundary layers with strong viscous-inviscid interaction", NLR TR 79023 U, (1979).

[8] STEWARTSON, K.:"Multistructured boundary layers of flat plates and related bodies", Advances in Appl. Mech., $\underline{14}$ (1974), pp.146–239.

[9] HERWIG, H.:"Wie entsteht die Potenz $Gr^{-9/28}$? Zum physikalischen Hintergrund mehrschichtiger asymptotischer Strukturen", Z. Flugwiss. Weltraumforsch. $\underline{15}$ (1991), pp. 185–191.

[10] VAN DYKE, M.:"Slow variations in continuum mechanics", Advances in Aplied Mechanics, $\underline{25}$ (1987), pp. 1–45.

[11] DRAZIN, P.G., REID, W.H.:"Hydrodynamic stability", Cambridge University Press, Cambridge, UK, (1981).

[12] HERWIG, H.:"Die Anwendung der asymptotischen Theorie auf laminare Strömungen mit endlichen Ablösegebieten", Z. Flugwiss. Weltraumforsch., $\underline{6}$ (1982) pp. 266–279.

[13] HERWIG, H.:"The asymptotic structure of laminar flow with finite regions of separation. In: Computational and asymptotic Methods for Boundary and Interior Layers", ed.: J.J.H. Miller. Proceedings of the Bail II Conf. (1982) pp. 278–284.

[14] NAYFEH, A.H., SARIC, W.S.:"Nonparallel stability of boundary layer flows". Phys. Fluids, vol. $\underline{18}$, No. 8 (1975).

[15] BRIGHTON, P.W.M.:"Boundary layer and stratified flows past obstacles". Dissertation. University of Cambridge, UK (1977).

[16] SMITH, F.T.:"Laminar flow over a small hump on a flat plate". JFM. $\underline{57}$ (1973), pp. 803–824.

[17] HERWIG, H.:"Die Anwendung der angepaßten asymptotischen Entwicklung auf laminare zweidimensionale Strömungen mit endlichen Ablösegebieten". Dissertation, Ruhr–Universität Bochum, Germany (1981).

[18] P. Schäfer, H. Herwig:"Drag and Heat Transfer in Laminar Boundary Layers With finite Regions of Separation", Proceedings of Eurotherm 31, Bochum, 17–22, (1993).

[19] SCHÄFER, P., HERWIG, H.:"Stability of plane Poiseuille flow with temperature dependent viscosity", Int. J. Heat Mass Transfer, $\underline{36}$, No. 9, 2441 - 2448, (1992).

[20] SAUNDERS, P.T.: "Katastrophentheorie",Vieweg, Braunschweig, (1986).

[21] HERWIG, H., SCHÄFER, P.: "A combined pertubation/finite difference procedure applied to temperature effects and stability in a laminar boundary layer",Arch. Appl. Mech, $\underline{66}$, (1996), pp. 264–272.

# TWO-DIMENSIONAL TURBULENT BOUNDARY LAYERS WITH SEPARATION AND REATTACHMENT INCLUDING HEAT TRANSFER

D. Vieth, R. Kiel, K. Gersten
Institut für Thermo- und Fluiddynamik, Ruhr-Universität Bochum
D-44780 Bochum, Germany

## SUMMARY

The essential problem calculating turbulent flows near solid walls with flow reversal is the physically correct modelling of the flow and the heat transfer adjacent to the wall. This includes the proper turbulence modelling as well as the use of correct boundary conditions. The results of an asymptotic approach to this problem for high Reynolds-numbers and the results of measurements are presented. The investigations were restricted to two-dimensional, incompressible and (in the mean) stationary flows.

The high Reynolds number asymptotic theory for turbulent wall-bounded flows leads to the typical three-layer structure involving the viscous wall layer, the fully turbulent part of the boundary layer, and the inviscid outer flow. It is shown that this structure also holds for boundary layers with local separation regions. The solutions in the viscous wall layer and in the overlap region between the wall layer and the fully turbulent flow turn out to be universal. New wall functions for the mean velocity, temperature, the turbulent stress components, the kinetic energy, and the dissipation rate are deduced. These wall functions evolve from a matching process of the viscous wall layer to the fully turbulent part of the boundary layer (in non-separated boundary layers: defect layer). They are clearly influenced by the wall pressure gradient and undergo drastic changes when pressure-induced separation occurs. It turned out that some common turbulence models are asymptotically correct in both attached and separated regions, but only if several model constants are suitably adjusted. As shown, the asymptotic theory leads to relations which can be used to determine those model constants.

A new experimental set-up was built that is capable of generating the specified flow and allows measurements of the flow field and the temperature field as well as the friction and the heat transfer at the wall. The most essential results of the measurements are presented. The analysis of the measurements leads to the empirical functions of universal nature, which are essential parts of the wall functions.

In contrast to the so-called low-Reynolds-number modelling the described theoretical and experimental results allow us to restrict the calculations of these kinds of flows on the fully turbulent part of the boundary layer, for attached as well as for reverse flow regions. The results of the new numerical prediction method applying the derived wall functions and the relations for the model constants agree favourably with the experimental results.

## INTRODUCTION

When flows at high Reynolds numbers are considered and boundary-layer separation occurs due to the pressure gradient, most prediction codes fail to give physically reasonable results in the separated region, especially if the heat transfer is considered. This can be explained by the large gradients of almost every function close to the wall and by the fact that the near-wall distributions of nearly all quantities show drastic changes when separation occurs, as shown in Figure 1 for the velocity. The figure illustrates the flow field under consideration. The theoretical and experimental investigations are restricted to fully turbulent, two-dimensional, and stationary flows. The wall heat flux is low enough to guarantee constant fluid properties. The separation of the boundary layer is not fixed by the geometry but occurs due to the pressure gradient in the inviscid outer region. This pressure gradient emerges in flows over curved surfaces or, as shown here, over plane surfaces with special arrangements at the opposite wall (suction, variable channel height). The separation vortex lies inside in the boundary layer.

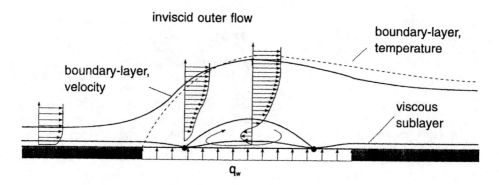

*Figure 1:* Flow field under consideration.

In Figure 1 the three-layer structure of the flow field at high Reynolds numbers is outlined. The flow field consists of the inviscid outer flow, the fully turbulent part of the boundary layer, and the viscous sublayer. The local separation region is located inside the boundary layer. The flow direction in the viscous sublayer of the separated region is reverse to the main flow direction.

The flow in the outer region is inviscid and thus can be described by the potential theory, which gives the streamwise pressure distribution. The viscous sublayer is characterised by the fact that the viscous and turbulent effects are of the same order of magnitude. The results in this layer are universal in that sense that they only depend on local parameters such as the local skin friction and streamwise pressure gradient. Hence the results can be described by analytical functions, e. g. see Gersten, Herwig [1] concerning the velocity and the temperature for $dp/dx = 0$.

At high Reynolds numbers, the flow is dominated by the fully turbulent region. This can be illustrated for an attached boundary layer with a typical Reynolds-number $Re_\delta = U\delta/\nu = 7 \cdot 10^4$: The ratio of the thickness of the viscous sublayer $\delta_\nu$ to the boundary layer thickness $\delta$ is equal to $\delta_\nu/\delta = 0.03\%$. What makes the predictions and the measurements of turbulent wall bounded flows so highly complicated is the unfortunate fact that the essential physical mechanisms take place in this small region near the wall. Here, nearly all gradients do have their maxima and the largest part of the energy losses of the whole boundary layer occurs in this region. The knowledge of the temperature distribution near the wall is necessary to determine the heat transfer from the wall into the flow field.

One major goal of the investigations was the development of a prediction method that describes correctly the aforementioned physical mechanisms. Since the flows under consideration are characterised by high Reynolds-numbers, asymptotic methods were used in connection with the dimensional analysis. This was done essentially for three purposes: to reduce the governing equations, so that they can be solved more easily, to choose and improve turbulence models and to develop appropriate boundary conditions near the wall.

The last point turned out to be the most essential. In principle, there are two strategies to treat the near-wall region numerically: the low-Reynolds-number modelling and the method of wall functions. Whereas the first method is applied to calculate the whole boundary layer, i. e. the fully turbulent part *and* the viscous sublayer, the latter allows one to calculate only the fully turbulent part. In that case the boundary conditions at the wall are replaced by wall functions. As it will be discussed, this method exhibits certain advantages in comparison to the low-Reynolds-number modelling.

The wall functions evolve from the matching of the two near-wall layers. They only depend on local values and are strongly dependent on the pressure gradient and the wall shear stress. They change their structure considerably at separation and reattachment. In this report it is shown for the first time, how generalised wall functions for all field quantities could be developed that are valid in all flow regions of attached or separating flows. These wall functions do have universal character in that sense that they are not restricted to boundary layer flows, but can also be applied to other flows, e. g. internal flows.

Prediction methods of turbulent flows always depend on empirical information. One essential characteristic of prediction methods is the way and the level, how this information is implemented into the mathematical description of the problem. In the method described here empirical functions and constants appear in the wall functions and in the turbulence models. So far, these functions and constants were known sufficiently well only for the case of zero pressure gradient. For all other cases, they have to be deduced from measurements.

The experimental investigations described here were conducted to determine these empirical functions and constants. Therefore detailed measurements of the distributions of the skin friction, the pressure, the heat flux, and the temperature at the wall as well as detailed three-component velocity field and temperature field measurements close to the wall were needed. The experimental apparatus, the measurement techniques, and the essential results will be described.

## THEORETICAL INVESTIGATIONS

### Asymptotic analysis

The method of matched asymptotic expansions has proven to be a very powerful tool in fluid mechanics, see Gersten, Herwig [1]. Undoubtedly the most popular example for this is the formulation of Prandtl's boundary layer theory with the help of the asymptotic high-Reynolds-number theory. see Schlichting, Gersten [40] and Gersten [2]. Asymptotic methods have also been successfully applied to turbulent boundary layer flows. The closed solution scheme formulated by Mellor [3] has been generally accepted to represent a physically and mathematically reasonable description of attached turbulent boundary layers. Extensions of this theory for curved flows have been given by Deriat [4] and Jeken [5]. In contrast to the laminar theory two small parameters arise, which are the reverse of the Reynolds number $\varepsilon$ and the wall shear stress at a reference point $\varepsilon_0$:

$$\varepsilon = \frac{1}{\text{Re}} = \frac{v}{U_{ref} L_{ref}} \quad , \qquad \varepsilon_0 = \frac{u_{\tau,ref}}{U_{ref}} = \frac{\sqrt{\tau_{w,ref}/\rho}}{U_{ref}} \quad . \tag{1}$$

A twofold singular perturbation problem follows leading to the 3-layer structure in the case of attached turbulent outer flows. As already outlined in Figure 1 this structure includes an inviscid outer region, a fully turbulent (defect-) region, and the viscous wall layer.

As a result of the matching procedure in the overlap region of the two near-wall layers it can be shown that $\varepsilon_0$ depends on $\varepsilon$ in the following manner (skin friction formula):

$$\varepsilon_0 \propto \frac{1}{\ln(\varepsilon_0^2/\varepsilon)} \qquad \Rightarrow \qquad \lim_{\varepsilon \to 0} \varepsilon_0 = 0 \quad . \tag{2}$$

Thus we do have a one-parameter problem for attached flows, in principle.

Mellor's scheme fails in the presence of very strong adverse pressure gradients, particularly when separation occurs. The difficulty to incorporate Mellor's theory into a more general high Reynolds number asymptotic theory, which is valid for both attached *and* separated boundary layers, can be traced back to the problem to find a physically reasonable single set of perturbation parameter. Klauer [6] presented an asymptotic solution scheme for the Stratford flow (boundary layer flow with $\tau_w = 0$ over a finite region) and for Couette-Poiseuille flows with $\tau_w = 0$. He found out that instead of $\varepsilon_0$ the parameter $\alpha$ is a suitable small parameter for the layer adjacent to the wall:

$$\alpha \propto \frac{\sqrt{-\overline{u'v'}\big|_{max,ref}}}{U_{ref}} \qquad \Rightarrow \qquad \lim_{\varepsilon \to 0} \alpha \neq 0 \quad . \tag{3}$$

This parameter does not vanish for $\varepsilon \to 0$, so that a real two-parameter exists.

Klauer's theory is partially influenced by the work of Melnik [7] and [8]. The parameter $\alpha$ can be identified either as a characteristic magnitude of the turbulent fluctuation velocity or as an appropriate turbulence model constant. Melnik ended up with a new 4-layer-structure equally valid for both the defect flow region and the separated region. However, according to the opinion of the present authors, the reasoning for introducing a new so-called equilibrium layer is mainly due to the chosen turbulence model. Secondly Melnik's work does not clearly show, how the matching conditions between the two near-wall layers - and thus the boundary conditions for the resulting differential equations in these layers - emerge *without* making any assumptions concerning the turbulence model. An asymptotic theory should - if possible - be independent of turbulence modelling. Turbulence models should be consistent with the conditions resulting from the asymptotic theory.

In this paper only Melnik's idea of the new perturbation parameter $\alpha$ is adopted. This parameter is already known from the analysis of free shear layers, see Schneider [9]. The parameter $\alpha$ characterises the slenderness of shear layers. Mainly experimental results, but also some numerical investigations show that turbulent free shear layers are slender in the sense of a low spreading rate. It is in fact very plausible to use $\alpha$ in this context, because we do have a free shear layer beyond the point of separation, see Figure 2.

The slenderness parameter $\alpha$ is a measure for the turbulent fluctuations and can be considered as a small parameter of the turbulence model. For example in the model of Cebeci and Smith [41] this parameter can be very easily identified. It is typical for free shear layers that $\alpha$ is independent of the Reynolds number.

In contrast to Klauer and Melnik it could be shown in the present study that a 3-layer structure suffices. This corresponds to the theory of Afzal [10], who used a so-called pressure velocity in the region of vanishing wall shear stress:

$$U_p = \left(\frac{\delta}{\rho}\frac{dp_w}{dx}\right)^{1/2}. \tag{4}$$

Relating the turbulent stresses in the governing equations (i. e. the Navier-Stokes equations and the Reynolds stress equations) to the pressure velocity at a reference point leads to the non-dimensional parameter

$$\varepsilon_\infty = \frac{U_{p,ref}}{U_{ref}} = \left(\frac{\delta_{ref}}{U_{ref}^2 \rho}\frac{dp_{w,ref}}{dx}\right)^{1/2}, \tag{5}$$

which is used in the present study as the perturbation parameter in the vicinity of vanishing wall shear stress. It can be shown that the parameter $\varepsilon_\infty$ and $\alpha$ are of the same order of magnitude.

The fundamental difference between the analysis for attached and separated flows is the fact that $\varepsilon_\infty$ (or $\alpha$) are not coupled with the parameter $\varepsilon$. Figure 2 illustrates the influence of $\varepsilon$ and $\alpha$ for a diffuser flow with separation. From laminar separating flows it is known that one of the limiting solutions for Re $\to \infty$ can be described by the free-streamline theory after Helmholtz and Kirchhoff, see Schlichting, Gersten [40]. The same can be applied to the turbulent case, if the turbulent activities vanish, that is for $\alpha \to 0$. Similar to the attached laminar boundary layers the attached turbulent boundary layers are reduced to lines of zero thickness due to the discussed coupling of the perturbation parameters. For $\alpha \neq 0$ one can assume that the limiting solution for Re $\to \infty$ leads to the flow structure shown in Figure 2. The developing turbulent free shear layers force the separation streamlines to reattach downstream.

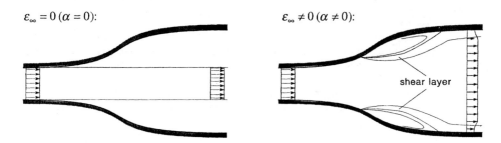

*Figure 2:* Limiting flow cases of the parameter $\varepsilon_\infty$ (or $\alpha$) for a separating diffuser flow at $\varepsilon = 1/\text{Re} = 0$.

For questions of heat transfer the additional parameter

$$\varepsilon_{T\infty} = \frac{T_{p,ref}}{\Delta T_{ref}} = \frac{q_{w,ref}}{\rho c_p U_{ref}\Delta T_{ref}} \tag{6}$$

enters the problem. which is then an additional perturbation parameter with $\varepsilon_{T\infty} = O(\varepsilon_\infty)$.

The asymptotic analysis finally leads to the governing equations in the different layers. For the fully turbulent region the governing equations of the first order are (for details see Vieth [11]):

$$\frac{\partial u}{\partial x} + \frac{\partial v}{\partial y} = 0 \, , \tag{7}$$

$$u\frac{\partial u}{\partial x} + v\frac{\partial u}{\partial y} = -\frac{1}{\rho}\frac{dp_w}{dx} + \frac{1}{\rho}\frac{\partial \tau_t}{\partial y} \, , \tag{8}$$

$$u\frac{\partial T}{\partial x} + v\frac{\partial T}{\partial y} = -\frac{1}{\rho c_p}\frac{\partial q_t}{\partial y} \, . \tag{9}$$

Equations (7) - (9) have to be closed by proper turbulence models. The model used here will be explained in detail in the chapter "numerical method". The boundary condition follow from the matching of the results in the overlap regions of the various layers:

$$y = \delta: \quad u(x,y) = U(x) \, , \quad T(x,y) = T_\infty \, , \quad \tau_t = q_t = 0 \, , \tag{10}$$

$$y \to 0: \quad \frac{\partial u}{\partial y} = \frac{\text{sign}(\tau_w)\sqrt{|\tau_w/\rho + y \cdot dp_w/dx/\rho|}}{\kappa(K) y} \, , \quad v(x,y) = 0 \, ,$$

$$\frac{\partial T}{\partial y} = \frac{-q_w/(\rho c_p)}{\kappa_\Theta(K) y \sqrt{|\tau_w/\rho + y \cdot dp_w/dx/\rho|}} \, , \tag{11}$$

$$\tau_t = \tau_w + y\frac{dp_w}{dx} \, , \quad q_t = q_w \, .$$

Equations (7) - (11) describe the flow in the attached *and* the separated region. For attached flows subjected to low pressure gradients Eq. (8) and (9) can be further simplified to the well known linearized equations for the defect, which can be solved, having carried out some transformations, independently from the Reynolds number, see Gersten, Vieth [12]. However, no way exists to solve the non-linear Eq. (7) - (9) without prescribing the Reynolds number, i. e. the viscosity, which influences the velocity and temperature field only through the near wall boundary conditions, Eq. (11). These boundary conditions are strongly dependent on the pressure gradient and the wall shear stress. Their structure changes considerably at separation. In the next chapters these boundary conditions and the proper boundary conditions for further turbulent quantities are derived and explained in detail.

## Matching conditions

**Viscous sublayer**  Figure 3 illustrates the different length scales of the various layers. The characteristic length of the fully turbulent outer layer is the boundary layer thickness $\delta$. The characteristic length and velocity of the sublayer result from a dimensional analysis. They differ for the attached and the separated case. In the vicinity of the wall the influence of the length scale of the outer region should vanish. Thus the following relation holds in the sublayer:

$$f_{sub}(u, y, \frac{\tau_w}{\rho}, \frac{1}{\rho}\frac{dp_w}{dx}, v) = 0 \, . \tag{12}$$

Two reference velocities can be deduced:

$$u_\tau = \sqrt{\frac{\tau_w}{\rho}} \quad \text{and} \quad u_s = \left(\frac{\nu}{\rho}\frac{dp_w}{dx}\right)^{1/3} . \tag{13}$$

Equation (12) can be reduced to

$$u^+ = F_{sub,0}(y^+, K) , \tag{14}$$

with

$$u^+ = \frac{u}{u_\tau} , \quad y^+ = \frac{y u_\tau}{\nu} , \tag{15}$$

and

$$\boxed{K = \frac{\nu/\rho}{(\tau_w/\rho)^{3/2}}\frac{dp_w}{dx} = \left(\frac{u_s}{u_\tau}\right)^3} . \tag{16}$$

Near separation, that is for low values of $\tau_w$ or large values of $K$, (14) and (15) are no longer useful. These relations are therefore replaced by

$$u^\times = F_{sub,\infty}(y^\times, K) \tag{17}$$

and

$$u^\times = \frac{u}{u_s} , \quad y^\times = \frac{y u_s}{\nu} . \tag{18}$$

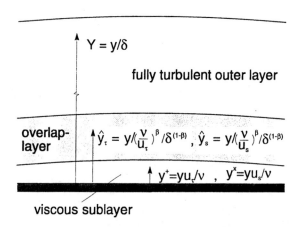

*Figure 3:* The coordinates in the different layers.

The parameter $K$ describes the transition from attached to separated flows. As it turns out later, $K$ is the crucial parameter for flows with separation.

An essential point of the derivation of the matching conditions is the existence of an overlap region between the fully turbulent and the viscous part, see Figure 3. The matching takes place in this overlap layer using a so-called intermediate coordinate $\hat{y}$. The stretching of the intermediate coordinate lies between the ones of the outer and the inner layer. This means that the exponent $\beta$ lies between 0 and 1, see Figure 3.

**Dimensional analysis in the overlap layer, momentum transfer** The dimensional analysis in the overlap layer leads to a functional dependency for the momentum transfer with the following quantities:

$$f(\frac{\partial u}{\partial \hat{y}}, \hat{y}, \frac{\tau_t}{\rho}, \frac{1}{\rho}\frac{dp_w}{dx}, \frac{\tau_w}{\rho}, v) = 0 \quad . \tag{19}$$

The essential hypothesis of the following derivations is that the quantities in the overlap region do not explicitly depend on $\delta$ but on the intermediate coordinate $\hat{y}$. Strictly speaking, we should again distinguish between the attached and the separating case. However, there is no difference for the matching procedure so that we will proceed with $\hat{y}$ rather than distinguishing between $\hat{y}_\tau$ and $\hat{y}_s$.

Equation (19) can be reduced to a relation between four non-dimensional parameters:

$$\lim_{\substack{\hat{y}=const \\ Re \to \infty}} \frac{\hat{y}}{|\tau_t/\rho|^{1/2}} \frac{\partial u}{\partial \hat{y}} = \Pi_1 = \frac{1}{\kappa} \quad , \tag{20}$$

$$\lim_{\substack{\hat{y}=const \\ Re \to \infty}} \frac{\hat{y}^3}{v^2} \frac{1}{\rho} \frac{dp_w}{dx} = \Pi_2 \quad , \tag{21}$$

$$\lim_{\substack{\hat{y}=const \\ Re \to \infty}} \left(\frac{\hat{y}}{\tau_w/\rho} \frac{dp_w}{dx}\right)^3 = \Pi_3 \quad , \tag{22}$$

$$\lim_{\substack{\hat{y}=const \\ Re \to \infty}} \frac{v}{(\partial u/\partial \hat{y})\hat{y}^2} = \Pi_4 \quad . \tag{23}$$

It should be noted that Eq. (20) - (23) and all the following non-dimensional quantities hold for the limit of high Reynolds numbers at a constant intermediate coordinate.

For $\tau_t = \tau_w$ the value $\Pi_1$ can be identified being the reverse of the well-known Karman constant. When the wall shear stress vanishes, the four parameters behave as follows for the limit $v \to 0$ (Re $\to \infty$):

$$\Pi_1 = O(1) \quad ,$$

$$\left.\begin{array}{l}\Pi_2 = O(v^{-2}) \to \infty \\ \Pi_3 = O(\tau_w^{-3}) \to \infty\end{array}\right\} \quad K = \left(\frac{\Pi_3}{\Pi_2}\right)^2 = O(1) \quad , \tag{24}$$

$$\Pi_4 = O(v) \to 0 \quad .$$

As indicated in Eq. (24), $\Pi_2$ and $\Pi_3$ can be combined such that we can formulate a distinguished limit. For Re $\to \infty$ and $\tau_w \to 0$ we assume that $\Pi_3/\Pi_2$ is $O(1)$. This reveals again the transition parameter $K$, see Eq. (16). Figure 4 summarises the various limits of this important quantity. In the literature, often $-p^+$ is used instead of $K$, see for example Bradshaw et al. [23]

Since $\Pi_4$ vanishes for Re $\to \infty$, (19) is reduced to

$$F(\kappa, K) = 0 \quad \text{or} \quad \kappa = f(K) \quad . \tag{25}$$

The matching condition concerning the velocity gradient for low values of $K$ finally reads:

$$\lim_{y^+ \to \infty} \frac{y^+}{|\tau_t^+|^{1/2}} \frac{\partial u^+}{\partial y^+} = \lim_{\substack{\hat{y}_\tau = \text{const} \\ \text{Re} \to \infty}} \frac{\hat{y}_\tau}{|\tau_t/\rho|^{1/2}} \frac{\partial u}{\partial \hat{y}_\tau} = \lim_{Y \to 0} \frac{Y}{|\tau_t/\rho|^{1/2}} \frac{\partial u}{\partial Y} = \frac{1}{\kappa(K)} \quad , \qquad (26)$$

and for high values of $K$:

$$\lim_{y^\times \to \infty} \frac{y^\times}{|\tau_t^\times|^{1/2}} \frac{\partial u^\times}{\partial y^\times} = \lim_{\substack{\hat{y}_s = \text{const} \\ \text{Re} \to \infty}} \frac{\hat{y}_s}{|\tau_t/\rho|^{1/2}} \frac{\partial u}{\partial \hat{y}_s} = \lim_{Y \to 0} \frac{Y}{|\tau_t/\rho|^{1/2}} \frac{\partial u}{\partial Y} = \frac{1}{\kappa(K)} \quad . \qquad (27)$$

Instead of the coordinate $Y$ the coordinate $y$ can be used also. The principle in Eq. (26) and (27) is the same: The outer solution of the inner layer - which is the viscous sublayer - has to correspond to the inner solution of the outer layer - which is the fully turbulent layer.

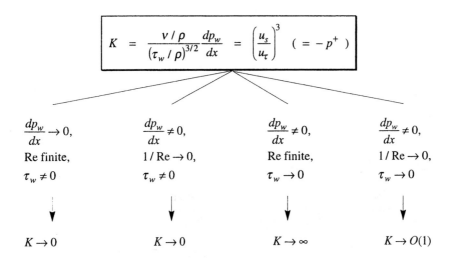

Figure 4: Various limits for the transition parameter $K$.

**Dimensional analysis in the overlap layer, heat transfer**  Similar to the investigations concerning the momentum transfer one can carry out a dimensional analysis for the transfer of thermal energy. In that case it follows:

$$h(\frac{\partial T}{\partial \hat{y}}, \hat{y}, \frac{q_t}{\rho c_p}, \frac{\tau_t}{\rho}, \frac{\partial u}{\partial \hat{y}}) = 0 \quad . \qquad (28)$$

In principle, the pressure gradient, the wall shear stress, and the viscosity should - in the same way as in Eq. (19) - also be taken into account. These quantities are not included in this analysis since they lead to the already known parameter $K$, see (20) - (24). The new non-dimensional parameters are

$$\lim_{\substack{\hat{y} = \text{const} \\ \text{Re} \to \infty}} \frac{\hat{y} \cdot |\tau_t/\rho|^{1/2}}{-q_t/(\rho c_p)} \frac{\partial T}{\partial \hat{y}} = \frac{1}{\kappa_\Theta(K)} \quad , \qquad (29)$$

$$\lim_{\substack{\hat{y}=const\\Re\to\infty}} \frac{(-\tau_t/\rho)\cdot\partial T/\partial\hat{y}}{\left[q_t/(\rho c_p)\right]\cdot\partial u/\partial\hat{y}} = \frac{\kappa(K)}{\kappa_\Theta(K)} = \Pr_t(K) \ . \tag{30}$$

Since Eq. (30) does not represent a new independent parameter, it can be shown that $\kappa_\Theta$ depends only on the transition parameter $K$:

$$H(\kappa_\Theta, K) = 0 \quad \text{or} \quad \kappa_\Theta = f(K) \ . \tag{31}$$

**Dimensional analysis in the overlap layer, transfer of turbulent quantities**  The dimensional analysis concerning the transfer of turbulent kinetic energy provides with

$$g(k, \hat{y}, \hat{\varepsilon}_u, \frac{\tau_t}{\rho}, \frac{\partial u}{\partial\hat{y}}) = 0 \tag{32}$$

the non-dimensional quantities

$$\lim_{\substack{\hat{y}=const\\Re\to\infty}} \frac{\hat{\varepsilon}_u\cdot\hat{y}}{|\tau_t/\rho|^{3/2}} = C_\varepsilon(K) \tag{33}$$

and

$$\lim_{\substack{\hat{y}=const\\Re\to\infty}} \frac{|\tau_t/\rho|}{k} = C_k(K) \ , \tag{34}$$

so that we finally end up with

$$G(C_k, C_\varepsilon, K) = 0 \ . \tag{35}$$

The quantity $C_k$ is equivalent to the well-known structural parameter. With the kinetic energy equation it can be shown that $C_k$ and $C_\varepsilon$ are independent of each other, see also (51) - (53). Thus these quantities depend only on $K$: $C_k = f(K)$, $C_\varepsilon = f(K)$. If the normal stresses are considered separately, one additionally gets

$$\lim_{\substack{\hat{y}=const\\Re\to\infty}}\frac{\overline{u'^2}}{k}=C_u(K) \ , \quad \lim_{\substack{\hat{y}=const\\Re\to\infty}}\frac{\overline{v'^2}}{k}=C_v(K) \ , \quad \lim_{\substack{\hat{y}=const\\Re\to\infty}}\frac{\overline{w'^2}}{k}=C_w(K) \ . \tag{36}$$

For the variance of the temperature fluctuations one gets with

$$j(\overline{T'^2}, \hat{y}, \frac{q_t}{\rho c_p}, \frac{\tau_t}{\rho}, \hat{\varepsilon}_\Theta) = 0 \tag{37}$$

the following quantities

$$\lim_{\substack{\hat{y}=const\\Re\to\infty}} \frac{\overline{T'^2}\cdot|\tau_t/\rho|}{\left[q_t/(\rho c_p)\right]^2} = C_T(K) \ , \tag{38}$$

$$\lim_{\substack{\hat{y}=const\\Re\to\infty}} \frac{\hat{\varepsilon}_\Theta\cdot\hat{y}\cdot|\tau_t/\rho|^{1/2}}{\left[q_t/(\rho c_p)\right]^2} = C_{\varepsilon\Theta}(K) \ . \tag{39}$$

It yields

$$J(C_T, C_{\varepsilon\Theta}, K) = 0 \ . \tag{40}$$

Again it results from the differential equation for $\overline{T'^2}$ that $C_T$ and $C_{\varepsilon\Theta}$ are independent, i.e. $C_T = f(K)$ and $C_{\varepsilon\Theta} = f(K)$.

To summarise we deduced 10 non-dimensional parameters ($\kappa, \kappa_\Theta, \text{Pr}_t, C_k, C_u, C_v, C_w, C_\varepsilon, C_T, C_{\varepsilon\Theta}$) and 10 matching conditions for the unknown quantities in the overlap layer. All these parameters depend on the transition parameter $K$. The only way to determine these functions is to analyse experimental data. Some results concerning this point will be discussed later.

## The Wall Functions

**Preliminary remarks** The term "wall function" is not uniformly used in the literature, see e. g. Wilcox [13]. Within the framework of the present report this term is defined as follows:

*"wall functions are universal, analytical relations for any flow, temperature or turbulence quantity in the overlap region between the viscous wall layer and the fully turbulent region of the boundary layer. They result from an analysis that is independent of turbulence modelling."*

The *method of wall functions* simply consists of the idea to evaluate the lower boundary conditions in the fully turbulent region for all unknown quantities with the help of the wall functions instead of calculating the viscous sublayer. As already mentioned, this method has some advantages in comparison to the low-Reynolds-number modelling. The essential items are:

- the number of grid points can be reduced,

- the damping of the turbulent momentum and heat transfer in the viscous sublayer does not need to be modelled. The choice of correct damping terms, especially near separation and for the heat flux, is still an unsolved problem,

- the turbulence models are less complex and thus the numerical effort is lower.

The strength of this method has been impressively shown by Wilcox [13]. In particular Wilcox indicated the usefulness of the method of wall functions for separated flows. Considering a simple flow case, he compared the solutions of both methods and did not find any essential differences, whereas the calculation time with the low-Reynolds-number model was notably higher.

Huang, Bradshaw [14] recently published a paper concerning the influence of the pressure gradient on the wall functions. Although they had started their derivation with arguments similar to those used in this study, they proceeded by using turbulence models to extend the laws of the wall. This is not the procedure of the present study.

Very often low-Reynolds-number models are employed with the argument, that there does not exist any correct wall function. But this argumentation is a fallacy, because mostly these turbulence models have to be changed also in those cases, when the wall functions change their structure. The results of this report can be used to validate low-Reynolds-number models.

The final step for the derivation of the wall functions is the evaluation of the turbulent shear stress and the heat transfer in the overlap region. At high Reynolds numbers the convective terms vanish in this region, so that the $x$-momentum and thermal energy equation are reduced to:

$$\tau_t = \tau_w + \frac{dp_w}{dx} y \quad , \qquad q_t = q_w \quad . \tag{41}$$

**Wall functions for the Velocity and temperature** The wall functions for the velocity and the temperature gradient emerge, if Eq. (41) is inserted into the matching conditions (20) and (29). They have already been presented in Eq. (11). Table 1 shows them in the scalings of the sublayer. These gradients integrated from the wall into the overlap region finally lead to the relations for the velocity and temperature as shown in Figure 5 for the scalings and the coordinates of the sublayer, see Table 1. The influence of the transition parameter $K$ now becomes very clear. For small values of $K$, that is for low pressure gradients or at high Reynolds numbers, the relations are reduced to the commonly known logarithmic laws of the wall. At the point of separation completely different relations hold, as shown on the right side. For the velocity the so called *square-root-law* holds and the temperature behaves proportional to $y^{-1/2}$ in the overlap layer. The limiting relation for the temperature follows with the rule of de l'Hospital after

*Table 1:* Scalings and limiting solutions for the velocity and the temperature.

| attached flows, $|K| \to 0$ | separation, reattachment, $|K| \to \infty$ |
|---|---|
| scalings $(\; y^\times = y^+ \cdot K^{1/3} \; , \quad u^\times = u^+ / K^{1/3} \; , \quad \theta^\times = \theta^+ \cdot K^{1/3} \;)$ ||
| $u_\tau = sign(\tau_w)\sqrt{|\tau_w/\rho|}, \quad T_\tau = \dfrac{q_w}{\rho c_p |u_\tau|}$, $y^+ = y \cdot |u_\tau|/\nu$ , $u^+ = \dfrac{u}{u_\tau}, \quad \Theta^+ = \dfrac{T_w - T}{T_\tau}$ | $u_s = \left[\dfrac{\nu}{\rho}\dfrac{dp_w}{dx}\right]^{1/3}, \quad T_s = \dfrac{q_w}{\rho c_p |u_s|}$, $y^\times = y \cdot |u_s|/\nu$ , $u^\times = \dfrac{u}{|u_s|}, \quad \Theta^\times = \dfrac{T_w - T}{T_s}$ |
| gradients ||
| $\dfrac{du^+}{dy^+} = \dfrac{\sqrt{1 + K y^+}}{\kappa(K) y^+}$ $\dfrac{d\Theta^+}{dy^+} = \dfrac{1}{\kappa_\Theta(K) y^+ \sqrt{1 + K y^+}}$ | $\dfrac{du^\times}{dy^\times} = \dfrac{sign(\tau_t^\times) \cdot \sqrt{|K^{-2/3} + y^\times|}}{\kappa(K) y^\times}$ $\dfrac{d\Theta^\times}{dy^\times} = \dfrac{1}{\kappa_\Theta(K) y^\times \sqrt{|K^{-2/3} + y^\times|}}$ |
| limiting solutions and integrals ||
| $\dfrac{du^+}{dy^+} = \dfrac{1}{\kappa y^+} \;\Rightarrow\; u^+ = \dfrac{1}{\kappa}\cdot \ln y^+ + C^+$ $\dfrac{d\Theta^-}{dy^-} = \dfrac{1}{\kappa_\Theta y^-} \;\Rightarrow\;$ $\Theta^+ = \dfrac{1}{\kappa_\Theta}\cdot \ln y^+ + C_\Theta^+(Pr)$ | $\dfrac{du^\times}{dy^\times} = \dfrac{1}{\kappa_\infty \sqrt{y^\times}} \;\Rightarrow\; u^\times = \dfrac{2}{\kappa_\infty}\cdot \sqrt{y^\times} + C^\times$ $\dfrac{d\Theta^\times}{dy^\times} = \dfrac{1}{\kappa_{\Theta\infty}(y^\times)^{3/2}} \;\Rightarrow\;$ $\Theta^\times = \dfrac{-2}{\kappa_{\Theta\infty}}\cdot \dfrac{1}{\sqrt{y^\times}} + C_\Theta^\times(Pr)$ |

$$\lim_{\substack{y^\times=const \\ K\to\infty}} \frac{2}{\kappa_\Theta(K)} \frac{\frac{d}{dK}\left[\ln\left(\left(\sqrt{y^\times + K^{-2/3}} - K^{-1/3}\right)/\sqrt{y^\times}\right)\right]}{\frac{d}{dK}\left[K^{-1/3}\right]} = \frac{-2}{\kappa_{\Theta\infty}} \frac{1}{\sqrt{y^\times}}$$

and
$$\lim_{K\to\infty} C_\Theta(K,\text{Pr}) = \frac{C_\Theta^\times}{|K|^{1/3}} - \frac{1}{\kappa_{\Theta\infty}} \ln\left(\frac{4}{|K|}\right).$$

Comparable relations concerning the velocity have already been derived and discussed by several authors. Some of the most important contributions are those of Townsend [15], Perry [16], Perry et al. [17], Kader, Yaglom [18], Afzal [10] and [19], Szablewski [20], and Klauer [6]. The methods of derivation and the results of these authors differ somewhat from the analysis of the present study. Less attention was paid to the temperature problem. The $y^{-1/2}$-behaviour was also deduced by Afzal [19], Perry et al. [17] and Szablewski [20]. The constants of integration $C$ and $C_\Theta$ also depend on the transition parameter $K$. Similar to the "constants" of the dimensional analysis these parameters have to be determined by using experimental results as shown later.

---

low values of K (e. g. $K \leq 1$):

$$u^+(y^+, K) = \frac{1}{\kappa(K)} \ln y^+ + \frac{2}{\kappa(K)} \ln\left|\frac{2}{\sqrt{1+Ky^+}+1}\right| + \frac{2}{\kappa(K)}\left(\sqrt{1+Ky^+}-1\right) + C(K)$$

$$\Theta^+(y^+, K) = \frac{1}{\kappa_\Theta(K)} \ln y^+ + \frac{2}{\kappa_\Theta(K)} \ln\left|\frac{2}{\sqrt{1+Ky^+}+1}\right| + C_\Theta(K,\text{Pr})$$

high values of K (e. g. $K > 1$):

$$u^\times(y^\times, K) = \frac{1}{K^{1/3}} \frac{2}{\kappa(K)} \ln\left|\frac{\sqrt{y^\times + K^{-2/3}} - K^{-1/3}}{\sqrt{y^\times}}\right| + \frac{1}{K^{1/3}} \frac{1}{\kappa(K)}\left(\ln\frac{4}{|K|} - 2\right)$$

$$+ \frac{2}{\kappa(K)} \sqrt{y^\times + K^{-2/3}} + \frac{1}{K^{1/3}} C(K)$$

$$\Theta^\times(y^\times, K) = 2\frac{K^{1/3}}{\kappa_\Theta(K)} \ln\left|\frac{\sqrt{y^\times + K^{-2/3}} - K^{-1/3}}{\sqrt{y^\times}}\right| \frac{2}{|K|^{1/2}} + K^{1/3} C_\Theta(K,\text{Pr})$$

---

$K \to 0$            $K \to \infty$

$$u^+ = \frac{1}{\kappa} \ln y^+ - C^+$$

$$\Theta^+ = \frac{1}{\kappa_\Theta} \ln y^- + C_\Theta^+(\text{Pr})$$

$$y^\times = y^+ K^{1/3}$$
$$u^\times = u^+ / K^{1/3}$$
$$\theta^\times = \theta^+ K^{1/3}$$

$$u^\times = \frac{2}{\kappa_\infty} \sqrt{y^\times} + C^\times$$

$$\theta^\times = \frac{-2}{\kappa_{\Theta\infty}} \frac{1}{\sqrt{y^\times}} + C_\Theta^\times(\text{Pr})$$

*Figure 5:* Wall functions for the velocity and the temperature and their limiting solutions.

**Wall functions for the turbulent quantities** The wall functions for the turbulent quantities can be deduced from the matching conditions (Eq. (33), (34), (36), (38) and (39)) and Eq. (41). These relations are also strongly dependent on the pressure gradient and thus on the parameter $K$. In non-stretched coordinates they read:

shear stress:
$$-\overline{\rho u'v'}\Big|_{y\to 0} = \tau_t(y \to 0) = \tau_w + y\frac{dp_w}{dx} \quad , \tag{42}$$

heat transfer:
$$\rho c_p \overline{v'T'}\Big|_{y\to 0} = q_t(y \to 0) = q_w \quad , \tag{43}$$

kinetic energy:
$$k(y \to 0) = \frac{1}{2}\overline{\left(u'^2 + v'^2 + w'^2\right)}\Big|_{y\to 0} = \frac{|\tau_t/\rho|}{C_k(K)} \quad , \tag{44}$$

normal stresses:
$$\overline{u'^2}(y \to 0) = C_u(K)\,k \quad , \quad \overline{v'^2}(y \to 0) = C_v(K)\,k \quad ,$$

$$\overline{w'^2}(y \to 0) = C_w(K)\,k \quad , \tag{45}$$

dissipation rate:
$$\varepsilon_u(y \to 0) = C_\varepsilon(K)|\tau_t/\rho|^{3/2}/y \quad , \tag{46}$$

specific dissipation rate (see Wilcox [13]):
$$\omega(y \to 0) = C_\omega(K)|\tau_t/\rho|^{1/2}/y \quad , \tag{47}$$

length scales:
$$l(y \to 0) = \kappa(K)\,y \quad , \tag{48}$$

variance of the temperature fluctuations:
$$\overline{T'^2}(y \to 0) = C_T(K) \cdot \frac{[q_w/(\rho c_p)]^2}{|\tau_t/\rho|} \quad , \tag{49}$$

dissipation rate of $\overline{T'^2}$:
$$\varepsilon_\Theta(y \to 0) = C_{\varepsilon\Theta}(K) \frac{[q_w/(\rho c_p)]^2}{y\,|\tau_t/\rho|^{1/2}} \quad . \tag{50}$$

With Figure 5 and Eq. (42) - (50) all near-wall boundary conditions for the calculation of turbulent boundary layers with heat transfer can be evaluated, if the empirical "constants" are known.

## The consequences for turbulence modelling

Up to this point, nothing was said about turbulence modelling. One can use the results of the matching procedure to check the turbulence models concerning their asymptotical correctness. If Eq. (42) - (50) and the results shown in Figure 5 are inserted into the model equations, relations for the model constants arise. It turns out that most common models have to be improved to satisfy the asymptotic results for attached *and* separating flows. This is shown in detail by Vieth [11] for Reynolds stress models. In the part that follows the results concerning the kinetic energy equation and an equation for the length scale are shown by example.

**The $k$ - equation**  At high Reynolds numbers the convective terms and the viscous diffusion term vanish in the overlap layer. It remains for $y \to 0$:

$$-\overline{u'v'}\frac{\partial u}{\partial y} + \frac{\partial}{\partial y}\left\{c_s \overline{v'^2}\frac{k}{\varepsilon_u}\frac{\partial k}{\partial y}\right\} - \varepsilon_u = 0 \quad . \tag{51}$$

By inserting the wall functions these relations evolve:

$$K \to 0: \quad \frac{1}{\kappa} - C_\varepsilon = 0 \qquad \text{(prod. = diss.)} \quad , \tag{52}$$

$$K \to \infty: \quad \frac{1}{\kappa_\infty} - C_\varepsilon + \frac{3}{2}\frac{c_{s\infty} C_v}{C_k^3 C_\varepsilon} = 0 \quad \text{(prod. + diff. = diss.)} \quad . \tag{53}$$

For attached flows and high Reynolds numbers the influence of turbulent diffusion is negligible in the overlap layer. The equilibrium of production and dissipation is called *structural equilibrium*. After Eq. (53) the diffusion term cannot be neglected near separation and reattachment. The quantities of the dimensional analysis and the Karman constant result from measurements, as discussed in the previous section. Thus the model constant $c_s$ has to be changed to satisfy Eq. (52) *and* (53). The transition can for example be described by:

$$c_s(K) = c_{s0} + (c_{s\infty} - c_{s0}) \exp(-0.1\, K^{-2/3}) \quad , \tag{54}$$

in which $c_{s0}$ is the value at $K = 0$ and $c_{s\infty}$ is the value at $K = \infty$ from Eq. (53).

**The $\varepsilon_u$ - equation**  In the overlap region the equation for the turbulent dissipation $\varepsilon_u$ is reduced to

$$-c_{\varepsilon 1}\frac{\varepsilon_u}{k}\overline{u'v'}\frac{\partial u}{\partial y} - c_{\varepsilon 2}\frac{\varepsilon_u^2}{k} + \frac{\partial}{\partial y}\left\{\sigma_\varepsilon \overline{v'^2}\frac{k}{\varepsilon_u}\frac{\partial \varepsilon_u}{\partial y}\right\} = 0 \quad . \tag{55}$$

With the wall functions it yields

$$K \to 0: \quad \frac{c_{\varepsilon 1}}{\kappa} - c_{\varepsilon 2} C_\varepsilon + \frac{\sigma_{\varepsilon 0} C_v}{C_k^3 C_\varepsilon} = 0 \quad , \tag{56}$$

$$K \to \infty: \quad \frac{c_{\varepsilon 1}}{\kappa_\infty} - c_{\varepsilon 2} C_\varepsilon + \frac{1}{2}\frac{\sigma_{\varepsilon \infty} C_v}{C_k^3 C_\varepsilon} = 0 \quad . \tag{57}$$

Again one of the model constants has to be chosen such that both equations are satisfied. One example would be

$$\sigma_\varepsilon(K) = \sigma_{\varepsilon 0} + (\sigma_{\varepsilon \infty} - \sigma_{\varepsilon 0}) \exp(-0.1 K^{-2/3}) \quad . \tag{58}$$

**Numerical method**

**Turbulence model**  The mathematical model thus far described is incomplete unless a proper turbulence model is chosen. The turbulence model used here is a 2-equation model with an analytic function for the integral length scale:

$$u\frac{\partial k}{\partial x} + v\frac{\partial k}{\partial y} = -\overline{u'v'}\frac{\partial u}{\partial y} + \frac{c_{s,k}C_v}{c}\frac{\partial}{\partial y}\left\{\sqrt{k}l\frac{\partial k}{\partial y}\right\} - c\frac{k^{3/2}}{l} \quad , \tag{59}$$

$$u\frac{\partial \overline{u'v'}}{\partial x} + v\frac{\partial \overline{u'v'}}{\partial y} = -C_v k\frac{\partial u}{\partial y}\left[1 - c_2 + \frac{3}{2}c_{2W}\frac{l}{c_L cy}\right]$$
$$- c\overline{u'v'}\frac{\sqrt{k}}{l}\left[c_1 + \frac{3}{2}c_{1W}\frac{l}{c_L cy}\right] + \frac{c_{s,uv}C_v}{c}\frac{\partial}{\partial y}\left\{\sqrt{k}l\frac{\partial \overline{u'v'}}{\partial y}\right\} \quad , \tag{60}$$

$$l(x,y) = L(x)c_l \tanh\left[\frac{\kappa(K)}{L(x)c_l}y\right] \quad . \tag{61}$$

Different equations for $L(x)$ were tested, but the best agreement with measurements gave the simple relation $L(x) = \delta(x)$. Equations (59) and (60) follow from the complete Reynolds-stress model, e. g. see Launder [21], with the following asymptotic correct assumptions:

$$\varepsilon_u = c\frac{k^{3/2}}{l} \quad , \quad \overline{v'^2} = C_v k \quad . \tag{62}$$

For the heat transfer an algebraic relation for $\overline{v'T'}$ was used:

$$\overline{v'T'} = -\frac{c_\Theta}{c}\frac{l}{\sqrt{k}}\left[C_v k\frac{\partial \Theta}{\partial y} - \overline{v'T'}\frac{\partial u}{\partial x}\right] \quad . \tag{63}$$

Equation (63) after Launder [21] follows from the complete Reynolds-heat flux relation neglecting the convective and diffusive terms.

The essential characteristics of the turbulence model (59) - (63) are:
- no eddy viscosity assumption is used,
- viscous effects do not appear,
- the equations are of parabolic nature similar to the momentum and heat flux equation,
- these equations are easier to solve than the full Reynolds stress equations,
- they are asymptotically correct for $y \to 0$, if the model constants are changed with respect to the wall functions (42) - (50) and Figure 5, for details see Vieth [11].

**Transformation of the governing equations** Turbulent boundary layers do have - in contrast to laminar boundary layers - a distinct thickness $\delta(x)$. The y-coordinate was related to $\delta(x)$ so that the calculation could be restricted for the whole field between 0 and 1. Unfortunately the wall functions for the velocity and temperature do have singularities for $y \to 0$. Because of this, the calculation was not started at the wall, but at a distance $y_u$ from the wall. The value $y_u$ was chosen such that the $u$-component after Figure 5 vanishes, see Rotta [22]. Hence the complete transformation reads

$$\eta = \frac{y - y_u(x)}{\delta(x) - y_u(x)} \quad , \quad \text{with} \quad 0 \le \eta \le 1 \quad .$$

The advantages of this rectangular calculation region is partially compensated by the drawback that the values of $y_u(x)$ and $\delta(x)$ - which in principle are parts of the solution - have to be calculated iteratively at each streamwise position. This sometimes leads to convergence problems.

For inverse boundary layer procedures it turned out to be most appropriate to use the stream function formulation in connection with the mechul-approach, see Bradshaw et al. [23].

**Boundary conditions**  The boundary conditions are already presented in Eq. (10), (11) and (42) - (50). For inverse calculations the outer boundary condition for $u$ is replaced by the definition of the displacement thickness $\delta_1(x)$ (see Eq. (65)) in order to avoid the singularity at the separation point:

$$\Psi(x,\eta=1) = \left.\frac{\partial \Psi}{\partial \eta}\right|_1 \cdot \frac{(\delta(x)-\delta_1(x))}{(\delta(x)-y_u(x))} . \tag{64}$$

The displacement thickness $\delta_1(x)$ is either prescribed (e. g. in case of calculating experimentally realised flows as reported here) or simultaneously determined with an interaction procedure. In both cases the interaction between the inviscid outer flow and the viscid boundary-layer flow is accounted for.

For the calculation of the heat transfer either a distribution of the wall heat flux $q_w(x)$ or the wall temperature distribution $T_w(x)$ has to be prescribed.

**Initial conditions**  Due to the parabolic nature of the equations, initial conditions are only needed for profiles at the up-stream boundary. It turned out that the results are very sensitive concerning these profiles. The essential items of the determination of the initial conditions are (for details see Vieth [11]):

- wall shear stress and boundary layer thickness: from self-similar solutions and skin friction law in case of prescribed Clauser parameter $\beta$ at the beginning,

- velocity profile: wake-function approach after Coles,

- turbulent shear stress profile: mixing-layer approach with Eq. (61),

- turbulent kinetic energy profile: assumption of structural equilibrium: $k(\eta) = \tau_t(\eta)/C_k$,

- temperature profile, fully developed case (the thickness of the temperature boundary-layer corresponds to the thickness of the velocity boundary-layer): wake function similar to the velocity and a proper analogy for the Stanton number,

- temperature profile, non-developed case (the temperature boundary-layer develops inside the velocity boundary-layer): self-similar solutions after Klick [24] for profile and Stanton number.

**Solution procedure**  Due to their parabolic nature, the boundary layer equations can be solved marching in the streamwise direction. An implicit finite-difference procedure after Keller, Cebeci [25] and Keller [26] was employed for solving the differential equations. The resulting algebraic equations are solved with the Newton-Raphson procedure. In the streamwise direction a uniform grid spacing was used, normal to the wall the grid spacing was varied proportional to a geometric series owing to the high gradients near the wall. If $\delta_1(x)$ is not known a priori, the flow field must be calculated several times until $\delta_1(x)$ converges. The implementation of the inverse boundary condition into the prediction code was done similar to Cebeci et al. [27]. For more details the reader may be referred to Vieth [11].

# EXPERIMENTAL INVESTIGATIONS

## Experimental set-up

A new experimental set-up was built to investigate the described flow field. The principle of generating the flow field is shown in Figure 6. The working fluid is air at a temperature of $T_\infty = 20°C$. The homogeneous oncoming flow has the velocity of $U_\infty = 18.5 m/s$. The entrance height of the single-sided diffuser is $H_1 = 50 mm$, the height at the end of the diffuser could be varied between $H_2 = 130 mm$ and $150 mm$. Due to the boundary-layer suction directly at the beginning of the diffuser the flow is forced to remain attached at the lower wall. In this way a closed separation region forms at the opposite plane wall. Inside the separation region a transverse vortex develops. In the region of separation a wall-heating segment could be implemented to provide a constant heat flux from the wall into the flow.

Due to the high channel width of $1800 mm$ (aspect ratio at the working section entrance: 36) and due to boundary-layer suction at the side walls a very good two-dimensionality concerning the mean flow field and the mean temperature field could be accomplished, see Kiel [28]. Without suction at the side walls longitudinal vortices form which disturb the flow field drastically.

The integration of the working section into the closed-loop wind tunnel is shown in Figure 7. With this construction the oncoming flow was independent from the ambient conditions and no pollution of the environment with seeding particles - necessary for the LDA-measurements - took place. The heat exchanger in connection with a temperature controller ensures a time-independent temperature of the oncoming flow.

Four metal screens (open-area ratio: 58%) in the wide-angle diffuser (diffuser angle: 40°) avoid separation. Two additional screens and a honeycomb (cell size: $9.5 mm$) in the settling chamber in connection with the nozzle (contraction ratio: 6) ensure a two-dimensional homogeneous flow at the test-section entrance. The open-area ratio of the screens in the settling chamber (chosen here: 69%) influences the flow conditions in the boundary layers developing at the channel walls considerably. In order to minimise irregularities in the oncoming boundary layer these screens had to be cleaned periodically from the soiling caused by the seeding particles.

In order to fix the transition line from a laminar to a turbulent boundary layer at the plane wall a tripping wire was placed at the nozzle exit. The optimum wire diameter turned out to be $0.7 mm$.

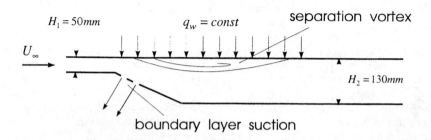

*Figure 6:* Principle of generating the flow field (working section).

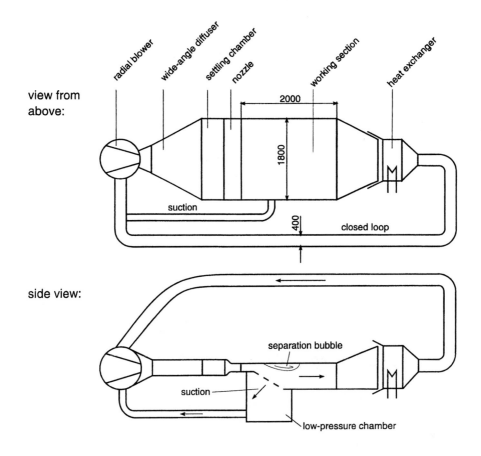

*Figure 7:* Closed-loop wind tunnel.

The set-up of the wall-heating segment becomes clear in the sectional drawing of Figure 8. The principle of generating the temperature field is based on the resistance heating of an extremely thin metal foil ($s = 0.25mm$) which is glued on a nearly adiabatic hard foam. The foil is made of Inconel 600 (NiCr15Fe), which allows the very low thickness. The specific electrical resistance of this foil is relatively high and nearly independent of the temperature, the thermal conductivity is very low. On this way the heat conduction inside the foil in flow direction was held less than 0.4% of the total heat flux, which was $q_w = 1570 W/m^2$. The thermal conductivity of the hard foam (PUR) and of the plastic base plate (Pertinax) was low enough so that the heat losses to the environment were less than 2.4% of the total heat flux.

The total heating area of $0.7 \times 0.7 m^2$ consists of 7 metal stripes with a width of $100mm$. These stripes are adjusted in flow direction. On this way the so-called "Hall effect" could be minimised. (The "Hall effect" describes the following mechanism: Due to the self-induced magnetic field the electrons in the foil are deflected in the direction of the middle of the foil, so that the electrical potential lines are confined and the temperature of the foils is no longer constant in spanwise direction.)

With this construction the heat losses due to conduction and radiation (the latter was estimated to be less than 1.6% of the total heat flux) were held at a minimum. Thus one can assume that the convective heat flux $q_w$ corresponds to the supplied electrical power and is constant over the whole area: $q_w = 1570 W/m^2$. The maximum wall temperature at this heat flux was $T_{w\,max} \approx 80°C$. The maximal possible heat flux was $q_{w\,max} \approx 3500 W/m^2$, where $T_w$ should not be higher than $T_{w\,max} \approx 150°C$.

The wall temperature distribution was measured with small thermocouples ($d \approx 0.5mm$, NiCr-Ni) directly underneath the foils. The accuracy of the wall-temperature measurements was better than $\pm 0.5°C$.

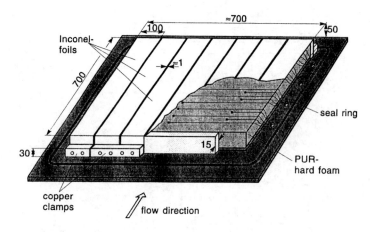

*Figure 8:* Total view and sectional view of the wall-heating segment.

## Measurement techniques

Table 2 summarises the applied measurement techniques and the measured quantities. In what follows the essential items will discussed, for more details see Kiel [28].

The instantaneous three velocity components were measured by using a Dantec LDA system based on two fibre probes. The system allows direct measurements of $u$ and $v$ components with one (four-beam) probe while the other (two-beam) probe is inclined at an angle of 30° to measure the $w$ component. Both optical probes have a focal length of $1200mm$ and a beam expansion ratio of $E = 2.9$, leading to a measuring volume with a maximum thickness in y-direction of $200\mu m$. They are inclined at a very small angle to allow near-wall measurements, the smallest distance to the wall being $y = 0.1mm$, corresponding to $y^+ \leq 7$ in wall layer coordinates.

The laser set-up is arranged with one probe on each side of the working section, so that all measurements presented here could be made in an "off-line" forward-scatter mode. In this way the data rates were 10 times higher than in the conventional back scatter mode. This set-up required a special traversing system that is stiff enough to keep the relative position between the probes exactly constant. The resolution of the traversing system in x- and z-direction is $10\mu m$, in the y-direction $1\mu m$. The accuracy of the spindle lead is better than $17\mu m$ traversing $100mm$.

Table 2: Measured quantities and measurement techniques.

| measuring techniques | measured quantities | | calculated quantities |
|---|---|---|---|
| 3D laser-Doppler-anemometer | velocities: | $\bar{u}(y),\bar{v}(y),\bar{w}(y)$ | $u,v$ - vectors, $\Psi$ <br> $\delta_1, \delta_2$ |
| | correlations: | $\overline{u'^2}(y),\overline{v'^2}(y),\overline{w'^2}(y),$ <br> $\overline{u'v'}(y),\overline{u'w'}(y),\overline{v'w'}(y)$ | turbulence intensity <br> $Tu(y)$ |
| | triple correlations: <br> $\overline{u'^3}(y),\overline{v'^3}(y),\overline{w'^3}(y),\overline{u'v'w'}(y),$ <br> $\overline{u'^2v'}(y),\overline{u'^2w'}(y),\overline{v'^2w'}(y),$ <br> $\overline{u'v'^2}(y),\overline{u'w'^2}(y),\overline{v'w'^2}(y)$ | | $\xrightarrow{Spline} f'(y)$ <br> $\rightarrow$ <br> terms of the governing equations |
| | reverse flow factors: | $\chi_u(y),\chi_v(y),\chi_w(y)$ | |
| sublayer fence | wall shear stress: | $\overline{\tau_w}(x)$ | $u_\tau$ |
| differential pressure transducer | wall pressure: | $\overline{p_w}(x)$ | $u_s$ |
| cold wire | temperature: | $\overline{T}(y),\overline{T'^2}(y)$ | |
| coincidental 3D-LDA and cold wire measurements | velocity-temperature correlations: <br> $\overline{vT'}(y),\overline{u'T'}(y),\overline{w'T'}(y)$ <br> $\overline{v'^2T'}(y),\overline{v'u'T'}(y),\overline{v'w'T'}(y),...$ | | |
| thermo couples | wall temperature: | $\overline{T_w}(x)$ | Stanton-number St |
| for reference used process data: | | | position of the profiles in flow direction: |
| pressure difference over the nozzle | reference velocity: <br> $U_\infty = 18.5\,m/s = const$ | | LDA-Measurements at 36 positions <br> $0.2m \leq x \leq 1.175m$ |
| Pt 100 in the settling chamber | reference temperature: <br> $T_\infty = 20°C = const$ | | cold wire at 23 positions |
| | wall heat flux: <br> $q_w = 1570\,W/m^2 = const$ | | $0.38m \leq x \leq 1.04m$ |

As seeding material DEHS ($C_{26}H_{50}O_4$) was used, the particle generator (type: Pallas AGF 5.0) gave poly-disperse particles with a mean diameter of $0.2-0.3\mu m$.

Signal processing was performed by Dantec burst spectrum analysers (BSAs). Prior to the measurements some examinations were conducted to find the best bias-elimination technique and to measure the integral time scale. It was found that *inter arrival time weighting* is the most appropriate technique whereas residence time weighting underestimates the results. The integral time scale is estimated from an auto correlation of a time series at high data rates. Both the integral time scale and number of particles lead to measuring times up to $25min$ (the longest times could be found in the shear layer of the separated region).

A differential pressure transducer (MKS-Baratron, $10mbar$) was used to measure the difference between the wall pressure (with pressure taps 0.8mm in diameter) and the reference pressure at $x = 50mm$. The error of the entire system is less than 1% for values higher than $1Pa$. The wall pressure was sampled 300 times at each position to ensure averaged values within the range $\pm 0.01Pa$ for a 90% confidence interval.

Skin friction measurements were performed with a sublayer fence (fence height: $0.1mm$) that had been calibrated with reference to the pressure loss in a fully developed flat channel flow. For measurements in the separated region the fence was also calibrated for the flow coming from the opposite side. The pressure difference was measured by a differential pressure transducer (MKS-Baratron, $1mbar$). For high skin friction values the skin friction was also deduced from the measured mean velocity profile with the help of an indirect turbulence model of Gersten, Herwig [1]. The difference between the skin friction values with the latter method and the sublayer fence were less than 1.3%, see also Figure 10.

The instantaneous temperature field was measured with a constant current anemometer (single cold-wire probe). The electric current through the Wollaston wire with $2.5\mu m$ diameter was chosen to $I_c = 200\mu A$. This value guaranteed a velocity independent signal. The DISA 55M20 temperature bridge was used, the resolution of the measuring range ($20\text{-}80°C$) amounted 0.25%. The probe was calibrated together with a reference resistance thermometer in a special calibration cell in a temperature controlled water bath. The cut-off frequency was determined to $265Hz$.

Much effort has been spent for coincidental 3-d-LDA and cold wire measurements. The two major problems to be solved were first the positioning of the wire relative to the measuring volume of the lasers and secondly to get signals from both measurement techniques at the same time with an acceptable high data rate.

The measuring volume and the cold-wire probe were positioned at the same distance to the wall very close to each other. The probe had to be positioned downstream of the measuring volume so that the flow through the measuring volume was not disturbed by the probe. Thus this arrangement had to be changed in the back flow region. Therefore the flow field had to be known a priori.

The second problem was overcome by triggering the constant-current anemometer with the LDA signals in a similar way as in Klein [29]. The measuring frequency was between 10 and $50Hz$, the measuring time was between 2 and $30min$ for one point of the profile depending on the position in the flow.

In this way it was possible for the first time to determine three-point correlations of the velocity and the temperature in all regions of a boundary layer with separation regions.

## Global results of the measurements

### Velocity field

The quality of the two-dimensionality of the flow field was investigated with flow visualisation by wall tufts and with wall pressure measurements as well as profile measurements at different $z$-positions. The flow turned out to be two-dimensional in a wide area of at least $1m$ span near the centre line, see Kiel [28].

Velocity profiles were measured at $z = 0$ at 36 positions in flow direction normal to the wall from $x = 0.2m$ to $x = 1.175m$. Figure 9 gives an overview of the measured flow field. The vector plot shows only a few of the measured profiles. The dividing streamline indicates the size of the separation vortex: The length is about $0.53m$, the maximum thickness $0.06m$.

Figure 10 shows the distributions of the wall shear stress and the wall pressure. Both quantities are non-dimensionalised with the dynamic pressure of the flow at the nozzle exit. The vertical lines indicate the location of the separation bubble. Inside the bubble, the skin friction coefficient has low negative values and is slightly wavy. This behaviour was also found by Patrik [30] and Dianat, Castro [31]. The $c_f$-distribution was measured several times during the whole measuring program showing that the size and the location of the separation bubble remained constant: The point of separation varied in the range $\pm 0.5mm$, the point of reattachment varied in the range $\pm 1.5mm$.

The negative values of the pressure distribution shows that the main flow is accelerated in the first part of the test section due to the boundary layers developing on both channel walls. In the region of the separation point the pressure increases sharply and becomes nearly constant in the region of the separation vortex. This plateau is typical for separating boundary layers. The value of $c_p \approx 0.36$ in this region was also measured by Patrick [30].

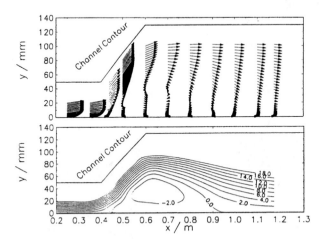

*Figure 9:* Overview of the measured flow field: mean velocity vectors and stream lines (the numbers indicate the constant values of the stream function $\psi(x,y) = \int_0^y u(x,y)/U(x)dy$).

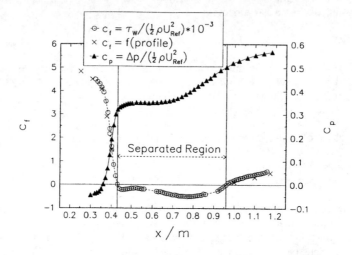

*Figure 10:* Non-dimensional skin friction and wall pressure coefficient.

The important integral boundary-layer parameter displacement thickness $\delta_1(x)$ and momentum thickness $\delta_2(x)$ are shown in Figure 11. These values result from integration of the measured mean velocity profiles:

$$\delta_1(x) = \int_0^\delta \left(1 - \frac{u(x,y)}{U(x)}\right) dy \quad , \quad \delta_2(x) = \int_0^\delta \left(1 - \frac{u(x,y)}{U(x)}\right) \frac{u(x,y)}{U(x)} dy \quad . \tag{65}$$

In contrast to the very large displacement effect of the boundary layer on the inviscid outer flow the loss of momentum inside the boundary layer is comparably low. This is also typical for separating boundary layers.

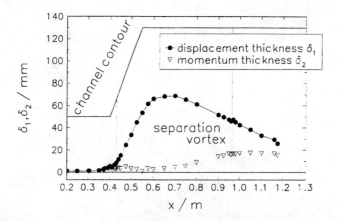

*Figure 11:* Boundary layer parameter.

Figure 12 shows the profiles of the mean velocity components. For clearness only a few profiles and only every second measuring point is shown. The $w$-component - not shown here - is nearly zero in the whole flow field. The largest value measured for $w$ is 4% of $U_\infty$. This confirms the good two-dimensionality of the flow field.

The normal Reynolds stresses and the turbulence intensity $Tu(y)$ (coupled with the turbulent kinetic energy by $Tu = \sqrt{2k/3}$) are shown in Figure 13 and 14. In general the behaviour of all normal stresses are comparable to each other. The largest values exist in the free shear layer between the back flow region and the outer flow, where $\overline{u'^2}$ is about 1/3 larger than the other components. In the separated region it can be seen that $\overline{u'^2}$ and $\overline{w'^2}$ approach constant values near the wall before dropping to zero at the wall with very steep gradients, while the fluctuations normal to the wall are damped more so that $\overline{v'^2}$ tends to zero with low gradients near the wall. The profiles for the turbulence intensity show that the turbulence level of the inviscid outer flow is very high even in the oncoming flow, where it reaches 3%. After the first half of the separation vortex the two boundary layers seem to be merged together. From this point on the turbulence intensity is about 10% in the "outer" flow.

The turbulent shear stress $-\overline{u'v'}$ is shown in Figure 15. The largest values do appear in the shear layer between the vortex and the outer flow. At the beginning of the second half of the separation region the turbulent shear stress begins to increase and is largest in the region of reattachment.

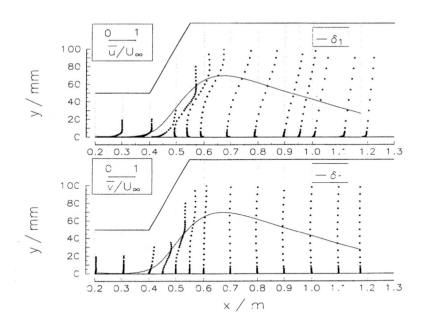

*Figure 12:* Profiles of the mean velocity $u(y)$ and $v(y)$.

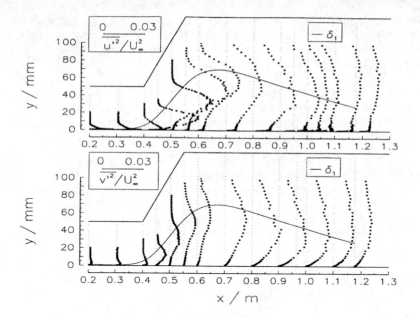

*Figure 13:* Profiles of the Reynolds stress components $\overline{u'^2}(y)$ and $\overline{v'^2}(y)$.

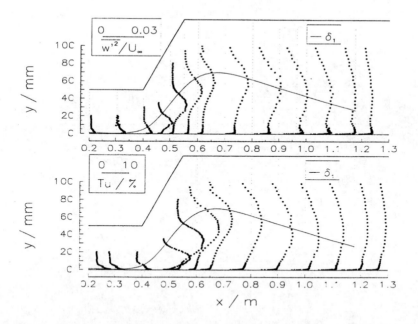

*Figure 14:* Profiles of the Reynolds stress component $\overline{w'^2}(y)$ and the turbulence intensity $Tu(y)$.

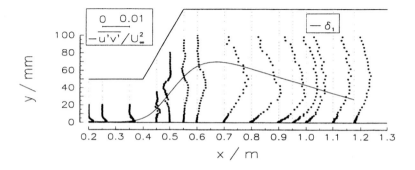

*Figure 15:* Profiles of the Reynolds stress component $-\overline{u'v'}(y)$.

**Temperature field**

Similar to the velocity field the temperature field was also investigated concerning its two-dimensionality. Measurements of the wall temperature at different z-positions as well as profile measurements on the centre line with different numbers of heated foils did not show any three-dimensional effects, see Kiel [28].

The distributions of the wall temperature and the Stanton number are shown in Figure 16. Due to the constant heat flux the latter, defined as

$$\text{St} = \frac{q_w}{\rho c_p U_\infty (T_w - T_\infty)} \quad , \tag{66}$$

behaves reciprocally to the wall temperature distribution. The vertical dashed lines indicate the heated region, the pointed lines indicate the location of the separation vortex.

The best heat transfer - and thus the lowest temperatures - are at the beginning of the heated region, where the temperature boundary layer is non-developed and the temperature gradients are highest. The local maximum concerning the heat transfer upstream of the reattachment point was also found by Vogel, Eaton [32] and Sparrow et al. [33]. This behaviour might be explained by the fact that the flow near the reattachment point can be compared with an impinging jet, where the wall heat flux is very high due to the large convective effects. The highest wall temperature - and thus the lowest heat transfer - can be observed 70$mm$ behind the separation point. In a comparable flow Rivir et al. [34] observed the same behaviour with nearly the same values for the maximum and the minimum Stanton number.

It is also worth mentioning that the Stanton number distribution does not show any distinct behaviour at the points of vanishing wall shear stress.

The accuracy of the local Stanton number was estimated using the root-sum-square method. Considering all possible inaccuracies - including the heat losses already discussed - the maximum error is about 15%.

For the discussion of the heat transfer results it is helpful to consider the distribution of the boundary-layer thicknesses in Figure 17 also. The thicknesses are defined as the positions, where 99% of the outer velocity is reached and the temperature deviates only 1% from the outer temperature, respectively. Downstream of $x = 0.8m$ the boundary layers at both walls have merged together so that the thicknesses could not be evaluated. It can be seen that the temperature boundary layer develops very fast inside the velocity boundary layer due to the high transport characteristics of turbulence. The thickness of the temperature boundary layer corresponds to the thickness of the velocity boundary layer downstream of $x = 0.45m$, that is shortly behind the separation point. The temperature field upstream of this point has to be considered as non-developed.

Some profiles of the non-dimensional temperature ($\Theta(y) = (T(y) - T_\infty)/(T_w - T_\infty)$) and the variance of the temperature fluctuations are shown in Figure 18. Remarkable are the very steep gradients near the wall in all flow regions: Over the first $2mm$ from the wall the temperature decreases to $\Theta \approx 0.4\Theta_w$. In the second half of the separation bubble a region exists with nearly constant temperature. The distribution of $\overline{T'^2}$ also demonstrates the importance of the region adjacent to the wall for the heat transfer. The highest values can be found near the wall. The variance of the temperature fluctuations is constant over the main part of

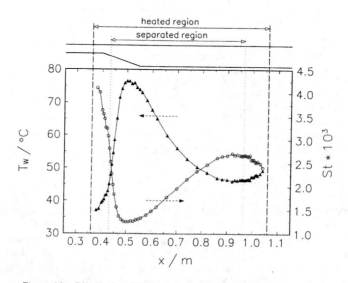

*Figure 16:* Distribution of the wall temperature and the Stanton number.

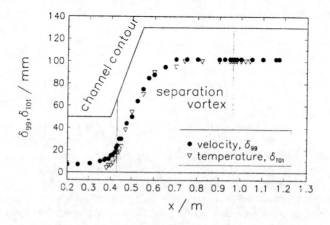

*Figure 17:* Thicknesses of the velocity and the temperature boundary layer.

the flow, except for the aforementioned region near the wall and in the shear layer at the beginning of the separation bubble.

The profiles of the turbulent heat flux $\overline{v'T'}(y)$ are shown in Figure 19. This quantity does not show the high gradients near the wall. In contrast to the turbulent shear stress no distinct maximum can be observed in the region of the free shear layer. At the end of the separation bubble the heat flux does not vanish for large y values because of the merging of the boundary layers.

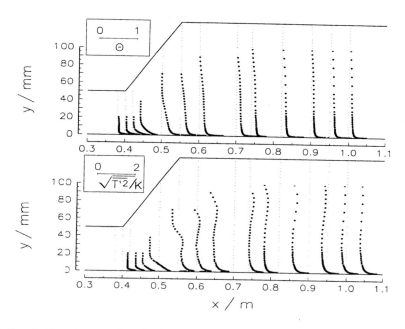

Figure 18: Profiles of the non-dimensional temperature $\Theta = (T(y) - T_\infty)/(T_w - T_\infty)$ and the variance of the temperature fluctuations $\overline{T'^2}(y)$.

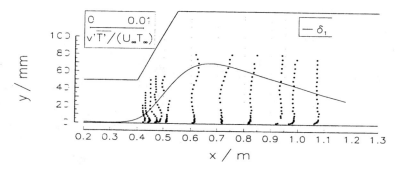

Figure 19: Profiles of the turbulent heat flux $\overline{v'T'}(y)$.

## The empirical functions of the wall functions

As frequently mentioned the non-dimensional empirical quantities $\kappa(K)$, $\kappa_\Theta(K,\text{Pr})$, $C_k(K)$, $C_u(K)$, $C_v(K)$, $C_w(K)$, $C_\varepsilon(K)$, $C_T(K,\text{Pr})$, $C_{\varepsilon\Theta}(K,\text{Pr})$, $C_\omega(K)$ and the constants of integration $C(K)$, $C_\Theta(K,\text{Pr})$ evolve from the described measurements. In what follows the results for some of these quantities are presented.

Figure 20 shows the quantities for the wall functions of the velocity and the temperature. The solid points indicate the results of the analysed velocity and temperature profiles, the lines represent the curve fits. The method of determining these values is described in detail in Vieth, Kiel [35] or in Kiel [28]. The following dependencies were found out (Eq. (69) and (70) are valid for $\text{Pr} = 0.72$):

$$\kappa = (\kappa_\infty - \kappa_0) \cdot \exp(-0.0045 \cdot K^{-4/3}) + \kappa_0 \quad , \tag{67}$$

$$C = \left( K^{1/3} \cdot C^\times - C^+ - \frac{1}{\kappa_\infty}\left(\ln(\frac{4}{|K|}) - 2\right)\right) \cdot \exp(-K^{-3.2/3}) + C^+ \quad , \tag{68}$$

$$\kappa_\Theta = (\kappa_{\Theta\infty} - \kappa_{\Theta 0}) \cdot \exp(-0.2 \cdot K^{-2/3}) + \kappa_{\Theta 0} \quad , \tag{69}$$

$$C_\Theta = \left( \frac{C_\Theta^\times}{|K|^{1/3}} - C_\Theta^+ - \frac{1}{\kappa_{\Theta\infty}} \ln\left(\frac{4}{|K|}\right)\right) \cdot \exp(-0.001 \cdot K^{-4/3}) + C_\Theta^+(\text{Pr}) \quad . \tag{70}$$

The limiting values turned out to be

$$\kappa_0 = 0.41 \quad , \quad \kappa_\infty = 0.59 \quad , \quad \kappa_{\Theta 0} = 0.47 \quad , \quad \kappa_{\Theta\infty} = 5 \quad , \tag{71}$$

$$C^\times = 0.0 \quad , \quad C^+ = 5.0 \quad , \quad C_\Theta^+(\text{Pr} = 0.72) = 3.5 \quad , \quad C_\Theta^\times(\text{Pr} = 0.72) = 1.8 \quad . \tag{72}$$

In Table 3 and in Figures 21 and 22 the results concerning the limiting values for the velocity are compared with the results of other authors (Afzal [10] and [19], Townsend [15], Perry et al. [17], Kader, Yaglom [18], Szablewski [20], El Telbany, Reynolds [36], Dengel, Fernholz [38]). In Kiel [28] a more detailed discussion can be found. It turns out that there is partially a good agreement concerning $\kappa_\infty$. This is not the case for $C$, $\kappa_\Theta$ and $C_\Theta$. As the scatter of the measured points indicates, there exists a relatively large uncertainty in the determination of these three functions. It is for example not clear if $C^\times$ has a positive or a negative value. Concerning the distributions of $\kappa_\Theta(K)$ and $C_\Theta(K)$ one has to consider that the temperature profiles upstream of the point of separation (i. e. $0.36m \leq x \leq 0.43m$) have not been developed and thus could not be taken into account.

As can be seen in Figure 20 all constants vary continuously through the point of separation. The curves seem to be symmetrical with respect to $K = \infty$. This means that the same values hold in the attached and separated region if the absolute value of $K$ is the same.

The functions $\kappa(K)$ und $C(K)$ have been determined experimentally also by Vieth [11] in turbulent, fully developed Couette-Poiseuille flows. In these experiments the hot wire anemometry was used to measure the mean velocity and the turbulent shear stress. While the limiting value $\kappa_\infty$ was measured slightly higher, the principle shape of $\kappa(K)$ is the same as may be seen in Figure 21. The results for $C(K)$ show the same uncertainty as the results measured by Kiel.

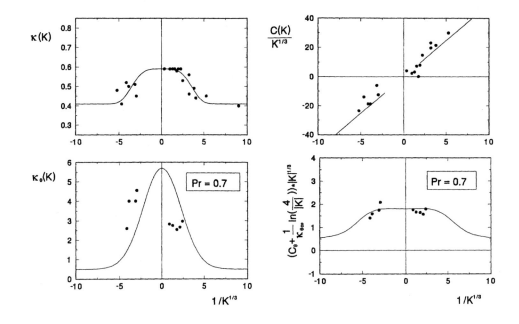

*Figure 20:* Empirical quantities of the wall functions for the velocity and the temperature.

*Table 3:* Constants of the square-root law after several authors.

| author | kind of investigation | $\kappa_\infty$ | $C^\times$ |
|---|---|---|---|
| Townsend (1960) | analytic | 0.5 ±0.05 | / |
| Townsend (1961) und (1976) | analytic | 0.48 ±0.03 | 2.2 |
| Mellor (1966) | analytic | 0.41 (0.44) | 1.33 |
| Perry et.al. (1966) | experimental | 0.48 | / |
| Szablewski (1972) | analytic | 0.41 | 2.23 |
| Kader/Yaglom (1978) | analytic | 0.45 | / |
| El Telbany/Reynolds (1980) | experimental | 0.8 | -3.2 |
| Nakoyama/Koyama (1984) | analytic | 0.5 | 0.0 |
| Spalart/Leonard (1987) | DNS | 0.6 | -3.36 |
| Vieth (1996) | experimental | 0.88 | -3 |
| present investigations | experimental | 0.59 | 0.0 |

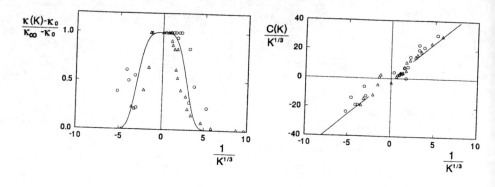

*Figure 21:* Comparison concerning $\kappa(K)$ und $C(K)$ of the present report (○) and the measurements of Vieth [11] (△).

The functions of the matching conditions (34) and (36) were also evaluated, see Figure 23. While the values for the $u$-component is a constant of about $C_u \approx 1.0$, no relations comparable to (67) - (70) could be found for the other quantities. In the region of vanishing wall shear stress and in the back flow region the Reynolds stress component $\overline{v'^2}/k$ becomes lower compared to the attached flow regions, while $\overline{w'^2}/k$ slightly increases towards separation. In the region of separation the values for $C_k = \overline{|u'v'|}/k$ nearly vanish, what can be explained with experimental uncertainties: in this region both values are very low. The mean values turned out to be $C_v \approx 0.3$, $C_w \approx 0.7$ and $C_k \approx 0.2$ with an accuracy of $\pm 0.1$.

It was also tried to determine the functions $C_\varepsilon(K)$, $C_T(K,\text{Pr}=0.72)$ and $C_{\varepsilon\Theta}(K,\text{Pr}=0.72)$ (see Eq. (33), (38) and (39)) from the measurements, but it was found that the experimental uncertainties are too high to allow definite statements concerning these points. This has to be seen under the consideration that much effort has been spent to enhance the accuracy of the measurements. As discussed in Vieth, Kiel [35] in detail for the function $C_\varepsilon(K)$ it seems to be questionable if these three functions can ever be derived reliably enough (e. g. with an accuracy better than 50%).

Concerning the experimental determination of the empirical quantities of the wall functions it can be concluded that some functions could be measured with a certain accuracy ($\kappa(K)$ with ±12%; $\kappa_\Theta(K,\text{Pr})$ with ±50%; $C(K)$ with ±20%; $C_\Theta(K,\text{Pr})$ with ±50%; $C_k(K)$, $C_u(K)$, $C_v(K)$ and $C_w(K)$ with ±20%) and some functions could not be measured ($C_\varepsilon(K)$, $C_T(K,\text{Pr})$ and $C_{\varepsilon\Theta}(K,\text{Pr})$).

One has to consider that even for the most simple flow case - that is fully developed pipe flow ($K$=0) - there exists a relatively high uncertainty concerning $\kappa$ and $C^+$ in the literature, i. e. $\kappa = 0.42 \pm 0.03$ and $C^+ = 5.4 \pm 0.8$. These values were recently measured with high accuracy by Zaragola et al. [37] as $\kappa = 0.44$ and $C^+ = 6.3$. The extreme experimental effort described in this report demonstrates the difficulty determining the wall function constants.

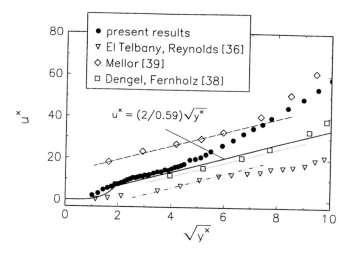

*Figure 22:* Comparison of different results concerning the velocity profile at $\tau_w = 0$.

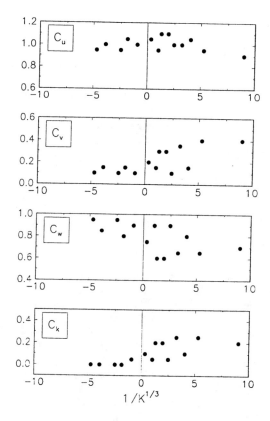

*Figure 23:* Universal constants of the wall functions for the turbulent stresses.

# COMPARISON OF EXPERIMENTAL AND NUMERICAL RESULTS

In what follows some results of the calculation method in comparison with the measured separated boundary layer are shown. At the starting point of the calculation the boundary layer was assumed to be fully turbulent, the flow outside the boundary layer was assumed to be inviscid and non-turbulent. All variables of the governing equations were non-dimensionalised with the following reference values, which are illustrated in Figure 24: The reference length $L_{ref}$ was chosen to be the distance of the starting point from the origin of a fictitious turbulent flat plate boundary layer and turned out to be $L_{ref} = 0.269m$. The reference velocity $U_{ref} = 18.5m/s$ is the velocity at the outer edge of the boundary layer at the position $x_c = 1$ ($x_c$: streamwise coordinate used in the calculation) or $x = 0.25m$ ($x$: streamwise coordinate used in the experiment). The Reynolds number of the problem thus turns out to be $Re = U_{ref} L_{ref} / \nu = 3.28 \cdot 10^5$. Concerning the heat transfer $\Delta T_{ref} = 20K$ and $q_{ref} = \Delta T_{ref} \lambda / L_{ref} = 1.881 W/m^2$ were chosen ($Pr = 0.72$). The independence of the calculations from the grid size was proven by doubling the number of grid points (e.g. from 80 to 320 points normal to the flow direction and 400 to 1600 grid points in flow direction).

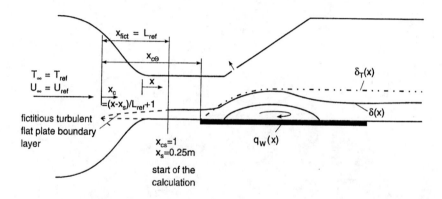

*Figure 24:* Reference values and coordinates used for the experiments ($x$) and the calculations ($x_c$).

The Figures 25 to 31 show the calculated results in comparison to the measurements. The displacement thickness, shown in Figures 11 and 25, was prescribed. Generally, the qualitative agreement between the calculations and the measurements is satisfactory. But nevertheless, one has to discuss some quantitative deviations from the measurements.

In general the typical behaviour for boundary layers under the influence of strong pressure gradients can be observed: increasing pressure causes a decrease of the skin-friction; the displacement thickness, boundary layer thickness, and the momentum thickness grow rapidly while the mean velocity profile loses fullness. The maximum shear stress moves away from the wall, the mean velocity gradient near the wall decreases and thus the production of Reynolds stresses decreases while the diffusion towards the wall increases. The separation bubble is characterised by a region of nearly constant pressure and constant negative wall shear stress, see Figure 26.

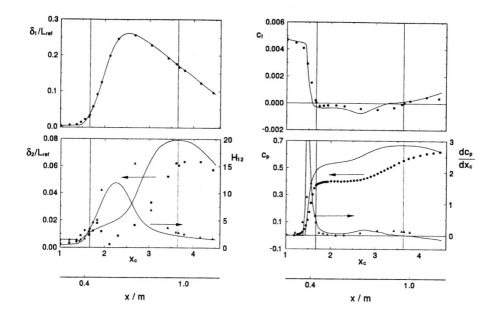

*Figure 25:* Measured (•, ▲) and calculated (—) boundary layer parameter $\delta_1(x)$, $\delta_2(x)$ and $H_{12}(x)$.

*Figure 26:* Measured (•, ▲) and calculated (—) wall shear stress, wall pressure and pressure gradient distribution.

Figure 27 shows some mean velocity profiles. The following seems to be noteworthy: The points of separation and reattachment were predicted slightly to early as may also be seen in Figure 26. Secondly, the measured velocities in the bubble are more flat than the calculated profiles. It should also be mentioned that the calculated maximum shear stress and kinetic energy (not presented here) markedly differ in the separation region from the measurements. This can be explained with the simple model for the integral length scale $l(x,y)$, Eq. (61). A differential equation for $l(x,y)$ (i. e. an equation for $\varepsilon_u$ or $\sqrt{k} \cdot l$) should give improvement concerning the last points.

Figure 28 shows the excellent agreement in the near wall region. A profile close to separation ($K \approx 30$ for both the calculated and measured profile) was chosen to illustrate the structure of the flow field. Additionally a profile from the measurements of Dengel, Fernholz [38] exactly at the point of vanishing wall shear stress ($K \approx \infty$) is shown. These data fit very well into the asymptotic theory except that the value of $C^x$ seems to be slightly lower.

One major aspect that cannot be accounted for within the boundary layer theory is the influence of the streamline curvature. This may be explained with Figure 29, which presents the calculated and measured free stream velocity. Additionally, the velocity derived from the measured wall pressure distribution and Bernoulli´s equation is shown (stream tube theory). The differences to the measured outer velocity are due to the streamline curvature (see also Figure 9). Under the effect of centrifugal forces the pressure decreases normal to the plane wall in the region of concave curvature and thus the outer velocity is larger than the one determined with the stream tube theory. The reverse holds in the region of convex curvature.

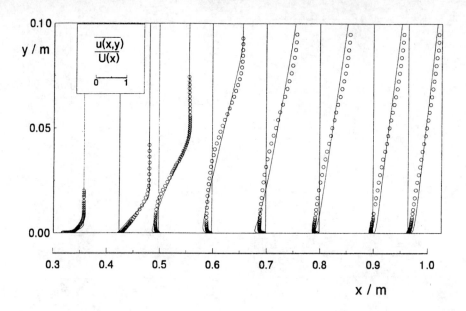

*Figure 27:* Comparison of the calculated (lines) and measured (symbols) velocity profiles.

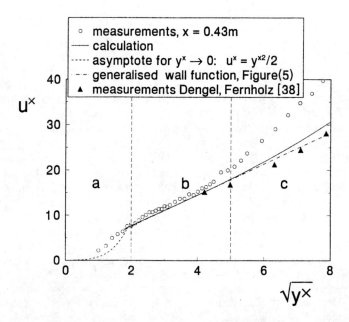

*Figure 28:* Velocity profile near the point of separation in wall layer scalings, comparison between measurements, calculation, the asymptotes and some values of the profile from Dengel, Fernholz [38]. Region a: viscous sublayer; b: overlap layer; c: fully turbulent region.

The deviations from the stream tube theory become also evident in the development of the momentum thickness. In the separation bubble the measured $\delta_2(x)$ nearly goes to zero and thus the shape parameter $H_{12}(x)$ gets markedly larger compared to the calculation, see Figure 25. The values of $H_{12}$ at the point of separation are 2.4 and 2.1 for the measured and calculated profile, respectively. These values are lower than those stated by Dengel, Fernholz [38] ($2.75 \leq H_{12,sep} \leq 2.95$) and are lower than those values calculated for the Stratford flow (a flow with $\tau_w(x) = 0$). The analysis concerning the Stratford flow excludes that this could be a Reynolds number effect. Small changes in the prescribed $\delta_1(x)$-distribution also did not influence the results in the separation region distinctly. Reasons for these deviations might be the influence of the streamline curvature and perhaps the equation for $l(x,y)$.

No reason could be found for the too high calculated pressure level in the separation region. The measured value $c_{p,sep} - c_{p,min} \approx 0.4$ corresponds to the measurements of other authors. However, the shape of $c_p(x)$ and thus the pressure gradient were predicted satisfactorily.

The influence of the parameters on the calculation were investigated in detail. It turned out that the results in the region of separation are only weakly dependent on moderate changes of the Reynolds number, the starting conditions, small changes in the prescribed $\delta_1(x)$-distribution and the number of grid points. The influence of the values of $C^x$ also seems to be negligible. The largest changes could be observed by varying the value of $\kappa_\infty$ as shown in Figure 30.

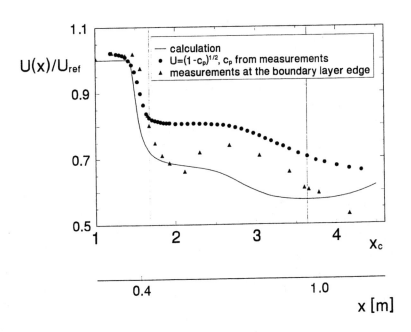

*Figure 29:* Measured and calculated free stream velocity.

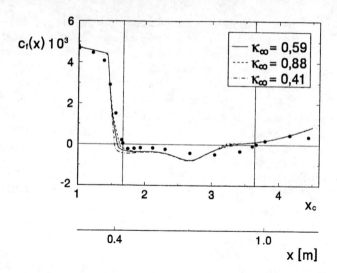

*Figure 30:* Influence of the value $\kappa_\infty$ on the skin friction, $C^x = 0$.

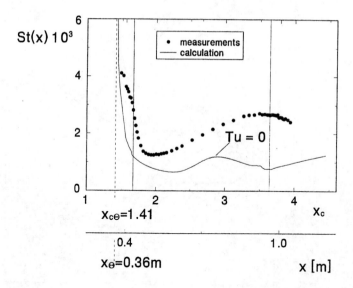

*Figure 31:* Measured (symbols) and calculated (line) distribution of St(x).

The results concerning the heat transfer are shown in Figure 31. As expected the heat transfer and thus the Stanton number is highest at the beginning of the heated region at $x_{c\Theta} = 1.41$ or $x_\Theta = 0.36 m$. From the calculations and the measurements one can see that the heat transfer is drastically reduced in the region of separation. The calculations do not confirm the local maximum of the measurements in the reattachment region. To the opinion of the authors the relative large differences between the experimental and the numerical results in this region can be traced back to two reasons: Probably the most essential point is the simple model for the turbulent heat flux (Eq. (62)) applied so far. It has to be expected that the use of a complete non-linear differential equation for $\overline{v'T'}$ would give a better description of the turbulent heat transfer in the separated region, e. g. see Launder [21]. Secondly the relatively high turbulence level in the free stream of the measurements might lead to higher values of the Stanton number - the calculations conducted so far are based on an outer flow without turbulence. A further improvement of the calculation method to take these two points into account is therefore needed. Such a method is currently under development by the authors. The reader may refer to Vieth [11] for the state of this work.

## SUMMARY AND CONCLUSION

Turbulent boundary layers with separation can be described in an asymptotically correct way by the classical concept of the three-layer structure. This structure also holds in the reverse flow region. Using the results of the dimensional analysis and the matching conditions between the viscous sublayer and the turbulent outer part, new wall functions for all unknown field quantities were derived, which are valid for attached and separating flows. The pressure gradient clearly influences these relations. A suitable parameter to describe this influence is the transition parameter $K$. Empirical relations for some non-dimensional quantities of the matching conditions and the constants of integration in the wall functions were given. It could be shown that most turbulence models are asymptotically correct, but only if certain model "constants" vary with $K$.

Based on this mathematical model a new prediction method has been presented. The new feature of this calculation method is the use of turbulence model constants that are adjusted according Eq. (54) in the way to satisfy the matching conditions, and the use of the new wall functions discussed here in detail, including the empirical relations for the functions (67) - (70).

An experimental set-up has been built to investigate the heat transfer. The flow field and the temperature field in the test section of this new wind-tunnel is characterised by an excellent two-dimensionality and a steady location of the separation bubble. Special care has been taken to assure steady flow and heat transfer conditions in order to establish reproducible measurements.

The experimental set-up and the three-dimensional LDA measurement technique in connection with a single cold-wire probe (the constant-current anemometer was triggered by the LDA signals) allow the coincidental measurement of the temperature and all three velocity components. For the first time it was possible to determine correlations of the velocity and the temperature in all regions of a boundary layer including a separation vortex.

The measured profiles of the mean velocity, the turbulent stresses and the temperature were used to determine the universal functions $\kappa(K)$, $C(K)$, $\kappa_\Theta(K)$, $C_\Theta(K)$, $C_k(K)$, $C_u(K)$, $C_v(K)$ and $C_w(K)$ of the laws of the wall.

The comparison of the experimental and numerical results confirm the statement that the method of wall functions is also successfully applicable for boundary layers with reverse flow and heat transfer. A detailed knowledge of the distributions of all quantities in the viscous sublayer is not necessary. Thus one may conclude that low-Reynolds-number modelling is - from an asymptotic point of view - not necessary to examine the flows considered in this study. Since the new wall functions are derived without any assumptions concerning turbulence modelling, these relations can be used to validate existing and future low- and high-Reynolds-number turbulence models.

## ACKNOWLEDGEMENT

The reported results were obtained within the frame of the project "Vortices and Heat Transfer". The authors highly appreciate the financial support of this project by the DFG.

## REFERENCES

[1]   Gersten, K.; Herwig, H. (1992):   *Strömungsmechanik: Impuls-, Wärme- und Stoffübertragung aus asymptotischer Sicht.* Verlag Vieweg & Sohn, Wiesbaden.

[2]   Gersten, K. (1989):   Die Bedeutung der Prandtlschen Grenzschichttheorie nach 85 Jahren. 32. Ludwig-Prandtl-Gedächtnisvorlesung, Karlsruhe, 28. März 1989, *Z. Flugwiss. Weltraumforsch.* **13**, 209-218.

[3]   Mellor, G.L. (1972):   The large Reynolds number asymptotic theory of turbulent boundary layers. *Int. J. Engng. Sci.* **10**, 851-873.

[4]   Deriat, E. (1987):   Asymptotic analysis of the k-$\varepsilon$-model for a turbulent boundary layer. *Rech. Aerosp.* no. 1987-5.

[5]   Jeken, B.H. (1992):   *Asymptotische Analyse ebener turbulenter Strömungen an gekrümmten Wänden bei hohen Reynolds-Zahlen mit einem Reynolds-Spannungs-Modell.* Dissertation, Ruhr-Universität Bochum, auch als VDI-Fortschrittbericht, Reihe 7, Nr. 215, VDI-Verlag, Düsseldorf.

[6]   Klauer, J. (1989):   *Berechnung ebener turbulenter Scherschichten mit Ablösung und Rückströmung bei hohen Reynolds-Zahlen.* VDI-Fortschrittberichte, Reihe 7, Nr. 155. VDI-Verlag Düsseldorf.

[7]   Melnik, R.E. (1987):   An asymptotic theory of turbulent separation. *Grumman Report RE-722.*

[8]   Melnik, R.E. (1989):   An asymptotic theory of turbulent separation. *Computers & Fluids* **17**, No. 1, 165-184.

[9]   Schneider, W. (1991):   Boundary-layer theory of free turbulent shear flows. *Z. f. Flugwiss. u. Weltraumforsch.* **15**, 143-158.

[10]   Afzal, N. (1980):   Asymptotic analysis of turbulent boundary layer near separation. *Proc. First Asian Congress on Fluid Mechanics,* Bangalore, Paper A-16, pp 1-7.

[11]   Vieth, D. (1996):   *Berechnung der Impuls- und Wärmeübertragung in ebenen turbulenten Strömungen mit Ablösung bei hohen Reynolds-Zahlen.* Ph. D. Thesis, Ruhr-Universität Bochum.

[12]   Gersten, K.; Vieth, D. (1995):   Berechnung turbulenter anliegender Grenzschichten bei hohen Reynolds-Zahlen. In: *Festschrift zum 70. Geburtstag von J. Siekmann,* Lehrstuhl für Mechanik, Universität Essen.

[13]   Wilcox, D.C. (1989):   Wall matching, a rational alternative to wall functions. *AIAA paper 89-611,* 27th Aerospace Sciences Meeting, Reno, Nevada.

[14]   Huang, P.G.; Bradshaw, P. (1995):   Law of the wall for turbulent flows in pressure gradients. *AIAA Journal* **33**, 624-632.

[15]   Townsend, A.A. (1961):   Equilibrium layers and wall turbulence. *J. Fluid Mech.* **11**, 97-120.

[16]   Perry, A.E. (1966):   Turbulent boundary layers in decreasing adverse pressure gradients. *J. Fluid Mech.* **26**, 481-506.

[17]   Perry, A.E.; Bell, J.B.; Joubert, P.N. (1966):   Velocity and temperature profiles in adverse pressure gradient turbulent boundary layers. *J. Fluid Mech.* **25**, 299-320.

[18]   Kader, B.A.; Yaglom, A.M. (1978):   Similarity treatment of moving-equilibrium turbulent boundary layers in adverse pressure gradients. *J. Fluid Mech.* **89**, 305-342.

[19]   Afzal, N. (1982):   Thermal turbulent boundary layer under strong adverse pressure gradient near separation. *J. of Heat Transfer* **104**, 397-402.

[20] Szablewski, W. (1972): Inkompressible turbulente Temperaturgrenzschichten mit konstanter Wandtemperatur. *Int. J. of Heat and Mass Transfer* **15**, 673-706.

[21] Launder, B.E. (1988): On the computation of convective heat transfer in complex turbulent flows. *J. of Heat Transfer* **110**, 1112-1128.

[22] Rotta, J.C. (1983): Einige Gesichtspunkte zu rationeller Berechnung turbulenter Grenzschichten. *Z. f. Flugwiss. u. Weltraumforsch.* **7**, Heft 6, 417-429.

[23] Bradshaw, P.; Cebeci, T.; Whitelaw, J.H. (1981): *Engineering calculation methods for turbulent flows.* Academic Press, London.

[24] Klick, H. (1992): *Einfluß variabler Stoffwerte bei der turbulenten Plattenströmung.* Ph. D. Thesis, Ruhr-Universität Bochum. VDI-Fortschrittberichte, Reihe 7, Nr. 213. VDI-Verlag Düsseldorf.

[25] Keller, H.B.; Cebeci, T. (1971): Accurate numerical methods for boundary layer flows. *Lecture Notes in Physics* **8**, Springer Verlag.

[26] Keller, H.B. (1978): Numerical methods in boundary-layer theory. *Ann. Rev. Fluid Mech.* **10**, 417-433.

[27] Cebeci, T.; Keller, H.B.; Williams, P.G. (1979): Separating boundary-layer flow calculations. *J. Comp. Phys.* **31**, 363-378.

[28] Kiel, R. (1995): *Experimentelle Untersuchung einer Strömung mit beheiztem lokalen Ablösewirbel an einer geraden Wand.* Ph. D. Thesis, Ruhr-Universität Bochum. VDI-Fortschrittberichte, Reihe 7, Nr. 281. VDI-Verlag Düsseldorf.

[29] Klein, B. (1993): *LDA- und HD-Meßsysteme für turbulente Wärmeströme in Grenzschicht- und rezirkulierenden Strömungen.* Ph. D. Thesis, Universität-Gesamthochschule Siegen. VDI-Fortschrittberichte, Reihe 7, Nr. 229. VDI-Verlag Düsseldorf.

[30] Patrick, W.P. (1987): Flowfield measurements in a separated and reattached flat plate turbulent boundary layer. *NASA Contractor Report 4052*, NASA Lewis Research Center.

[31] Dianat, M.; Castro, P. (1991): Turbulence in a separated boundary layer. *J. Fluid Mech.* **226**, 91-123.

[32] Vogel, J.C.; Eaton, J.K. (1985): Combined heat transfer and fluid dynamic measurements downstream of a backward facing step. *J. of Heat Transfer* **107**, 922-929.

[33] Sparrow, E.M.; Kang, S.S.; Chuck, W. (1987): Relation between the points of flow reattachment and maximum heat transfer for regions of flow separation. *Int. J. of Heat and Mass Transfer* **30**, 1237-1246.

[34] Rivir, R.B.; Johnston, J.P.; Eaton, J.K. (1992): Heat Transfer on a flat surface under a region of turbulent separation. *37th ASME International Gas Turbine and Aeroengine Congress, 92-GT-198*, Köln.

[35] Vieth, D.; Kiel, R. (1995): Experimental and Theoretical Investigations of the Near-Wall Region in a Turbulent Separated and Reattached Flow. *Exp. Thermal and Fluid Sci.* **11**, 243-254.

[36] El Telbany, M.M.M.; Reynolds, A.J. (1980): Velocity distribution in plane turbulent channel flows. *J. Fluid Mech.* **100**, 1-29.

[37] Zaragola, M.V.; Smits, A.J.; Orszag, S.A.; Yakhot, V. (1996): Experiments in high Reynolds number turbulent pipe flow. *AIAA-Paper 96-0654*, 34th Aerospace Science Meeting & Exhibit, January 15-18, 1996, Reno.

[38] Dengel, P.; Fernholz, H.H. (1989b): An experimental investigation of an incompressible turbulent boundary layer in the vicinity of separation. *J. Fluid Mech.* **212**, 615-639.

[39] Mellor, G.L. (1966): The effects of pressure gradients on turbulent flow near a smooth wall. *J. Fluid Mech.* **24**, 255-274.

[40] Schlichting, H; Gersten, K. (1997): *Grenzschicht-Theorie.* Springer-Verlag, Berlin, Heidelberg.

[41] Cebeci, T.; Smith, A.M.O. (1974): *Analysis of turbulent boundary layers.* Academic Press, New York.

# Turbulent Flow Structure and Local Heat Transfer in Asymmetrically Ribbed Channels

W. Leiner, S. Lorenz, M. Dierich, J. Torkar

Ruhr-Universität Bochum, Institut für Thermo- und Fluiddynamik, 44780 Bochum, Germany

## SUMMARY

Turbulent flow structure and local heat transfer in asymmetrically ribbed channels are investigated experimentally. The Reynolds number $Re_{dh}$ is varied between 10,000 and 160,000. Starting from a basic configuration of rectangular ribs (P/e = 4, e/H = 1/4) the relative rib height e/H, the pitch ratio P/e and the rib shape, rectangular and semi-cylindrical respectively, are varied. The flow structure in the free cross section and in the grooves between the ribs, the wall pressure distribution, the local distribution of convective heat transfer coefficients, the global pressure loss and the mean heat transfer coefficient are measured. Global results are compared to values of fully developed plane channel flow. The ratio of heat transfer augmentation to the increase of pressure loss as compared to a plane channel is $(St_m/St_{plane})/(f_{app}/f_{plane}) > 1$ for both the configuration with rectangular and semi-circular ribs and pitch ratio P/e = 4. Highest pressure drop and highest local and mean Stanton numbers are measured ($f_{app,max}$=0.042, $St_{m,max}$=0.015) for the long pitch ratio P/e = 10, and rectangular ribs for $Re_{dh}$.

## NOMENCLATURE

| | | |
|---|---|---|
| $c_p$ | J/(kg K) | specific heat capacity |
| $d_h$ | m | hydraulic diameter = $4 V_{fluid}/A$ |
| d | m | diameter of semi-cylindrical ribs, d = s |
| e | m | rib height |
| i | - | integer number, number of considered geometric period |
| m | kg/(m²s) | mass flow |
| q | W/m² | heat flux |
| s | m | rib width |
| u | m/s | time averaged velocity component in main flow (x-) direction |
| v | m/s | time averaged velocity component in y- direction (perpendicular to plane wall) |
| x | m | coordinate in main flow direction |
| y | m | coordinate perpendicular to the plane wall |
| z | m | coordinate parallel to ribs |

| | | |
|---|---|---|
| A | - | constant |
| A | m² | wetted surface of one geometric period |
| B | m | total width of channel (in z-direction) |
| C | - | constant in eq. 1 |
| H | m | clear channel height (in y-direction) |
| L | m | deployed wrapping length of one rib period |
| P | m | length of one period in main flow (x-) direction |
| T | K | temperature |
| $T_o$ | K | air temperature at the entrance of the heated module |
| V | m³ | fluid volume inside one geometric period |
| $\Delta p$ | N/m² | pressure loss through one geometric period |
| $\alpha$ | W/(m² K) | heat transfer coefficient (in general plotted as $\alpha = \alpha(\zeta)$, mean values are based on the deployed wrapping surface of one period) |
| $\delta$ | m | thickness of heated metal foil |
| $\varepsilon$ | - | emissivity |
| $\zeta$ | m | coordinate following the contour of the ribbed wall |
| $\kappa$ | - | Kármán's constant |
| $\Lambda$ | m | position of origin of the logarithmic velocity profile from the bottom of the groove |
| $\lambda$ | W/(m K) | heat conductivity |
| $\nu$ | m²/s | cinematic viscosity |
| $\rho$ | kg/m³ | density |
| $\tau$ | - | transmissivity |

**SUBSCRIPTS**

| | |
|---|---|
| 2H | based on twice the clear channel height or hydraulic diameter of a plane channel instead of the hydraulic diameter $d_h$ |
| air | air |
| conv | convective: corrected for both, tangential conduction and radiation |
| cond | conductive: corrected for tangential conduction only |
| e | based on the rib height e instead of the hydraulic diameter $d_h$ |
| foil | heated foil of heated module |
| l | local value, smoothened by spline functions |
| loss | losses |
| m | mean value based on the deployed wrapping area of one period |
| max | maximum value |
| o | total local heat transfer: including effects of tangential conduction and radiation |
| plane | plane channel |
| r | reattachment |
| rad | corrected for radiative flux only |
| ref | reference position at plane wall, opposite to the front edge of the rib |
| wall | wall |
| A | mean correction for radiation, particular for each wall surface segment |
| N | groove ($x_N = x + s$; $y_N = H + e - y$) |
| R | rib ($y_R = H - y$) |
| $\infty$ | mean value of main flow |

# CHARACTERISTIC NUMBERS

$$C_p = \frac{2 \cdot (p(\zeta) - p_{ref})}{\rho_{air} \cdot u_\infty^2}$$ local wall-pressure coefficient

$$f_{app} = \frac{2 \cdot \Delta p}{\rho_{air} \cdot u_\infty^2} \frac{d_h}{4 \cdot P}$$ apparent friction factor, total pressure loss

$$Nu = \frac{\alpha \cdot d_h}{\lambda}$$ Nusselt number

$$Re = \frac{u_\infty \cdot d_h}{\nu}$$ Reynolds number, defined with $d_h$

$$St = \frac{\alpha}{\rho_{air} \cdot c_{p,air} \cdot u_\infty} = \frac{Nu}{Re \cdot Pr}$$ Stanton number

# PROJECT AND AIMS OF INVESTIGATION

Heat transfer enhancement in fluid flows is often obtained by geometric modifications or structures of the wall. These modifications induce vortices to augment transverse fluid mixing and to reduce the thermal boundary layer thickness. Particular types of modifications allow to control generation and orientation of vortices.

The effect of vortices on heat transfer has been investigated intensively by the research group „Vortices and Heat Transfer", guided by *Prof. Fiebig* [1], Bochum. Transverse vortices in 2D-flow with their main axis perpendicular to the main flow direction and longitudinal vortices with their main axis in flow direction are investigated in several particular projects of the research group. Effects of <u>transverse vortices</u> induced by periodically arranged transverse ribs in essentially <u>2-dimensional</u>, <u>turbulent</u> flows are investigated <u>experimentally</u> in the present project. The effect of transverse vortices in turbulent flows on heat transfer is additionally investigated by *Braun et al.* [2], where transverse ribs are one limiting case of inclined winglets, and by *Gersten et al.* [3], where transverse vortices are induced by adverse pressure gradients. *Grosse-Gorgemann et al.* [4] and *Hahne et al.* [5] investigate numerically the effect of transverse vortices induced by periodically arranged transverse ribs in laminar and transient flows.

The present investigations are concerned with asymmetrically ribbed channels with essentially 2-dimensional boundaries as shown in Fig. 1. The flow structure and the local convective heat transfer are measured. Basic investigations are made for one periodic rectangular-rib configuration (basic configuration P/e = 4, e/H = 1/4) at one Reynolds number. Further investigations consider the variation of the Reynolds number, of several geometric parameters and of the rib shape. Particular interests of this project are
- to investigate the time averaged flow field and local turbulent fluctuations in the free cross section and in the grooves between the ribs,
- to measure wall pressure distribution and flow losses,

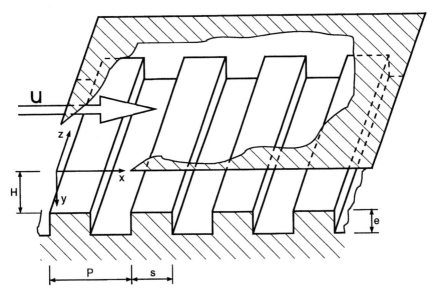

**Fig. 1**: 2-D-asymmetrically ribbed channel with geometric measures and coordinates

- to measure the temperature distribution along the ribbed wall and to evaluate local convective heat transfer coefficients,
- to investigate effects of tangential conduction and of radiation on measured local heat transfer coefficients for the boundary condition of constant heat flux, and to identify effect of neglecting radiation and conduction in the evaluation of local convective heat transfer coefficients from measured temperature data,
- to investigate effects of geometric parameters, of the relative rib height e/H, the pitch ratio P/e and the rib shape (rectangular ribs and semi-cylindrical ribs respectively) on the flow field, wall pressure distribution and on local convective heat transfer coefficients,
- to investigate the validity of the logarithmic law of wall for flows over structured surfaces (especially transversely ribbed walls),
- to give a detailed documentation of results for one basic configuration valid to serve as a benchmark for numerical simulations of turbulent flow and heat transfer in channels with periodic boundary conditions and partially simplified thermic boundary conditions (neglecting tangential conduction and/or radiative heat transfer).

## BACKGROUND

Detailed literature reviews have been given by *Lorenz* [6] for turbulent flow and by *Grosse-Gorgemann* [7] for laminar and transient flow through asymmetrically ribbed channels. The background is presented here only as far as particularly relevant for this project.

Asymmetrically ribbed channels with rectangular ribs are geometrically simple to describe (2-dimensional, cartesian). They are easy to manufacture for experimental investigations and grids are easy to discretise for numerical simulations. Nevertheless flow through

asymmetrically ribbed channels may be rather complex; for this reason flow through and heat transfer in ribbed channels have been widely investigated earlier (see [6]) and yield the following findings:
- The velocity profile is asymmetrical, the location of $y(u_{max})$ of the peak of the velocity profile is situated closer to the plane than to the ribbed wall.
- The position of zero turbulent shear stress $y(u'v' = 0)$ is even closer to the plane wall and does not coincide with the position of peak velocity, $y(u_{max}) \neq y(u'v' = 0)$, see *Hanjalic and Launder* [8], *Meyer* [9], and *Lorenz* [6]. This indicates that turbulence models based on the eddy-viscosity as the k-ε-model do not describe properly the shear stress distribution, and that experimental data are particularly required to verify the quality of applied turbulence models.
- Asymmetrically ribbed channels are the only type of configuration for which investigations of the validity of the logarithmic law of wall are kown (see literature review by *Meyer* [9]). *Meyer* showed for the flow over ribbed surfaces that the constant A figuring in the logarithmic law of wall

$$u^+ = A \cdot \ln(y^+) + C \tag{1}$$

differs from the reciprocal value of Kármáns constant $A \neq 1/\kappa \approx 0{,}41$, but depends on the rib configuration, on the relative rib height e/H, and on the appropriate choice of the origin Λ of the logarithmic velocity profile, $A = f(s/e, P/e, e/H, \Lambda) \neq 1/\kappa$.

Local heat transfer along ribbed surfaces has been much less investigated than the flow field. Only two investigations exist where the local heat transfer coefficients along the complete contour of the ribbed surface have been measured by infrared-thermography (*Aliaga et al.* [10, 11]). Viewing through an opening in the wall instead of a solid IR-transparent window *Aliaga et al.* did not generate channel-flow boundary conditions. Furthermore they made no corrections for radiation and tangential conduction to obtain net convective heat transfer coefficients.

Combined investigations of both, local heat transfer and flow structure in the same ribbed channels are not known yet. The present reference case seems suitable for better physical understanding of the effect of vortex flow structures on heat transfer and for verification of numerical simulations of flow and heat transfer in ribbed channels.

Values of global pressure loss $f_{app}$ and heat transfer $Nu_m$ and $St_m$ respectively in ribbed channels are known to be highest for pitch ratios of $7 \leq P/e \leq 10$ (*Wilkie* [12]), where the shear layer reattaches and again separates at the bottom of the groove between the ribs. The ratio $St_m/f_{app}$ seems to be highest for relative rib distances in the order of $P/e \cong 4$ (cf. *Grosse-Gorgemann* [7]) for which the shear layer stretches plainly over the groove between the rib edges and reattaches at the front edge of the next rib.

## CONFIGURATIONS, BOUNDARY CONDITIONS AND REFERENCE CASES

The investigated basic configuration of an asymmetrically ribbed channel with rectangular ribs inducing transverse vortices in the grooves between is shown in Fig. 2a. Keeping the width-to-height ratio s/e = 2 of the ribs constant the following variations are made:
1. Variation of the Reynolds number Re in the turbulent domain ($10{,}000 \leq Re_{dh} \leq 160{,}000$).

2. Variation of the relative rib height e/H and the channel height H respectively in the range $1/6 \leq e/H \leq 1/2$ with constant pitch ratio P/e = 4, Fig. 2a. The hydraulic diameter $d_h = 4 V_{fluid}/A$ of the configuration with the highest value of relative rib height e/H = 1/2 is equal to the twice clear channel height H, $d_h = 2H$, which allows for particularly easy comparison with the plane channel for this case (e/H = 1/2).
3. Variation of the pitch ratio P/e and the width of grooves (P-s) respectively in the range 4 (highest $St_m/f_{app}$ expected) $\leq P/e \leq 10$ (highest $St_m$ and highest $f_{app}$ expected) with constant relative rib height e/H = 1/4, Fig. 2a and 2b.
4. Variation of the rib shape including rectangular ribs with s/e = 2, Fig. 2a, and semi-cylindrical ribs with d/e = 2, Fig. 2c, with equal pitch ratio P/e = 4 and equal relative rib height e/H = 1/4.

The main parameters of geometric proportions and the hydraulic diameters of all investigated configurations are presented in Table 1. The boundary conditions are approximately:
- essentially 2-D wall configuration with time averaged 2-D turbulent flow,
- hydrodynamically and thermally periodic turbulent flow,
- constant heat flux along the contour of the ribbed wall, $q_{el}(\zeta, z)$ = const.,
- adiabatic plane wall.

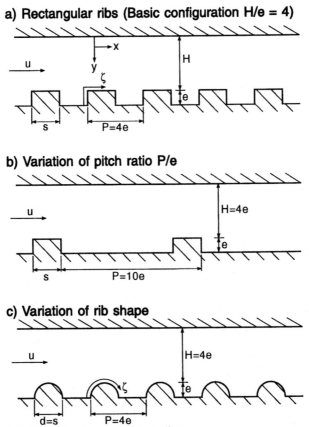

**Fig. 2:** Investigated rib configurations with coordinates

**Table 1:** Geometric variations for the investigations according to Fig. 2 (s/e = 2)

| configuration | rib shape | P/e | e/H | $d_r/H$ | Fig. |
|---|---|---|---|---|---|
| basic configuration | rectangular | 4 | 1/4 | 1.80 | 2a |
| e/H varied | rectangular | 4 | 1/2; 1/6 | 2.0; 52/30 | 2a |
| P/e varied | rectangular | 10 | 1/4 | 2.1825 | 2b |
| rib-shape varied | semi-cylindrical | 4 | 1/4 | 2.106 | 2c |

Hydrodynamically and thermally fully developed flow through a plane channel heated asymmetrically from one side with constant heat flux is used as a reference case of periodic flow and heat transfer in asymmetrically ribbed and heated channels. The friction factor of a plane channel is calculated from a correlation of *Dean* [13]:

$$f_{plane} = 0.0868 \cdot Re_{2H}^{-1/4}. \tag{2}$$

Data of heat transfer coefficients of fully developed flow through a plane channel for asymmetric heating with constant heat flux were measured earlier by *Sparrow et al.* [14] and calculated by *Sakakibara* [15] and *Kays and Leung* [16]. These data have been correlated in a range of $10,000 \leq Re_{2H} \leq 150,000$ yielding

$$St_{plane} = 0.04 \cdot Pr^{-0.6} \cdot Re_{2H}^{-0.274}. \tag{3}$$

The local heat transfer coefficient on the top of the rectangular ribs is compared to the local heat transfer coefficient of a plane turbulent boundary layer at an external velocity equal to the mean velocity in the channel according to *Kays and Crawford* [17]. Inserting the definition of Re (based on the hydraulic diameter) into the equation for the local Stanton number of the boundary layer yields:

$$St(x) = 0.03 \cdot \left(\frac{x}{d_h}\right)^{-0.2} \cdot Pr^{-0.4} \cdot Re_{dh}^{-0.2}. \tag{4}$$

## EXPERIMENTAL TECHNIQUES AND EVALUATION

The wind tunnel serving for the present investigations is shown in Fig. 3. The flow is hydrodynamically periodic in the test section (6), Fig. 3. The 3-dimensional positioning mechanism (7) allows to position an IR-camera or various probes for flow measurements accurately inside the channel. Details of experimental setup, measuring procedure and data evaluation are described by *Lorenz* [6].

The flow through these asymmetrically ribbed channels has essentially 2-dimensional boundary conditions. Both u and v-velocity components in the x-y-plane (see Fig. 1) are measured by 2D-Laser-Doppler-Anemometry (LDA), Hot-Wire-Anemometry (HWA, x-wire) and Pitot-probes.

The wall pressure distribution is measured by an array of wall holes distributed along the contour of the ribbed wall and connected through a scanning valve to a pressure gauge with a sensitivity of 0,1 Pa, see also *Lorenz et al.* [18]. The position for measuring the reference pressure is at the plane wall, at the origin of the x-coordinate, just opposite to the front edge of the rib.

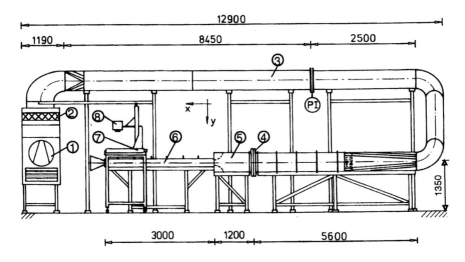

**Fig. 3:** Wind tunnel with test section and traverse mechanism

Flow has been visualized by Helium-filled soap bubbles of nearly the same global mass density as air, *Lorenz et al.* [19]. Such He-bubbles are able to follow the main time-averaged flow but are insensitive to small-scale turbulent fluctuations because of their size.

Temperature distributions along the contour of the ribbed wall are measured by Infrared-Thermography (IRT) with the boundary condition of constant heat flux at the ribbed wall and evaluated to yield local heat transfer coefficients. The experimental setup is shown in Fig. 4. The main items are:

- <u>IR-Camera</u> AVIO TVS 200 System with a spectral sensitivity range of 3 - 5.4 μm and a temperature resolution of 0.01 K,
- <u>Infrared-transparent window</u> made of $CaF_2$ with a transmission coefficient of $\tau = 0.918$ in the sensitive spectral range of the IR-camera [20],
- Electrically <u>heated module</u> providing constant heat flux along the contour of the ribbed wall. This heated module consists of an electrically heated metal foil which forms the ribbed surface and is backed a homogenous and adiabatic foam for stabilisation and insulation. The metal foil is coated by Nextel-Velvet-Coating 2010 [21] to obtain a high and constant emissivity of $\varepsilon = 0.95$ in the sensitivity range of the IR-camera.

The local convective heat transfer coefficient $\alpha_{conv}(\zeta, z)$ is defined as

$$\alpha_{conv}(\zeta, z) = \frac{q_{conv}(\zeta, z)}{T_{wall}(\zeta, z) - T_{air}}. \tag{5}$$

The temperatures indicated by the IR- camera were calibrated (see [22]) for measurements through windows. To determine the object coordinates belonging to each pixel of the IR image, a nonlinear coordinate-transformation is applied, see *Lorenz et al.* [23] and *Leiner et al.* [24].

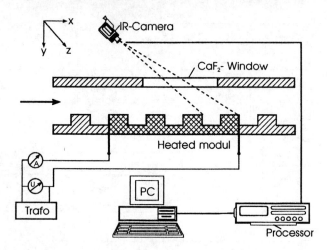

**Fig. 4:** Experimental setup for measuring local temperature distributions by IR-thermography for evaluation of local heat transfer coefficients

The bulk temperature $T_{air}$ is assumed to be the mean air temperature at the entrance cross section of the rib period considered. It is calculated for the i-th period of the module by an energy balance from the temperature $T_0$ of the air flow at the front edge of the heated module (measured by thermocouples) and from the electric power as

$$T_{air} = T_0 + (i-1) \cdot q_{el} \cdot L \cdot B / (c_{P,air} \cdot m_{air}). \tag{6}$$

The convective heat flux $q_{conv}$ in eq.(5) is evaluated from the energy balance of a surface element for steady state flow and electric heating correcting for conduction in the foil and for radiation

$$q_{conv} = q_{el} - q_{cond} - q_{rad} - q_{loss}. \tag{7}$$

The heat flux by tangential conduction in the metal foil is evaluated according to

$$q_{cond} = -\lambda_{foil} \cdot \delta_{foil} \cdot \left( \frac{\partial^2 T_{wall}}{\partial \zeta^2} \right). \tag{8}$$

The one-dimensional temperature distribution $T(\zeta)$ corresponding to a 2-dimensional time averaged flow is obtained by averaging in z-direction. The second derivative for evaluating conductive heat flux from eq.(8) is determined from the measured data by means of cubic spline functions (see e.g. *Fleischer* [25]). Tangential conduction in the foam back of the heating module is neglected.

The heat flux $q_{rad}$ by radiation to and from the surrounding walls is calculated by means of the enclosure method (see e.g. *Siegel et al.* [26]). This method yields averaged values for the heat flux by radiation of each (plane) wall element of the geometric period.

Heat losses $q_{loss}$ through the back side of the heated wall module are estimated to be below 2.6% of the electric heat flux $q_{el}$ and are neglected.

## EFFECTS OF TANGENTIAL CONDUCTION AND RADIATION ON HEAT TRANSFER

The effect of tangential conduction and of radiation on local heat transfer coefficients is shown in Fig. 5 for the basic configuration (P/e = 4, e/H = 1/4, rectangular ribs). The tangential conduction $q_{cond}$ in the metal foil affects the local Stanton number by up to 10%, Fig. 5b). Because of the thermally periodic condition tangential conduction has nearly no effect on the mean Stanton number $St_m$ averaged along one geometric period.

The correction applied here for heat flux by radiation yields average values of this heat flux for each plane wall element. Consequently calculated effects of radiation do not affect local distribution within a surface element but shifts the heat transfer coefficient distribution of the element by up to 13% (at the rib head) and yield nonphysical discontinuities of the evaluated convective Stanton number distribution at the edges. These discontinuities are smoothened here by using cubic spline functions. Radiative heat transfer changes the value of the mean heat transfer coefficient or the mean Stanton number $St_m$.

The peak difference between heat transfer coefficients from both evaluations with and without corrections for both conduction and radiation is 18% at $\zeta/P \approx 0.77$ at the rear flank of the rib. The mean purely convective Stanton number $St_m$ is approximately 8% lower than the mean Stanton number evaluated without correcting for tangential conduction and for radiation (basic configuration, Re = 90 000).

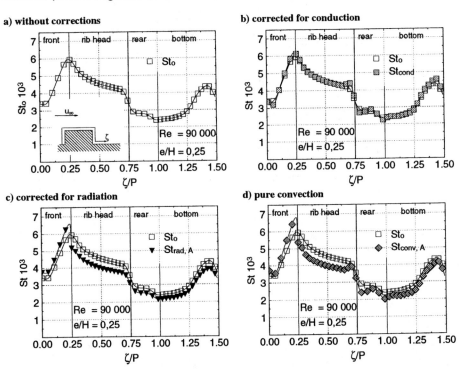

**Fig. 5:** Stanton number distribution along the ribbed contour of the basic configuration (Table 1, P/e = 4)
a) <u>without corrections: $St_o$</u>,  b) <u>corrected for conduction only: $St_{cond}$</u>
c) <u>corrected for radiation only: $St_{rad}$</u>,  d) <u>purely convective: $St_{conv, A}$</u>: corrected for both conduction and radiation.

113

# RESULTS

Velocity fields of the basic configuration (P/e = 4, e/H = 1/4, s/e = 2, rectangular ribs) and the analysis of the validity of the logarithmic law of wall are described in detail by *Lorenz* [6]. Results for the configuration of Fig. 2a (short pitch ratio P/e = 4, rectangular ribs) with a relative rib height of e/H = 1/2 are described by *Lorenz et al.* [27] (local heat transfer coefficient), in [18] (flow structure, wall pressure distribution and numerical simulation) and in [19] (flow visualisation).

Validation of the experimental results by different experimental techniques and by numerical simulations are described in detail by *Lorenz* [6]. The phenomenon of permanent three-dimensional flow patterns in turbulent flow through two dimensional ribbed channels is described and validated by other experimental investigations, numerical simulations and by comparison with similar flow phenomena observed by other researchers by *Lorenz et al.* [28]. Flow structure, wall pressure distribution and local heat transfer coefficients of all investigated configurations are presented in comparison with results from the basic configuration (e/H = 1/4, P/e = 4, s/e = 2, rectangular ribs), Fig. 2a for better clearness of presentation. The presentation of the flow structure, the wall pressure distribution and the local heat transfer coefficient for all investigated configurations is presented as follows:
- Comparison of results for the relative rib height values e/H = 1/2 and 1/6 respectively with results for the basic configuration (e/H = 1/4, symbol: solid circle),
- Comparison of results for the pitch ratio P/e = 10 with results for the basic configuration,
- Comparison of results for semi-cylindrical ribs with results for the basic configuration.

## TURBULENT FLOW STRUCTURE

### VELOCITY PROFILES IN THE FREE CROSS SECTION

The relative velocity profiles $u/u_\infty$ measured at different Reynolds numbers $10\,000 \leq Re \leq 140\,000$ coincide within the range of accuracy of measurement, for details see *Lorenz* [6]. The plane of zero shear stress $y(u'v'=0)$ is located closer to the plane wall than the peak of velocity $y(u_{max})/H$ in all investigated configurations, see table 2. Both relative positions of peak velocity $y(u_{max})/H$ and of zero shear stress $y(u'v'=0)/H$ shift towards the plane wall with increasing relative rib height e/H, Fig. 6. The distance $y(u_{max})/H - y(u'v'=0)/H$ between the locations of peak velocity and of zero shear stress decreases proportionally with increasing e/H, see table 2.

**Table 2:** Relative positions $y(u_{max})/H$ of peak velocity and $y(u'v'=0)/H$ of zero shear stress of investigated rib configurations according Fig. 2 at Re = 90 000

| rib shape | P/e | e/H | $y(u_{max})/H$ | $y(u'v'=0)/H$ | $(y(u_{max})-y(u'v'=0))/H$ |
|---|---|---|---|---|---|
| rectangular | 4 | 1/2 | 0.325 | 0.28 | 0.045 |
| rectangular (basic conf) | 4 | 1/4 | 0.37 | 0.31 | 0.06 |
| rectangular | 4 | 1/6 | 0.41 | 0.34 | 0.07 |
| rectangular | 10 | 1/4 | $0.225^1$ - $0.258^2$ | 0.15 | $0.075^1$ - $0.108^2$ |
| semi-cylindrical | 4 | 1/4 | 0.25 | 0.2 | 0.05 |

[1] above the rib at x/P = 0.1
[2] above the middle of the groove at x/P = 0.773

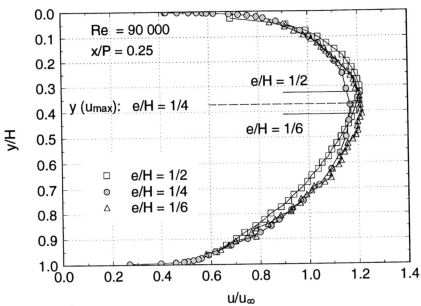

**Fig. 6:** Velocity profiles in the clear cross section of asymmetric channels with rectangular ribs (P/e = 4, Fig. 2a) above the rib at x/P = 0.25 and x/s = 0.5 respectively for different relative rib heights e/H with indicated position of peak velocity $y(u_{max})/H$

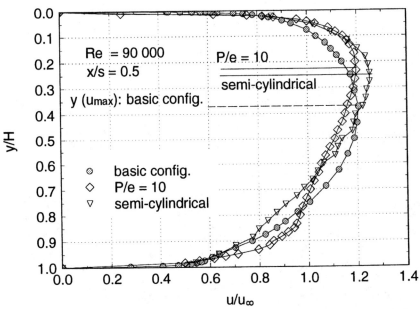

**Fig. 7:** Velocity profiles in the clear cross section above the rib; asymmetric ribbed channels with the basic configuration (P/e = 4, rectangular ribs), rectangular ribs with pitch ratio P/e =10, and semi-cylindrical ribs and positions of peak velocity $y(u_{max})/H$; H/e = 4, Re = 90 000

Velocity profiles in the clear cross section above the ribs (x/s = 0.5) are shown in Fig. 7 for rectangular ribs with a pitch ratio of P/e = 10, for semi-cylindrical ribs p/e = 4 and for the basic configuration. For both, increased pitch ratio p/e = 10 and for semi-cylindrical ribs, the position of peak velocity is closer to the plane wall than for the basic configuration. The relative change of position is in both cases stronger than the relative change due to variation of the relative rib height e/H, see table 2.

Velocity profiles above rectangular and semi-cylindrical ribs respectively with pitch ratio P/e = 4 do not change in flow direction with in the clear cross section $0 \leq y/H \leq 0.95$. For the pitch ratio P/e = 10 the position of peak velocity changes through the pitch length P according to table 2 and the peak velocity value consequently changes periodically in flow direction by up to 20% of the mean value $u_\infty$.

## FLOW STRUCTURE BETWEEN AND ABOVE THE RIBS

The time averaged flow velocity field in the groove between the ribs of the basic configuration (P/e = 4, e/H = 1/4, rectangular ribs) measured by LDA is presented in Fig. 8 as a vector plot. u-velocity-components in different longitudinal planes (y = const.) above one rib are shown in Fig. 9. The shear layer above the groove between two neighboured ribs is almost plane. The terrain of recirculation fills the groove completely. One principal transverse vortex develops with its centre at $x_N/s = 0.7$ in front of the front flank of the following rib, Fig. 8. The u-velocity component in the clear cross section is positive along the complete geometric rib period even at a small distance just above the rib, which indicates that no separation is detected on the top of the rib for P/e = 4, see Fig. 9.

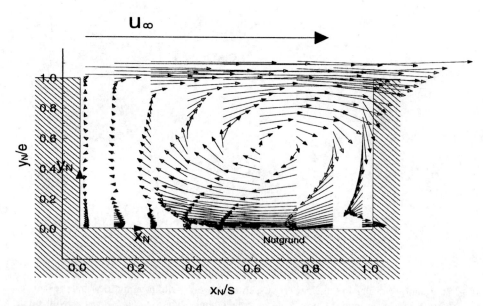

**Fig. 8:** Plot of the time averaged velocity field between two subsequent ribs of basic configuration measured by 2D-LDA (Re = 90 000, e/H = 1/4, P/e = 4, s/e = 2)

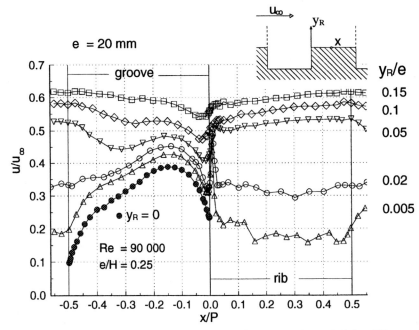

**Fig. 9:** Time averaged u-velocity plotted along the geometric period in different planes $y_R$ = const. for the basic configuration (P/e = 4, e/H = 1/4, rectangular ribs) measured by 1D-LDA

The flow structure between the ribs and the u-velocity components in longitudinal sections in the clear cross section are presented <u>for the large pitch ratio P/e = 10</u> in Fig. 10 and 11. The flow separates from the rear top edge of a rib and reattaches at the bottom of the groove at $x_{N,r} \approx 4e$. This distance $x_{N,r}$ of reattachment agrees well with results of *Martin and Bates* [29] ($x_{N,r} \approx 4e$, for s/e = 1, P/e = 7.2) and *Okamoto and Nakaso* [30] ($x_{N,r} \approx 4e$, for s/e = 1, 6.6 ≤ P/e ≤ 12). A big transverse vortex having its centre at 1.9 e is located behind the rear flank of the rib. In front of the front edge of the following rib high values of transverse v-velocity components occur. In the region from $x_N/e \approx 5$ to $x_N/e \approx 7,5$ the flow is almost parallel (v=0) through the whole cross section, Fig. 10. A second vortex in front of the following rib is hardly to detect. The flow separates from the top of the rib behind the front edge, Fig. 11. Velocity gradients ∂u/∂x and the change of flow direction respectively in front of the rib edge have opposite signs for the long pitch ratio P/e = 10 (Fig. 11) as compared to the short pitch ratio P/e = 4 (Fig. 9).

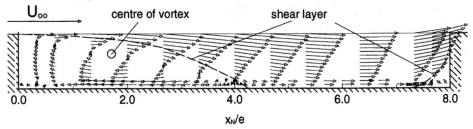

**Fig. 10:** Vector plot of the time averaged velocity field between two following ribs for pitch ratio P/e = 10 from 2D-LDA measurements (Re = 90 000, e/H = 1/4, s/e = 2)

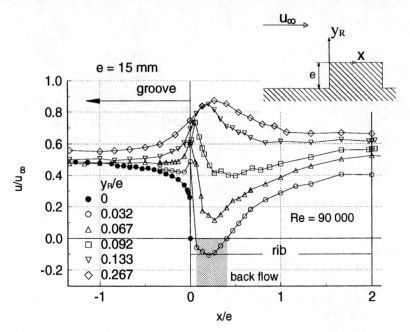

**Fig. 11:** Time averaged u-velocity component in front of the rib and along the rib head in different distances $y_R$ from the rib for $P/e = 10$ ($e/H = 1/4$, rectangular ribs) from 1D-LDA-measurements

The flow structure between and above <u>semi-cylindrical ribs</u> is presented in Fig. 12. One large vortex having its centre in the middle of the groove is generated between the ribs. The flow reattaches at approximately 45° of the rib contour angle, where the $C_p$-distribution has a maximum value, compare Fig. 15, rhs.

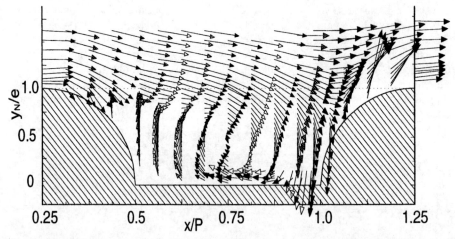

**Fig. 12:** Plot of the time averaged velocity vectors between and above two following ribs for semi-cylindrical ribs from 2D-LDA measurements (Re = 90 000, $e/H = 0.25$, $P/e = 4$)

## WALL PRESSURE DISTRIBUTIONS

The wall-pressure-coefficient ($C_p$)-distributions are nearly independent from the Reynolds number in the investigated turbulent range in all investigated rib configurations, c.f. *Lorenz* [6].

The wall-pressure-coefficient ($C_p$)-distributions along rectangular-rib configurations of short pitch ratio P/e = 4 with <u>different relative rib heights e/H</u> are presented in Fig. 13. The $C_p$-distribution is almost independent from e/H. The relatively largest differences occur at the front flank of the rib ($0.0 \leq \zeta/P \leq 0.25$) and at the bottom of the groove close to the rear flank at $1.4 \leq \zeta/P \leq 1.50$. The largest gradients $\partial C_p/\partial \zeta$ of the $C_p$-distribution occur at the greatest relative rib height examined e/H = 0.5. The changes of $c_p$ through one geometriv period of short pitch ratio p/e = 4 are of the order of only

$$c_{pmax} - c_{pmin} \approx 0{,}25$$

which may be explained by the fact that the shear sheet between the ribs is nearly plane and no strong changes of flow direction are imposed to the main flow.

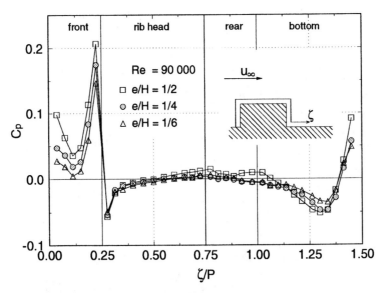

**Fig. 13:** Wall-pressure-coefficient ($C_p$)-distribution along the contour of the ribbed wall for rectangular ribs of different relative rib heights e/H, P/e = 4, Re = 90 000

Wall pressure distributions along the contours of both rectangular-rib configurations, pitch ratio <u>P/e = 10</u> and P/e = 4 (basic configuration), are presented in Fig. 14. $C_p$-values and gradients $\partial C_p/\partial \zeta$ are much higher for P/e = 10 than for the basic configuration. For p/e = 10 the flow reattaches in the groove between two ribs and the main flow is subject to essential periodic changes of velocity profile linked with strong differences in the wall pressure distribution in the order of

$$c_{pmax} - c_{pmin} \approx 1{,}3 \; .$$

Wall pressure distributions in the basic configuration with rectangular ribs and in the configuration with semi-cylindrical ribs, both with p/e = 4, are compared in Fig. 15. Above cylindrical ribs the near-all main flow accelerates from the point of reattachment to the top and then slows down until the point of separation similar to the flow in a Venturi nozzle. Strong differences of wall pressure result along the ribbed contour:

$$c_{pmax} - c_{pmin} \approx 0{,}5.$$

The shapes of the $c_p$ plots for both configurations (maximum near the stagnation point, minimum in the region of highest tangential velocities at the edge of the rib for rectangular ribs and on the top of the cylinder for semi-cylindrical ribs) have some similar features. However the jump in the $C_p$-distribution caused by the front edge of the rectangular ribs is sharp whereas this jump extends over almost 45° for semi-cylindrical ribs. Both, the $C_p$-values and the gradients $\partial C_p/\partial \zeta$ are higher for the semi-cylindrical ribs than for the basic configuration (both of same pitch ratio P/e = 4), but even smaller than for the large pitch ratio P/e = 10. At the position $x_{N,r}/e \approx 4 \Leftrightarrow \zeta/H \approx 2$ where the flow reattaches at the bottom of the groove for P/e = 10, the $C_p$-value is zero ($C_p(x_{N,r}/e \approx 4) = 0$).

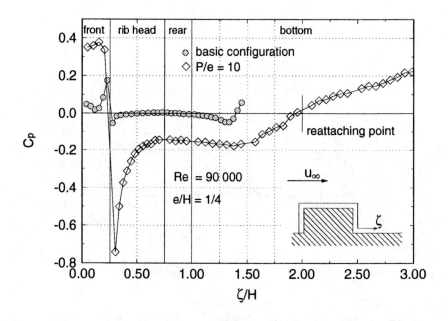

**Fig. 14:** Distribution of wall-pressure-coefficient ($C_p$) along the contours of ribbed walls with rectangular ribs with pitch ratio P/e = 10 and of the basic configuration (P/e = 4) respectivly, Re = 90 000

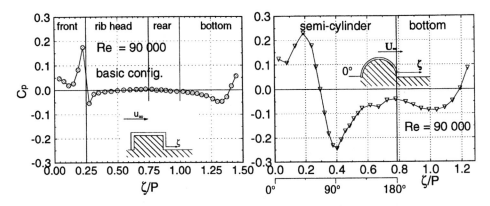

**Fig. 15:** Distribution of wall-pressure-coefficient ($C_p$) along the contours of ribbed walls with semi-cylindrical and rectangular ribs (P/e = 4, e/H = 1/4) respectively, Re = 90 000

## LOCAL CONVECTIVE HEAT TRANSFER

Local distributions of Stanton number and of Nusselt number respectively in the basic configuration are shown for different Reynolds numbers in Figs. 16 and 17 respectively. The shapes of the plots are qualitatively similar for all investigated Reynolds numbers. The mean values St and gradients $\partial St/\partial \zeta$ decrease whereas Nu and $\partial Nu/\partial \zeta$ respectively increase with Re. The position of the local relative maximum of St and Nu respectively at the bottom of the groove, $1.4 \leq \zeta/P \leq 1.5$, shifts slightly to the rear flank of the rib with increasing Re.

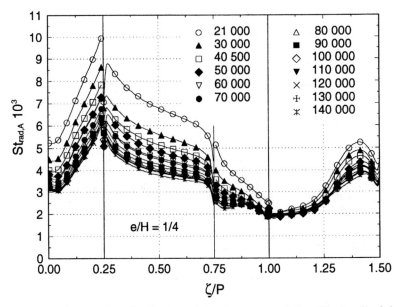

**Fig. 16:** Local Stanton number distributions along the contour of the ribbed wall of the basic configuration (P/e = 4, e/H = 1/4, rectangular ribs) for different Reynolds numbers

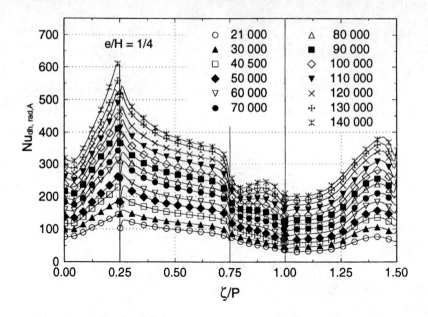

**Fig. 17:** Local Nusselt number distributions along the contour of the ribbed wall of the basic configuration (P/e = 4, e/H = 1/4, rectangular ribs) for different Reynolds numbers

The Stanton number distributions are shown in Fig. 18 for different relative rib heights $1/6 \leq e/H \leq 1/2$. The rib-Reynolds number $Re_e$ given in Fig. 18 is defined with the rib height e

$$Re_e = \frac{u_\infty \cdot e}{\nu}. \qquad (9)$$

In Fig. 18 the local Stanton number is related to the mean Stanton-number $St_m$ averaged along one geometric period. The shapes of plot for same rib-Reynolds number $Re_e$ are almost equal for all relative rib heights e/H. Greatest differences occur at the front flank, as they do in the (unscaled) $C_p$-distributions.

Stanton number distributions for the large pitch ratio P/e = 10 and for the basic configuration (P/e = 4) are compared in Fig. 19. The local Stanton number is much higher along the whole contour with the large pitch ratio P/e = 10 than it is in the basic configuration (P/e = 4).

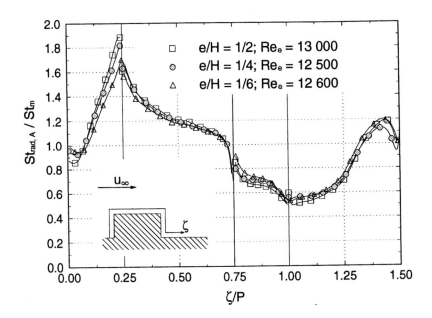

**Fig. 18:** Relative Stanton-number distributions $St/St_m$ along the contour of the ribbed wall for rectangular ribs with $P/e = 4$ and different relative rib heights $e/H$ at constant rib-Reynolds-number $Re_e$ (basic configuration: $e/H = 1/4$).

For the small pitch ratio $P/e = 4$ the maximum of local Stanton-number appears just at the upper edge of the front flank, whereas for the long pitch ratio $P/e = 10$ the maximum of the Stanton number is located on the top of the rib downstream from the edge. For the long pitch ratio $P/e = 10$ the Stanton number has a local maximum value near the position $x_{N,r}/e \approx 4$, $\zeta/H \approx 2$ where the flow reattaches at the bottom of the groove.

Stanton number distributions in configurations of semi-cylindrical ribs and rectangular ribs (basic configuration) are compared in Fig. 20. The Stanton number in an asymmetrically heated plane channel according to eq.(3) and of a turbulent boundary layer according to eq.(4) are shown additionally as reference cases in the diagram for the basic configuration, Fig. 20, lhs. The differences between both plots, Fig. 20, are similar to those of the Cp-distribution, Fig. 15. The points of reattachment (edge of the front rib for rectangular ribs and at approximately 45° of the rib contour angle for semi-cylindrical ribs) coincide in both cases with maxima of local heat transfer coefficient and Stanton number respectively.

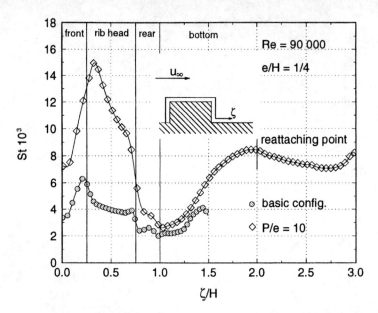

**Fig. 19:** Stanton-number distribution along the contour of the ribbed wall for rectangular ribs with pitch ratio P/e = 10, compared to the basic configuration.

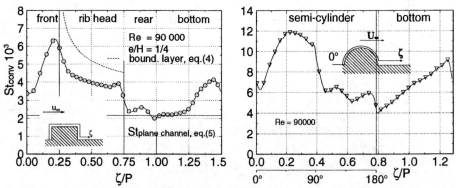

**Fig. 20:** Stanton-number distribution along the contours of the ribbed wall for semi-cylindrical ribs, rhs, compared to that of the basic configuration with rectangular ribs, lhs, (e/H = 1/4, P/e = 4)

## GLOBAL HEAT TRANSFER AND PRESSURE LOSSES

Average values of apparent friction factors $f_{app}$ and of Stanton numbers $St_m$ are plottet versus $Re_{dh}$ in Fig. 21 and 22 respectively for all investigated configurations. The reference values of the plane channel according to equations (2) and (3) are presented additionally for comparison. In Fig. 23 the ratio $(St_m/St_{plane})/(f_{app}/f_{plane})$ describing the proportion of heat transfer augmentation and increase of pressure loss relative to the plane channel is presented.

Friction factors decrease slightly with decreasing relative rib height e/H, but the mean Stanton number is almost independent from e/H. For the long pitch ratio P/e = 10 the apparent friction factor and mean heat transfer is up to 2.5 times higher than for the small pitch ratio P/e = 4, both with rectangular ribs. This is in good agreement with results of other researchers (c.f. *Wilkie* [12], see also *Lorenz* [6]). The change of $f_{app}$ resulting from variation of relative rib height e/H is negligible as compared to the change in $f_{app}$ due to variation of pitch ratio P/e. For semi-cylindrical ribs, $f_{app}$ is just between the values for rectangular ribs of the small pitch ratio P/e = 4 and of the long pitch ratio P/e =10, whereas the mean Stanton number $St_m$ is of the order of the values for the pitch ratio P/e = 10 and rectangular ribs.

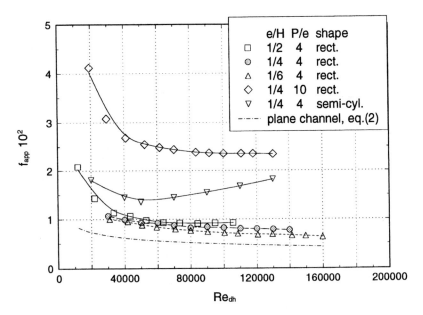

**Fig. 21:** Apparent friction factor $f_{app}$ plotted versus Re for investigated rib configurations and plane channel

The ratio of heat transfer augmentation to increase of pressure loss is $(St_m/St_{plane})/(f_{app}/f_{plane}) > 1$ for the small pitch ratio P/e = 4 and both semi-circular (e/H = 1/4) and rectangular ribs (e/H =1/6), which means that the heat transfer is more increased than the global pressure loss as compared to a plane channel of the same hydraulic diameter $d_h$. The long pitch ration P/e = 10 with highest heat transfer and pressure loss yields up to 50% lower values of $(St_m/St_{plane})/(f_{app}/f_{plane})$ than the small pitch ratio P/e = 4.

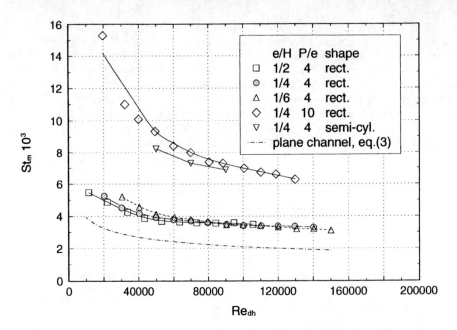

**Fig. 22:** Mean Stanton number for different rib configurations and for a plane channel plotted versus Re

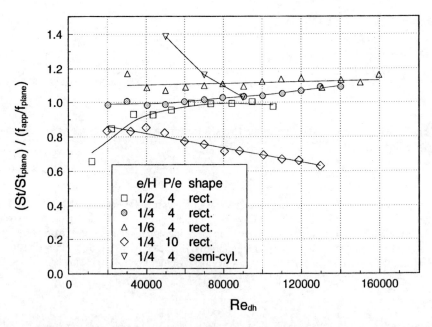

**Fig. 23:** Ratio $(St_m/St_{plane})/(f_{app}/f_{plane})$ for different rib configurations related to the value of the plane channel, plotted versus Re

# SUMMARISED RESULTS

The velocity field, the wall pressure distribution and the local heat transfer coefficients are measured for turbulent flow in asymmetrically ribbed channels along the contour of the ribbed wall. The relative rib height e/H, the pitch ratio P/e and the shape of the ribs are varied.

## FLOW STRUCTURE

The positions of peak velocity $y(u_{max})/H$ of the velocity profiles and of zero shear stress $y(u'v' = 0)/H$ in the free cross section are located closer to the plane wall for increasing relative rib height e/H. The dislocation of the peak velocity is even stronger for the variation of the pitch ratio from P/e = 4 to P/e = 10 and for variation of the rib shape from rectangular to semi-cylindrical ribs. The positions of peak velocity in the free cross section are almost equal $(y(u_{max})/H \approx 0,2)$ for rectangular ribs with long pitch ratio (P/e = 10) and for semi-cylindrical ribs with short pitch ratio (P/e = 4). The flow structure between and above the ribs depends strongly on both the pitch ratio P/e and on the rib shape.

## WALL PRESSURE DISTRIBUTION

The relative rib height e/H has only little effect on the local wall pressure distribution. Absolute values of $C_p$ and gradients $\partial C_p/\partial \zeta$ are up to 10 times higher for rectangular ribs with long pitch ratio P/e = 10 than for small pitch ratio P/e = 4. The absolute values and the local gradients $\partial C_p/\partial \zeta$ are greater for semi-cylindrical ribs than for rectangular ribs of equal pitch ratio P/e = 4, but smaller than for rectangular ribs of long pitch ratio P/e = 10.

## LOCAL HEAT TRANSFER

The distribution of relative local Stanton number scaled by the mean Stanton number $St_m$ along one rib period, $St/St_m$, is almost equal for different relative rib heights e/H and for equal values of rib Reynolds number $Re_e$ or equal rib size e and flow velocity u, respectively. The local heat transfer coefficient is much higher along the whole rib contour and the maximum of local heat transfer is displaced from the front edge of the rib (at small pitch ratio P/e = 4) to a position behind the edge on the top of the rib for long pitch ratio P/e = 10. The reason seems to be a flow separation on the top of the rectangular rib for the long pitch ratio P/e = 10, whereas for the small pitch ratio P/e = 4 no separation on top of the rectangular rib was observed. The absolute values of St are in the same order of magnitude for semi-cylindrical ribs (P/e = 4) as for rectangular ones of large pitch ratio P/e = 10. Rectangular ribs of long pitch ratio P/e = 10 have highest values of St and the gradients $\partial St/\partial \zeta$. The relative differences between rectangular ribs and semi-cylindrical ribs in both, the wall pressure distribution and in the heat transfer coefficient distribution, are qualitatively similar. The stronger effect of the semicylindric rib shape as compared to rectangular ribs, both of short pitch ratio p/e = 4, is evidently due to the fact that the flow attaches along a good part of the curved rib contour and induces locally a strong acceleration and subsequent deceleration, linked with strong changes of wall pressure and a pressure minimum on the top of the rib. By this way relatively strong disturbances are induced into the main flow, increasing strongly heat transfer and pressure loss.

# GLOBAL HEAT TRANSFER AND PRESSURE LOSS

The channel height and relative rib height respectively have almost no effect on the mean Stanton number $St_m$ and only little effect on the apparent friction factor. Highest mean Stanton number and pressure loss occur in the rectangular-rib configuration with the long pitch ratio P/e = 10 ($f_{app,max}$ = 0.042, $St_{m,max}$ = 0.015, $Re_{dh}$ = 30,000). The highest values for the relation $(St_m/St_{plane})/(f_{app}/f_{plane})$ occur for semi-circular ribs with pitch ratio P/e = 4. The Reynolds number has a great influence on the ratio $(St_m/St_{plane})/(f_{app}/f_{plane})$. The greatest value $(St_m/St_{plane})/(f_{app}/f_{plane}) \approx 1.38$ is obtained for Re = 50,000.

# REFERENCES

[1] Fiebig, M.: *Vortices as Tools to Influence Heat Transfer - Overview of the Results of the DFG-Research Group 'Vortices and Heat Transfer'*, in Vortices and Heat Transfer, Closing Seminar of the DFG-Research Group „Wirbel- und Wärmeübertragung", Proceedings of Advance Program Seminar , 28/29 November, Bochum, Ruhr-Universität, Germany, 1996.

[2] Braun, H., Neumann, H., Fiebig, M.: *Vortex Structure, Heat Transfer and Flow Losses in Turbulent Channel Flow with Periodic Longitudinal Vortex Generators*, in Vortices and Heat Transfer, Closing Seminar of the DFG-Research Group „Wirbel- und Wärmeübertragung", Proceedings of Advance Program Seminar , 28/29 November, Bochum, Ruhr-Universität, Germany, 1996.

[3] Gersten. K., Herwig, H., Kiel, R., Vieth, D.: *Two Dimensional Turbulent Boundary Layers with Separation and Reattachment Including Heat Transfer*, in Vortices and Heat Transfer, Closing Seminar of the DFG-Research Group „Wirbel- und Wärmeübertragung", Proceedings of Advance Program Seminar , 28/29 November, Bochum, Ruhr-Universität, Germany, 1996.

[4] Grosse-Gorgemann, Weber, D., Fiebig, M.: *Heat Transfer and Flow Losses in Transition from Longitudinal to Transverse Vortices in Steady and Oscillating Channel Flow*, in Vortices and Heat Transfer, Closing Seminar of the DFG-Research Group „Wirbel- und Wärmeübertragung", Proceedings of Advance Program Seminar , 28/29 November, Bochum, Ruhr-Universität, Germany, 1996.

[5] Hahne, W., Weber, D., Fiebig, M.: *Flow and Heat Transfer in Ribbed Channels with Self-sustained Oscillating Transverse Vortices*, in Vortices and Heat Transfer, Closing Seminar of the DFG-Research Group „Wirbel- und Wärmeübertragung", Proceedings of Advance Program Seminar , 28/29 November, Bochum, Ruhr-Universität, Germany, 1996.

[6] Lorenz, S.: *Lokaler Wärmeübergang und Strömungsstruktur bei turbulenter Strömung in einseitig querberippten Kanälen*, Bochum, Univ., Diss., 1995, zgl. VDI Fortschrittberichte, Reihe 7, Nr. 285, VDI Verlag Düsseldorf, 1996.

[7] Grosse-Gorgemann, A. *Numerische Untersuchung der laminaren Strömung und des Wärmeübergangs in Kanälen mit rippenförmigen Einbauten*, Bochum, Univ., Diss., 1995, zgl. VDI Fortschrittberichte, Reihe 19, Nr. 87, VDI Verlag Düsseldorf, 1996.

[8] Hanjalic, K., and Launder, B.E. *Fully developed asymmetric flow in a plane channel*, J. Fluid Mech., Vol. 51, Part 2, pp. 301-335, 1972.

[9] Meyer, L. *Turbulente Strömung an Einzel- und Mehrfachrauhigkeiten am Plattenkanal*, Karlsruhe, Univ., Diss., 1978, Vgl.: *Turbulent flow in a plane channel having one or two rough walls*, Int. J. Heat Mass Transfer, Vol. 23, pp. 591- 608, 1980.

[10] Aliaga, D.A., Lamb, J.P., Klein, D.E. *Convective heat transfer distribution over plates with square ribs from infrared thermography measurements*, Int. J. Heat Mass Transfer, Vol. 37, No. 3, pp. 363 - 374, 1994.

[11] Aliaga, D.A., Lamb, J.P., Klein, D.E. *Heat transfer measurements on a ribbed surface at constant heat flux using infrared-thermography*, Experimental Heat Transfer, Vol. 6, No.1, pp 17-34, 1993.

[12] Wilkie, D. *Forced Convection Heat Transfer from Surfaces Roughened by Transverse Ribs*, in: Heat Transfer 1966: proceedings of the Third International Heat Transfer Conference, Chicago, USA, Vol. 1, pp. 1-19, 1966.

[13] Dean, R.B. *Reynolds number dependence of skin friction and other bulk flow variables in two dimensional rectangular duct flow*, J. Fluids Engng., Vol. 100, pp. 215-223, 1978.

[14] Sparrow, E.M., Lloyd, J.R., Hixon, C.W. *Experiments on turbulent heat transfer in an asymmetrically heated rectangular duct*, J. of Heat Transfer, Vol. 88, pp. 170-174, 1966.

[15] Sakakibara, M. *Analysis of heat transfer in the entrance region with fully developed turbulent flow between parallel plates - The case of uniform heat flux*, Memoirs of the Faculty of Engineering, Fukui University, Vol. 30, pp. 107-120, 1982.

[16] Kays, W.M., Leung, E.Y. *Heat transfer in annular passages: hydro-dynamically developed turbulent flow with arbitrary prescribed heat flux*, Int. J. Heat Mass Transfer, Vol. 6, pp. 537 - 557, 1963.

[17] Kays, W.M., Crawford, M.E. *Convective Heat and Mass Transfer*, McGraw-Hill, New York, 2nd Edition, 1980.

[18] Lorenz, S., Braun, H., Bai Li, Leiner, W. *Wall pressure distribution in a channel with periodic transverse grooves - comparison of experimental and numerical results for turbulent flows*, in: Vortices and Heat Transfer: Proceedings of Eurotherm Seminar No. 31, 24-26 May, 1993, Bochum, Ruhr-Universität Bochum, pp. 87 - 92, 1993.

[19] Lorenz, S., Neumann, H., Schulz, K., Leiner, W. *Evaluation of streamlines and velocities of turbulent air-flow by visualization with helium-bubbles*, in "Optical Methods and Data Processing in Heat and Fluid Flow", IMechE Seminar Papers, City-University, London, pp. 111 - 115, April 1994.

[20] Zeiss: Produktinformationen Optische Kristalle, Kalziumfluorid ($CaF_2$), 1992.

[21] Lohrengel, J. *Gesamtemmissionsgrad von Schwärzen*, Wärme- und Stoffübertragung, Bd. 21, S. 311 - 315, 1987.

[22] Neumann, H., Lorenz, S., Leiner, W. *Infrarot-Thermographie durch Fenster bei niedrigen Temperaturen*, Wärme- und Stoffübertragung, Bd. 29, S. 219 - 225, 1994.

[23] Lorenz, S., Velesco, N., Leiner, W. *Verfahren zur Koordinatenbestimmung (Entzerrung) aus Infrarot-Aufnahmen von strukturierten Oberflächen*, Institutsbericht ITF-WST 95-02, Ruhr-Universität Bochum, 1995.

[24] Leiner, W., Schulz, K., Behle, M., Lorenz, S. *Imaging techniques to measure local heat and mass transfer* IMechE Conference Transactions 1996 - 3, in: Optical Methods and Data Processing in Heat and Fluid Flow, ISSN 1356-1448, ISBN 0 85298 992 X.

[25] Fleischer, P. *Optimale ausgleichende Spline-Funktionen: Berechnung, Eigenschaften und analytische Anwendung*, Würzburg, Univ., Diss., 1989.

[26] Siegel, R., Howell, J., Lohrengel, J. *Wärmeübertragung durch Strahlung*, Teil 2: Strahlungsaustausch zwischen Oberflächen und in Umhüllungen, Springer, Berlin (u.a.), 1991.

[27] Lorenz, S., Mukomilow, D., Leiner, W. *Distribution of the heat transfer coefficient in a channel with periodic transverse grooves*, Experimental Thermal and Fluid Science, Vol. 11, No. 3, pp. 234 - 242, Oct. 1995.

[28] Lorenz, S., Nachtigall, C., Leiner, W. *Permanent tree-dimensional patterns in turbulent flows with essentially two-dimensional wall configurations*, Int. J. Heat Mass Transfer, Vol. 39, No. 2, pp. 373 - 382, 1996.

[29] Martin, S.R., Bates, C.J. *Small-probe-volume laser doppler anemometry measurements of turbulent flow near the wall of a rib-roughened channel*, Flow Meas. Instrum., Vol. 3, No 2, pp. 81 - 88, 1992.

[30] Okamoto, S., Nakaso, K. *Turbulent Shear Flow Over Rows of Two-dimensional Square Ribs On Ground Plane*, in: Eighth Symposium on Turbulent Shear Flows, 9-11, Sept. 1991, Tech. Univ. of Munich, Germany, Vol. 1, Session 14-3.

# FLOW AND HEAT TRANSFER IN RIBBED CHANNELS WITH SELF-SUSTAINED OSCILLATING TRANSVERSE VORTICES

A. Grosse-Gorgemann, H.-W. Hahne, H. Neumann, D. Weber and M. Fiebig
Lehrstuhl für Wärme und Stoffübertragung,
Institut für Thermo- und Fluiddynamik
Ruhr-Universität Bochum, 44801 Bochum, Germany

## SUMMARY

Numerical and experimental investigations of flow structure and heat transfer in a channel with periodically mounted transverse vortex generators (ribs) have been conducted in the Reynolds number range of steady laminar to oscillatory transitional flow. The unsteady Navier-Stokes equations and the energy equation have been solved by the finite volume code FIVO developed by the research group 'Vortices and Heat Transfer' at Bochum. Two component hot wire anemometry and the ammonia absorption technique have been used to deduce the velocity field and the wall heat transfer respectively. The numerical investigations encompass the Reynolds number range from steady flow to the lower transition region. In the upper Reynolds number range, Re>1000, experimental investigations have been carried out. Numerical and experimental results have been compared for validation in the Reynolds number range where transition sets in. Due to the periodic geometry instability leads at relatively low Reynolds numbers to self-sustained oscillations and vortex shedding from the ribs. With increasing Reynolds number their amplitudes and number of frequencies increase until a nearly continuous spectrum is obtained. Data for heat transfer and flow losses are presented for different rib heights as well as for different longitudinal pitch in a Reynolds number range of Re=100 to Re=3000.

## 1. INTRODUCTION

Induced flow oscillations will always lead to enhanced heat transfer, but also to additional pressure losses. Pioneering investigations on self-oscillating flows over cavities started in the seventies. Sarohia 1977 [1] investigated flow oscillations over shallow cavities experimentally. He distinguished between „open" and „closed" cavities. For open cavities where the cavity lenght to depth ratio is smaller than 8 the separating shear layer reattaches at the front edge of the next protrusion. For closed cavities the shear layer reattaches at the base of the cavity. Here the lenght to depth ratio is greater than 8. Closed cavities became unsteady at lower Reynolds numbers than the corresponding plane channel flow.

Open cavities were considered by Ghaddar et al. [2,3] and Amon and Mikic [4,5]. They investigated open cavities in grooved channels numerically and determined the flow field and temperature distribution, flow losses and heat transfer. They observed self-sustained oscillations at a critical Reynolds number which was lower than the critical Reynolds number of the correponding plane channel flow. The detected frequency was identical with the frequency of the Tollmien-Schlichting wave of minimal damping for plane channel flow at the investigated Reynolds number. The global Nusselt number was considerably enhanced by the self-sustained oscillations. Greiner et al. [6] and Greiner [7] investigated resonant heat transfer in grooved channels. Flow pulsations where actively imposed on the flow. The frequency of the pulsations was tuned to the natural frequency of the grooved channel flow and a further increase in heat transfer was observed.

Closed cavities were investigated by Tropea and Gackstalter [8] and Durst et al. [9]. They studied the flow field over two obstacles in tandem and found lower critical Reynolds numbers than for open cavities. No heat transfer measurements were reported. Rowley and Patankar [10] investigated tubes with internal circumferential fins, and Webb and Ramadhyani [11] studied heat transfer in a channel with staggered ribs. In both cases no heat transfer enhancement was reported for steady flow.

Up to now systematic studies for periodically ribbed channels have not been reported for open cavities which include heat transfer. For this purpose a channel with periodic transverse ribs attached to one channel wall is considered here, see Fig. 1.1. As base geometry a rib with height (h) of 0.5 the channel height (H), a longitudinal pitch ($L_p$) of 10 times the rib height, and a rib thickness ($\delta$) of 10% the rib height is chosen. Usually cavities are closed for $(L_p - \delta)/h \geq 8$ and opened for $(L_p - \delta)/h \leq 8$, so the base geometry represents a closed cavity.

Questions of major interest are the following:
- What is the critical Reynolds number for the onset of self-sustained large, amplitude oscillations?
- How do these amplitudes and frequencies change with Reynolds number?
- Does the flow become three dimensional for this two dimensional geometry?
- What is the influence of Reynolds number, rib height ratio and longitudinal pitch ratio on the oscillations, heat transfer and flow losses?

Here an experimental and numerical study is performed. For the detailed study of the self-sustained oscillatory flow and temperature field a three dimensional unsteady finite volume method has been developed by our group for the solution of the continuity equation, Navier-Stokes equations and energy equation for a constant property fluid. The code is validated by grid independence studies and by comparision with experimental results. The most severe test for a code is probably the detailed comparison of self sustained amplitudes and frequencies determined from experimental results. Here, for a reference point in the flow field, the computed and measured mean velocity components, their frequencies and amplitudes are compared. The measurements were performed with a special hot wire probe in a dedicated wind tunnel for low velocities, see section 3.1.2.

For Reynolds numbers above $10^3$ local Nusselt numbers and pressure drop are determined experimentally, see sections 3.1.3 and 3.1.4. The ammonia absorption method (AAM) was further developed to yield a high local resolution which could be analysed digitally.

The numerical and experimental results are discussed in section 4. The numerical and experimental local velocities and their fluctuations are in key with the experimental accuracy.

Pressure loss and heat transfer data for low Reynolds numbers were only determined numerically. The good agreement between hot wire measurements and numerical calculations at low Reynolds numbers gave sufficient confidence into the numerical values. This was substantiated by the grid independent studies, see section 2.5.

For Reynolds numbers between 1000 and 3000 experimental results are presented with two different rib heights of h/H = 0.5 and 0.25. For Re = 3000 the results at h/H = 0.25 are compared with the large eddy simulation of Braun [12]. The presented results draw heavily on the theses of A. Grosse-Gorgemann [13] and D. Weber [14].

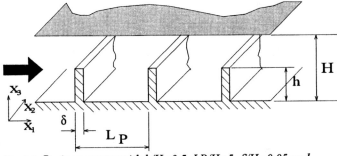

Fig. 1.1: *Basic geometry with h/H=0.5, LP/H=5, δ/H=0.05 and* $Re = 2H\bar{u}/\nu$.

## 2. MATHEMATICAL AND NUMERICAL MODEL

### 2.1 BASIC EQUATIONS

The flow and temperature fields are governed by the continuity, nonsteady three dimensional Navier-Stokes and energy equation. The fluid properties are assumed to be constant and the dissipation terms in the energy equation are neglected. The equations then read in nondimensional cartesian tensor notation[1]:

Continuity equation

$$\frac{\partial u_i}{\partial x_i} = 0 \qquad (2.1)$$

Momentum equation

$$\frac{\partial u_i}{\partial t} + \frac{\partial u_i u_j}{\partial x_j} = -\frac{\partial p}{\partial x_i} + \frac{2}{Re}\frac{\partial^2 u_i}{\partial x_j^2} \qquad (2.2)$$

Energy equation

$$\frac{\partial T}{\partial t} + \frac{\partial T u_i}{\partial x_i} = \frac{2}{Re\,Pr}\frac{\partial^2 T}{\partial x_i^2} \cdot \qquad (2.3)$$

The velocity components have been nondimensionalised with the average velocity $\bar{u}$ and the length with the channel height H. The Reynolds number is $Re=\bar{u}2H/\nu$. The temperature T is nondimensionalized with the reference value $T_\infty$ and the pressure with $\rho * \bar{u}^2$ where $\rho$ is the density. Pr is the Prandtl number given by $\nu/a$.

---

[1] In the further progress of the articel the cartesian notation for the axis is used likewise (x=$x_1$, y=$x_2$, z=$x_3$).

## 2.2 BOUNDARY CONDITIONS

The solutions depend on the hydrodynamic and thermal boundary conditions. With the notation of Fig. 1.1 the hydrodynamic and thermal boundary conditions for the computational domain are:

Solid walls:

- no slip: $u_i = 0$ (2.4)

- constant wall temperature: $T_w = \text{const.}$ (2.5)

Side boundaries: Two different conditions are considered, (a) symmetry and (b) periodicity.

- (a) symmetry:
$$\Psi = \{u_i, p, T\}$$

$$\left.\frac{\partial \Psi}{\partial x_2}\right|_{x_2=0} = \left.\frac{\partial \Psi}{\partial x_2}\right|_{x_2=Bp} = 0 \qquad (2.6)$$

$$u_2|_{x_2=0} = 0 \;;\; u_2|_{x_2=Bp} = 0$$

- (b) periodicity:
$$\Psi = \{u_i, p, T\} \qquad (2.7)$$

$$\Psi(x_2 = 0) = \Psi(x_2 = Bp) \qquad (2.8)$$

$$\left.\frac{\partial \Psi}{\partial x_2}\right|_{x_2=0} = \left.\frac{\partial \Psi}{\partial x_2}\right|_{x_2=Bp} \qquad (2.9)$$

Inlet and outlet:

- periodicity with $p = p_p - \frac{\partial p}{\partial x_i} \cdot x_i$ and $\theta = \frac{T_W - T}{T_W - T_B}$

$$\Psi = \{u_i, p_p, \theta\}$$

$$\Psi(x_1 = 0) = \Psi(x_1 = Lp) \qquad (2.10)$$

$$\left.\frac{\partial \Psi}{\partial x_1}\right|_{x_1=0} = \left.\frac{\partial \Psi}{\partial x_1}\right|_{x_1=Lp} \qquad (2.11)$$

Due to the assumption of incompressibility the mass conservation law is decoupled from the pressure. This leads to difficulties in finding the pressure field that satisfies the momentum and continuity equation. To remove these difficulties an equation for the pressure is derived with the use of the continuity equation. This equation leads to a pressure field that fulfills the momentum and continuity equation. The result is that the equation system to solve consists of the momentum equations, a Poisson equation for the pressure and the energy equation. The applied algorithm for coupling velocity and pressure is the SIMPLE-C algorithm. It is described in detail in [15].

## 2.3 DISCRETIZATION

A finite-volume method on a co-located grid is used for discretizing the equations (2.1)-(2.3). For the time dependent terms a backward Euler scheme with first order accuracy is used. The viscous and the pressure terms are discretized with central differences with second order accuracy. The non-linear convective terms (CT) are determined through the 'deferred-correction approach', described in [13]. According to this technique, the convective flux is split into an implicit part, expressed through the upwind differencing scheme (UDS) and an explicit part containing the differences between the central scheme (CDS) and the UDS approximations. The explicit part is weighted with a factor ($\Gamma$-factor) so that the discretization of the convective terms leads to the following expression:

$$CT = UDS + \Gamma (CDS - UDS) . \qquad (2.15)$$

This procedure results for $\Gamma=1$ in a second order difference scheme and for $\Gamma=0$ in a first order difference scheme for the convective terms.

## 2.4 NUMERICAL METHOD

The discretization results in a system of three non-linear algebraic equations for the three velocities and in two linear algebraic equation systems for the pressure and temperature. The nonlinearity is caused by the convective terms of the momentum equation. For solving the linearized and linear algebraic equation systems the SIP-algorithm [16] is used.
To handle the problem of non-linearity an iterative solution method is used. The computation cycle starts with guessed fields for velocity, pressure and temperature. From these fields, corrected pressure and velocity fields are interactively calculated with the modified SIMPLE-C algorithm [13]. From the initial temperature field and the afore mentioned corrected velocity field, a temperature distribution for the first cycle is calculated from the energy equation. This new temperature field and the forgoing corrected pressure and velocity fields are used to solve Eq. (2.2) to evaluate velocities and temperatures for the following increment. In subsequent cycles the solution proceeds in three steps. In the first step, the momentum equations are solved with the help of converged velocity and temperature fields of the previous time step.

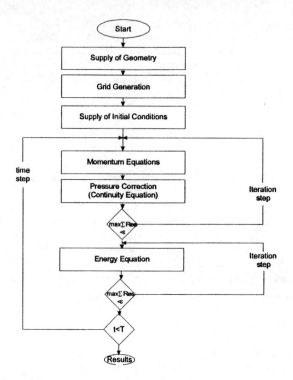

**Fig. 2.4.1** *Flow chart for the numerical solution algorithm.*

In the second step, a pressure velocity correction equation is solved, following the modified SIMPLE-C algorithm which determines the correct pressure and a divergence-free velocity field. In the third step, the solution of the energy equation is accomplished in order to obtain a temperature field for the next cycle. This algorithm is illustrated in the flow chart for the described method shown in Fig. 2.4.1.

## 2.5 VALIDATION

The transformation of the basic differential equations (Eqs. 2.1-2.4) into an algebraic equation system gives rise to errors. These errors are uncertainties that depend on the precision of the differential scheme in space and time and on the number of volumes in the computational domain.

The aim of this investigation is the error estimation as a function of the grid refinement and the applied discretiziation scheme for the convective terms.
As reference values for the grid refinement study values for the grid size equal zero are used. The reference values (global Nusselt number, apparent friction factor and a reference velocity are determined with a Romberg extrapolation. The importance of the $\Gamma$-factor (Eq.2.15) was examined by setting the factor to $\Gamma = 0; 0.5; 1.0$ in Eq. (2.15). This leads to different influences of the central and up-wind terms on the discretiziation.

The flow for the basic geometry was computed with four grids and two Reynolds numbers. The grids are obtained through grid doubling, beginning with the coarse grid of 100 grid points in x-direction and 20 grid points in y-direction, see Tab. 2.1. The Reynolds numbers are Re=100 for steady flow and Re=300 for unsteady flow.

**Tab. 2.1** *Used grids*

| grid no. | nx | ny |
|---|---|---|
| 8 | 102 | 22 |
| 4 | 202 | 42 |
| 2 | 402 | 82 |
| 1 | 802 | 162 |

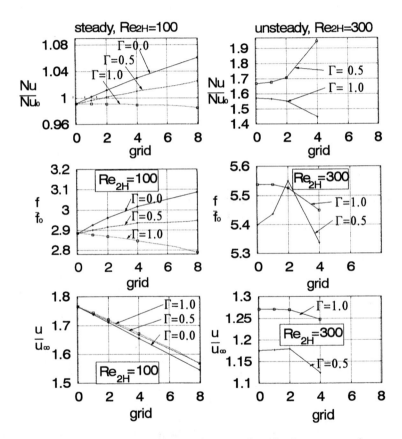

**Fig. 2.5.1:** *Nusselt number, apparent friction factor and axial velocity in a reference point as a function of grid spacing*

As shown in Fig. 2.5.1 the calculated values of normalized Nusselt number and friction factor for steady flow (Re=100) have a monotonous dependence on the grid size; i.e. with decreasing grid size they converge to the grid independent solution without changing slope for constant $\Gamma$. Furthermore the values for the grid independent solution achieved with the extrapolation are independent of the $\Gamma$-factor. Computations with up-wind terms ($\Gamma \neq 1$) result in an overestimation of the global Nusselt numbers and the apparent friction factors. The velocity is always underestimated; independend of the Γ-factor.

Calculation of the unsteady flow (Re=300) leads to a more complex situation. The time averaged global Nusselt number and reference velocity are still monotonous functions of the grid size, the values converge continuously to the grid independent solution. But the values for a grid size equal zero are a function of the difference scheme (Γ-factor). This is due to the higher damping through the UDS, that occurs because of its lower precision. The time averaged apparent friction factor is only for central differences a monotonous function of the grid size.

Summarizing it can be stated that the finer the grid the better are the results. The linearity of the functions for steady flow in fig. 2.5.1 implies a quadratic error reduction with doubling the number of grid points. In contrast the functions for the unsteady flow are not linear. This implies a non quadratic error reduction with doubling the number of grid points in case of unsteady flow. The error reduction is probably linear due to the first order time discretization. It can be concluded that in case of steady flow the numerical algorithm has a second order accuracy. In case of unsteady flow the accuracy is lower than for steady flow and it decreses with increasing Renolds number, see section 4.3.

# 3. EXPERIMENTAL METHODS

## 3.1 EXPERIMENTAL SETUP

### 3.1.1 Hot-wire measurements

For the investigation in the transition region measurements with a two-component hot-wire anemometer are performed in an open wind tunnel. The test section has a cross sectional area of height H = 20mm and width B = 22.5·H. The overall length of the test section is 150·H.

The hot-wire equipment consists of a DISA type 55P61, 5-µ, tungsten-wire, X-wire probe, and a PSI model 6100 constant temperature anemometer. The data were sampled via a fast Keithley-Metrabyte DAS-20 analog/digital converter with a Keithley-Metrabyte simultaneous-sample-and-hold SSH-4 module. The data were directly stored on a PC hard disc at a sample frequency of 2048 Hz for each channel. The reference velocity was monitored via an orifice meter that was connected to an MKS Baratron 100 mbar pressure gauge. The signal was also stored on the PC hard disc.

The hot-wire probe is positioned by a computer-controlled traversing mechanism that is described in detail in [13]. Its design principle is based on three circular discs mounted eccentrically within each other. The discs are free to rotate in the x-y plane which is the plane of the wall, see Fig. 1. The probe can be positioned in the x-y plane within a circular area of 340mm and in any desired z position. A rotation of the probe around the z-axis is also possible.

Because of the small flow velocities of $0.1 \text{m/s} < \overline{u} < 1 \text{m/s}$, an *in-situ* calibration of the probe is performed before and after each run. For calibration, only the ribbed wall is exchanged for a plane channel wall, so the hot-wire equipment is kept turned on and stays at the same position during the calibration and the actual measurement. Fully developed laminar flow is established then, since the length of the test section upstream of the probe is larger than 200·H.

Jörgensen's directional sensitivity equation [14] describes the cooling of each wire i of the probe

a) velocity calibration
(and measurement)

b) yaw calibration

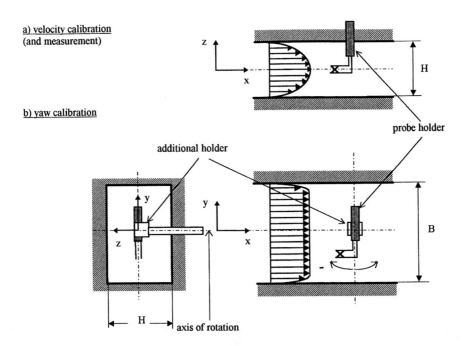

*Fig. 3.1: Test section and probe arrangement during calibration and measurement.
a) sideview on velocity calibration, b) side and top view on yaw calibration*

$$U_{eff,i}^2 = U_{n,i}^2 + k_{t,i} U_{t,i}^2 + k_{b,i} U_{b,i}^2 \qquad (i=1,2) \qquad (3.1)$$

where $U_{eff}$ is the effective cooling velocity and $U_n$, $U_t$, $U_b$ are the components of the velocity vector U, normal, tangential and binormal to the wire i. If one assumes that the flow vector is two-dimensional, and that the instantaneous velocity field is uniform over the sensing volume of the probe, the binormal velocity component $U_b$ can be neglected. Introducing the angle θ between flow direction and probe axis, Eq. (1) can be expressed as

$$f_i(E_i) = U_{eff,i}^2 = U^2 g_i(\theta) \qquad i=1,2 \qquad (3.2),$$

$$g_i(\theta) = \left(\cos^2\theta + k_i^2 \sin^2\theta\right). \qquad (3.3).$$

$g_i(\theta)$ and $f_i(E_i)$ are the yaw and the speed function, respectively. These functions must be determined by calibration. The calibration procedure is performed as follows: The probe is positioned in the plane channel as shown in Fig. 2 (a). Its wires are positioned in the x-z plane.

The probe is rotated around the y-axis first, while the flow velocity U is kept constant. A step motor controlled rotation of the probe in the interval of -40° < θ < 40° is performed in a single run at constant angular velocity within approximately 25 seconds. From this we get the dependence of the effective cooling velocity on the flow angle θ as

$$\frac{U_{\text{eff},i}^2}{U^2} = g_i(\theta).\tag{3.4}$$

Following Döbbeling et al. [19], it is assumed that $g_i(\theta)$ is not a function of U. This assumption is reasonable for laminar flow and, especially when the Reynolds number interval that is investigated is small, and the calibration is performed within this interval. In order to prove the validity of this assumption, the yaw angle calibration was repeated for several U. The probe is rotated into the x-y plane next, see Fig. 2 (b). It is worth mentioning that this rotation does not require the disconnection of the probe cables. For $\theta = 0°$ we defined $U_{\text{eff}} = U$ (Eqs. (2) and (3)), i.e. $g_i(\theta) = 1$ (i=1,2). Hence we can perform the speed calibration, simplifying Eq. (2) to

$$f_i(E_i) = U_{\text{eff},i}^2 = U^2 \qquad i=1,2.\tag{3.5}$$

The flow speed U of the fully developed laminar flow in the plane channel is varied continuously during the speed calibration. Mass flow rate and the two anemometer offset voltages are recorded simultaneously with the A/D-converter and stored at the hard disk of the PC with a frequency up to 12500 Hz. The entire Reynolds number interval is covered by the calibration procedure within several minutes. It was shown at an early stage of the investigation, that the continuous variation of U, e.g. decrease of U from 1 m/s to 0.1m/s within 3 minutes, leads to the same speed function as a fit based on several steady state calibration points, each with a constant U. Approximately 20000 samples are recorded during the speed calibration.

The speed calibration functions $f_i(E_i)$ for the two wires (Eq. (4)) are determined by use of a 4th order polynomial best fit approximation to the recorded data. The functions $f_i(E_i)$ are used to calculate the effective velocities $U_{\text{eff},i} = f_i(E_i)$ from the $E_i$ that were sampled in the yaw calibration. Following Eq. (3), the yaw calibration functions $g_i(\theta)$ are also determined by a 4th order polynomial best fit approximation.

After calibration the plane channel wall is exchanged for a ribbed wall and the measurements are performed. Up to 170000 samples are recorded at typical sampling frequencies between 100 and 1000 Hz. U and $\theta$ are calculated from the sampled $E_i$, by solving

$$\sum_{i=1}^{2}\left(\frac{f_i(E_i)}{U^2} - g_i(\theta)\right) = 0, \qquad i=1,2.\tag{3.6}$$

Eq. (6) is solved iteratively by a conventional Newton-Raphson algorithm. The computation time on a PC is less than 15 minutes for 170000 samples.

The frequencies of the fluctuations are read from the power density spectra of the flow velocity. Those were calculated by transforming the calculated velocity components via a discrete Fast-Fourier-Transform- (FFT-) algorithm.

### 3.1.2 Pressure drop measurements

For the investigation of the flow losses and the heat transfer a open wind tunnel is used. The flow losses are determined from the measured static pressure drop. For this purpose three pressure taps are positioned laterally at the plane wall opposite of each of the first ten ribs. Three pressure taps at each cross section increase the statistical precision of the results.

Following Kakac et al. [20] the apparent friction factor $f_{app}$ is defined from the pressure drop as:

$$f_{app} = \frac{2\Delta p}{\rho \bar{u}^2} \frac{A_f}{A} . \tag{3.7}$$

For the case of periodically fully developed flow the apparent friction factor is the same for each period. The constancy of the apparent friction factor for the successive periods in the flow direction is a necessary criterion of periodically fully developed flow.

### 3.1.3 Heat/Mass transfer measurements

The analogy between mass and heat transfer is used to determine the local heat transfer coefficients. The optical *Ammonia-Absorption Method* (AAM) is used that yields the local surface mass transfer distribution for constant wall concentration. The colour distribution of a filter paper caused by the chemical reaction is analysed photometrically. Because of the fast chemical reaction at the surface the ammonia concentration on the wall is equal to zero. This corresponds to the thermal boundary condition of constant wall temperature. The evaluation of local Nusselt numbers using the analogy between mass and heat transfer and using the periodically constant mass transfer coefficient is described in detail by Weber [14].

For the mass transfer experiments a small amount of reaction gas ($NH_3$) is added to the air stream in a short gas pulse. The ammonia is partially absorbed by the filter paper, which is soaked with a reactive solution of $H_2O$, $MnCl_2$ and $H_2O_2$. A colour reaction takes place with the ammonia. The locally transferred $NH_3$ mass becomes visible as a colour density distribution. This colour intensity is digitally determined by a commercial flat bed scanner (HP scanner IIc).

The mass transfer coefficient is defined as:

$$\beta_{NH_3-air} = \frac{\dot{m}_{NH_3}}{\left(\rho_{NH_3\infty} - \rho_{NH_3 wall}\right)} . \tag{3.8}$$

Here $\left(\rho_{NH_3\infty} - \rho_{NH_3,wall}\right)$ describes the difference of the Ammonia density between the wall and far away from it. Because of the chemical reaction at the wall the density of ammonia at the wall $\rho_{NH_3,wall}$ is equal to zero. The absorbed ammonia during the experimental time $t_m$ can be determined by the following integration of Eq. (3.8):

$$\int_{t=0}^{t=t_m} \dot{m}_{NH_3} dt = \int_{t=0}^{t=t_m} \beta_{NH_3-air}\, \rho_{NH_3\infty} dt . \tag{3.9}$$

The density of the ammonia in the volume flux is equal to the ratio of the added mass flow of ammonia and the volume flux:

$$\rho_{NH_3\infty} = \frac{\dot{M}_{NH_3}}{\dot{V}}. \tag{3.10}$$

With Eq. (3.9) and (3.10) the locally absorbed mass density of ammonia is equal to:

$$b = \frac{M_{NH_3\,local}}{A} = \beta_{NH_3-air} \int_{t=0}^{t=t_m} \rho_{NH_3\infty} dt = \beta_{NH_3-air} \frac{M_{NH_3}}{\dot{V}}. \tag{3.11}$$

In Eq. (3.11) $M_{NH_3\,local}$ is the locally absorbed ammonia and $M_{NH_3}$ is the mean mass of ammonia at the investigated cross section. The mass of reacted ammonia per unit area b corresponds to the colour intensity determined by a flat bed scanner. The quantitative determination of the mass transfer coefficient requires a thorough calibration with a surface object of well-known mass transfer distribution. The calibration surface is a circular disc in cross flow.

With the knowledge of the local mass transfer density it is possible to determine the local mass transfer coeffiecient by the equation

$$\beta_{NH_3-air}(x,y) = b(x,y) \frac{\dot{V}}{M_{NH_3}(x)}. \tag{3.12}$$

Here $M_{NH_3}(x)$ describes the mean mass of ammonia in the air flow at the cross section x. The so determined mass transfer coefficient implies the assumption that the density gradient of ammonia in the air flow over the span is neglectable $\rho_{NH_3\infty}(x,y) = \rho_{NH_3\infty}(x) \neq f(y)$.

With the assumption of a power law dependance on Prandtl and Schmidt number the analogy between heat and mass transfer yields the following relation between the local Nusselt number and the local Sherwood number, see Eckert [24]

$$\frac{Nu}{Sh} = \left(\frac{Pr}{Sc}\right)^n = \left(\frac{D_{NH_3-air}}{a_{air}}\right)^n = Le^{-n}. \tag{3.13}$$

In Eq. (3.13) D is the ammonia-air diffusion coefficient, a the thermal diffusivity, and n the exponent of the Lewis-number. The diffusion coefficient and its dependence on pressure and temperature are well known quantities, but n depends on the flow situation. For Pr between 0.5 and 2.5 Mitzushima and Nakajima [25] show that the value of n=0.5 is a good choice. For the used ammonia-air mixtures ($0.9 \leq Le \leq 1.1$), the variation of n between 0.37 (turbulent boundary layer) and 0.6 (laminar boundary layer) is not critical, the uncertainty is below 2%. In addition the determination of the local mass transfer rate depends on the measured grey scale value because of the non linear calibration curve (s. Zhang [26]), but for spanwise averaged Nusselt distributions Zhang [26] shows that the statistical error is not more than 4%.

Kottke et al. [22] and Schulz and Fiebig. [23] determine the amount of ammonia in equation (12) with the known injected amount of ammonia. For a low volume flux as in the present investigations the ammonia losses in the upstream parts of the test section cannot be neglected. For periodically developed flow, the periodically constant mass transfer coefficient is used by Weber [14] to determine the amount of ammonia remaining at one cross section. With the periodic boundary condition:

$$\beta(x,y) = \beta(x+n \cdot L_P, y). \tag{3.14}$$

The change of the mean mass of ammonia in the air flow $\Delta M_{NH_3}(x; x + \Delta x)$ between two cross sections x and x+Δx is determined by the mass balance.

$$M_{NH_3}(x) - M_{NH_3}(x + \Delta x) = \int_{X=x}^{X=x+\Delta x} \left( \int_y b(X,y) dy \right) dX \equiv \Delta M_{NH_3}(x; x + \Delta x) \tag{3.15}.$$

With the local mass transfer coefficient (Eq. (3.12)) for points (x,y) and (x + n·L_P, y) and equations (3.14) and (3.15) it follows for the local mass transfer coefficient in the periodically developed section.

$$\beta(x,y) = \left(b(x,y) - b(x+n \cdot L_P, y)\right) \frac{\dot{V}}{\Delta M_{NH_3}(x; x + n \cdot L_P)}. \tag{3.16}$$

With Eq. (3.16) the determination of the local Sherwood number distribution is independent of the injected amount of ammonia. However the practicability of Eq. (3.16) is restricted by the noise level of the measured mass transfer density. Instead of using Eq. (3.16) alone for the determination of the Sherwood number one can equate Eqs. (3.12) and (3.16) to determine the amount of ammonia at one cross section. So the amount of ammonia at one cross section x = A is determined with the span averaged mass transfer densities in the periodically ribbed section.

$$M_{NH_3}(A) = \Delta M_{NH_3}(A; A + \Delta x;) + \frac{\overline{b}(A + \Delta x) \Delta M_{NH_3}(A + \Delta x; A + \Delta x + n \cdot L_P)}{\overline{b}(A + \Delta x) - \overline{b}(A + \Delta x + n \cdot L_P)} \tag{3.17}$$

with: $\overline{b}(x) = \int_y b(x,y) dy$.

$\Delta M_{NH_3}$ describes the absorbed ammonia between two cross sections determined by Eq. (3.15). With eq. (3.17) and the here realized resolution it is possible to get nearly 1000 values for the determination of the amount of ammonia at one cross section. With the knowledge of this amount $M_{NH_3}(A)$ the span averaged mass transfer coefficient is:

$$\beta(x,y) = \frac{b(x,y) \dot{V}}{M_{NH_3}(A) - \left( \int_{X=A}^{X=x} \overline{b}(X) dX \right)}. \tag{3.18}$$

The Nusselt number follows from Eq. (3.13).

## 3.2 VERIFICATION

Ghaddar, Korszak, Mikic and Patera [2] investigated the transition in a grooved channel flow with constant mass flow, shown in Fig. 3.2 The flow was periodically fully developed. With a two-dimensional numerical simulation of the flow Ghaddar et al. [2] found that the critical Reynolds number is reduced to $Re_{crit.}$ = 4080 compared with the critical Reynolds numbner of the plane channel flow estimated via linear stability theory $Re_{0,crit}$=15392, Drazin et al. [35] .

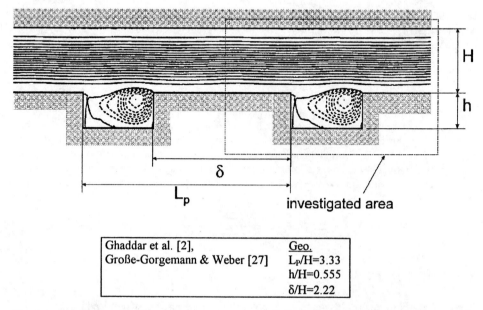

| Ghaddar et al. [2], Große-Gorgemann & Weber [27] | Geo. $L_P/H$=3.33 $h/H$=0.555 $\delta/H$=2.22 |
|---|---|

**Fig. 3.2**: *Streamlines in a flat channel with periodically arranged cavaties, Ghaddar et al.[2], Re=4080*

Futhermore they determined a self-sustained oscillation of the flow field for Re above $Re_{crit.}$ with only one frequency.
In this investigation experiments and numerical calculations were made. The experiments with hotwire probes between the eighth and the ninth groove in a range of $1865 \leq Re \leq 2611$ show self-sustained oscillations with a dominant Strouhal number of S = 0.9. This result differs from the value calculated in [29] by 4%. Furthermore with increasing Re more frequencies were detected experimentally and numerically. In Fig. 3.3 three frequencies are dominant at Re = 2419, at Re = 2746 a nearly continous spectrum exists.

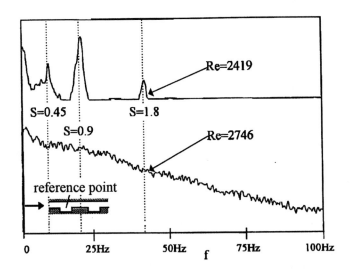

**Fig. 3.3**: *Measurement of frequency spectrum of self-sustained oscillations in a one-sided grooved channel at two Reynolds numbers, geometry parameters as in Ghaddar et al. [2], Fig. 3.3, ordinate without unit, linear*

# 4. RESULTS

First numerical data are shown to give a detailed insight into the flow structure, flow losses and heat transfer of the investigated base configuration (BC) in dependence on Reynolds number and geometry parameters. Then experimental results are shown and compared with the numerical results as well as with data from other authors.

## 4.1 TWO DIMENSIONAL CALCULATIONS

In a first step the flow and heat transfer for the two dimensional flow have been investigated. In a second step the assumption of two dimensionality is relaxed.

### 4.1.1 Reynolds Number Variation: Flow Structure

The results show the flow structure for the base configuration from steady flow to time periodic flow up to the beginning of turbulent flow.

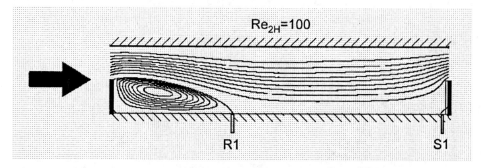

**Fig. 4.1.1** *Streamlines for Re=100 with reattachement point $R_1$ and separation point $S_1$ for the base configuration.*

At Reynolds number Re=100 the flow is steady. The flow field is characterised by acceleration above the rib and separation behind and in front of the rib, see Fig. 4.1.1. The flow has a reattachment point at $R_1$=1.75H behind the rib and a separation point at $S_1$=0.13H in front of the rib.

At Reynolds number Re=200 the flow structure is already time periodic. The time dependent stream- lines for one period are shown in Fig. 4.1.2 . Two transverse vortices appear in the separation zone behind the rib, they grow together during one time period. The reattachment point $R_1$ oscillates around a mean value of 2.9H within a range of 10.5%. The separation point $S_1$ in front of the rib lies in a range of 4.3H to 4.95H. A secondary transverse vortex is induced at the upper wall of the channel.

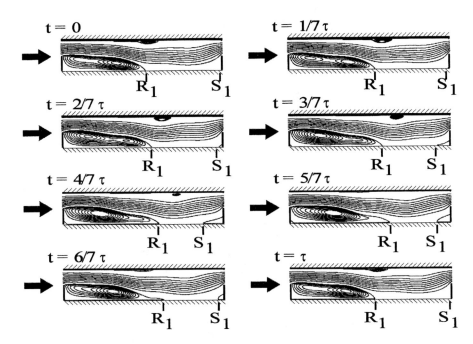

**Fig. 4.1.2** *Instantaneous streamlines for one time period $\tau$ at Re=200, with the oscillating reattachement points $R_1$ and separation points $S_1$*

For Reynolds number Re=300 and 350 the number of frequencies and their amplitudes increase. The flow is still time periodic. During one time period two transverse vortices appear behind the rib. The first generated vortex separates from the rib and moves through the channel. In the meantime a second vortex is generated. The two vortices grow together and separate from the rib. This vortex moves through the channel until it dissipates in front of the next rib. Two secondary vortices are induced at the upper wall in one time period they are dissipated above the rib. Stream lines for one time period are shown in Fig. 4.1.3. for Re = 350.

At Reynolds number Re=500 the flow is no more time periodic, as can be seen in the power density spectra over the dimensionless frequencies of Fig. 4.1.4

The dimensionless frequencies obtained (via fast fourier transformation) for different Reynolds numbers are shown in figure 4.1.4. The frequency spectrum shows the transition process from a laminar flow to a nearly turbulent flow. At Re=200 only one frequency is present (oscillation of the reattachment point) this indicates a pure periodic solution. With increasing Reynolds number more and more frequencies can be detected (vortex shedding), at Re=300 four and at Re=350 six frequencies. At Re=500 a nearly continuous spectrum can be seen, the flow is on the way to turbulence.

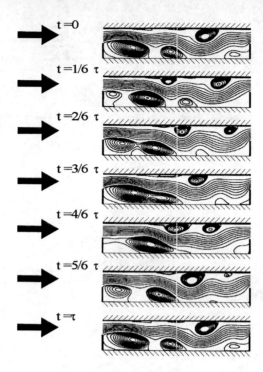

**Fig. 4.1.3** *Instantaneous streamlines for one time period $\tau$ at Re=350.*

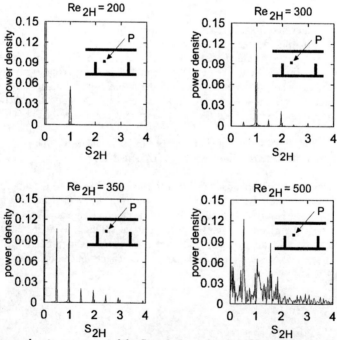

**Fig. 4.1.4** *Power density spectrum of the Strouhal number S for the axial velocity component at the reference point (x:Lp/4; y:middel of channel height). Number of grid points 400x80.*

## 4.1.2 Reynolds Number Variation: Heat Transfer and Friction

Figure 4.1.5 shows the friction coefficients and Nusselt numbers relative to the plane channel values for the upper and lower wall and the rib sides. On the upper wall the friction coefficient and the Nusselt number distribution are similar. The application of the extended Reynolds analogy (Colburn analogy: $Nu = c_f/2 \, RePr^{1/3}$) gives a deviation of 12.7% in the Nusselt number and 15% for the friction coefficient. This shows that the analogy holds qualitatively but not quantitatively. The average value of the friction coefficient is about 2.0% lower and the average Nusselt number is 12% higher than the corresponding plane channel value. For the lower wall a correspondence between momentum and heat transfer does not exit because the analogy between momentum and heat transfer is not valid for separated flow.

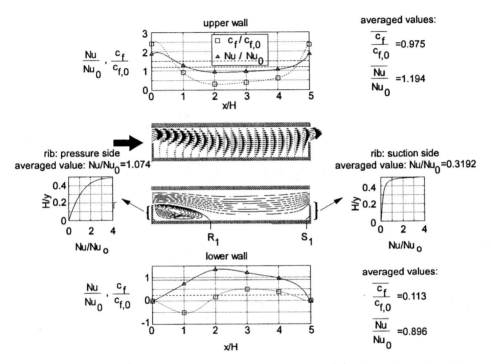

**Fig. 4.1.5** *Velocity profile, streamlines, friction coefficients and Nusselt number at Re=100.*

The heat transfer has a maximum near the reattachment point and approaches zero near the junction of wall and rib. The average friction coefficient amounts to only 10% while the average Nusselt number is 84% of the corresponding plane channel value. On the pressure and suction side of the rib the Nusselt number increases from the root to the top. On the pressure side its average value is equal to the plane channel value while it amounts to only 32% on the suction side. The rib drag coefficient is 4.29, based on the rib area and the mean velocity. The global result is an increase in the apparent friction factor by 287% and an increase in Nusselt number by 1%. In general no substantial increase in global heat transfer could be observed for steady flow.

Figure 4.1.6 shows the time averaged Nusselt number for the upper and lower wall. The global Nusselt number at the upper wall is for all four Reynolds numbers (Re=100, 200, 300, 350) nearly 29% higher than the Nusselt numbers for the lower wall e.g. the increase for the global

Nusselt number is equal for both walls with increasing Reynolds number. At the upper wall the increase in Nusselt number stems from the induced oscillating secondary vortices and the stronger flow acceleration above the rib. At the lower wall the heat transfer increases due to the oscillatory or shedding vortices.

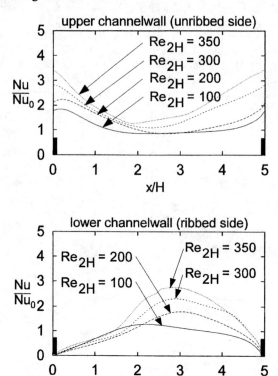

**Fig. 4.1.6** *Time averaged Nusselt number distribution on the lower and upper wall for different Reynolds numbers.*

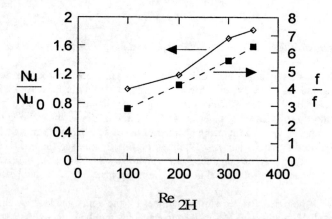

**Fig. 4.1.7** *Global time average Nusselt number and apparent friction factor for different Reynolds numbers.*

Figure 4.1.7 shows the global normalised Nusselt number and normalised apparent friction factor for the calculated Reynolds numbers. The apparent friction factor is a linear function of the Reynolds number in the whole Reynolds number range. For the Nusselt number an increase above the plane channel value occurs with the onset of oscillatory flow.

4.1.3 Rib Height Variation

The rib height is varried between h=0.1H to h=0.55H for constant Reynolds number (Re=350). Figure 4.1.8 shows a dominant Strouhal number of S≈1 until the rib height is greater than 0.5H. With increasing rib height the frequency spectra contain more harmonic Strouhal numbers. For a rib height of 0.55H the frequency spectrum is noisy with a dominant Strouhal number at S=0.5. The power density spectra show an increasing number of frequencies with increasing rib height analogous to the Reynolds number variation for constant rib height.

To find the rib height where the flow becomes unsteady the amplitudes of the v-components in the reference point are used. For the rib height of 0.35H, 0.4H and 0.45H their amplitudes grow linear, for larger rib height a digressive dependence exists, Fig. 4.1.9. The extrapolation from this data to a rib height with an amplitude of zero gives the critical rib height as 0.34.H.

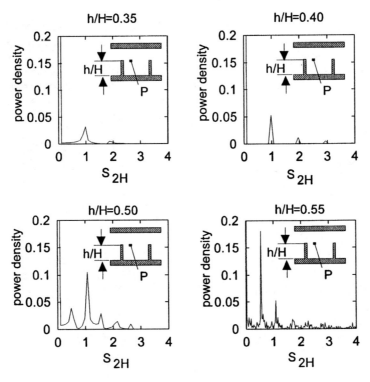

**Fig. 4.1.8** *Power density spectrum of the Strouhal number S for different rib heights at Re=350.*

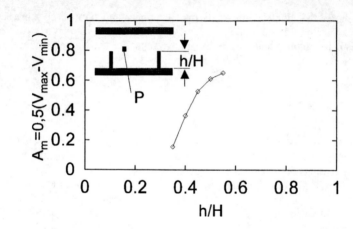

**Fig. 4.1.9** *Amplitude of the axial velocity in a reference point as a function of rib height for Re=350.*

The global Nusselt number and the apparent friction factor as a function of rib height for Re = 350 are shown in figure 4.1.10. The Nusselt number has a nearly constant value very close to unity for steady flow with small rib heights (h=0.1-0.3H). An increase of the heat transfer over the value of plane channel flow takes place, analogous to the Reynolds number variation, when the flow becomes unsteady at a rib height of 0.34H. For higher ribs the Nusselt number becomes a nearly linear function of rib height. The critical Reynolds number based on rib height remains nearly constant for varying rib heights.

Comparing the critical Reynolds number of the Reynolds number variation for the base configuration and the critical Reynolds number from the rib height variation shows that they have nearly the same value if the Reynolds number is based on the rib height.

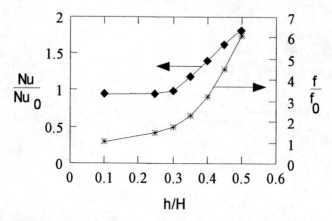

**Fig. 4.1.10** *Time averaged global Nusselt number and apparent friction factor for different rib height, Re = 350.*

### 4.1.4 Logitudinal Pitch Variation

Figure 4.1.11 shows that maximum amplitudes occur where the transition from open to closed cavity flow takes place (($L_p-\delta$)/h $\cong 7$, here $L_p/H \cong 4$). Here the configuration has the lowest damping. This observation is independent of Reynolds number.

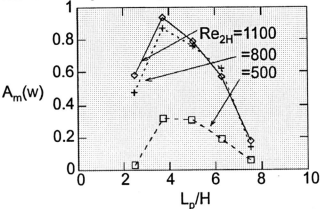

**Fig. 4.1.11** *Amplitudes at the reference point as function of longitudinal pitch for different Reynolds numbers.*

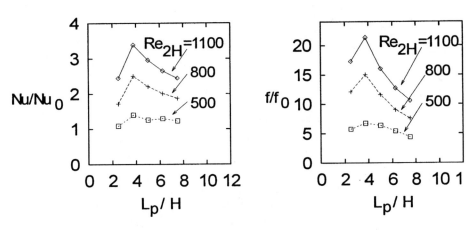

**Fig. 4.1.12** *Time averaged global Nusselt number and apparent friction factor as function of longitudinal pitch for different Reynolds numbers.*

The configuration with minimum damping is also the configuration with maximum heat transfer and pressure loss, shown in fig. 4.1.12. This is due to the high amplitude fluctuations that cause an additional convective energy transport to the wall but also additional Reynolds stresses. The former increase the heat transfer the latter the pressure loss.

## 4.2. THREE DIMENSIONAL CALCULATIONS

For the Reynolds number Re=350 a three dimensional domain was investigated with the periodic width of $B_P$=4H and periodic side boundary conditions. The calculations indicated a non periodic behaviour in time. Figure 4.2.1 shows the time averaged streamlines on the upper and lower wall. A two dimensional distribution exits with a symmetry line in the middle of the channel width (y=$B_P$/2). The two dimensional structure of the time averaged streamlines at the walls results from a three dimensional time averaged flow field. An analysis of the flow field shows that the fluctuations transverse to the main flow possess a maximum amplitude of 13% (at the location: x=2.65H, y=0.45H and z=0.5H) of the mean velocity.

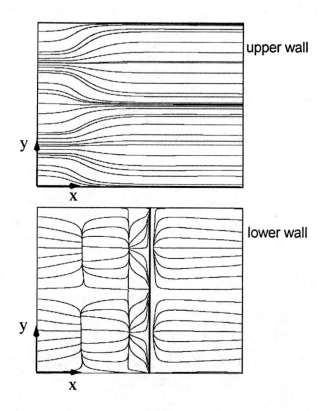

**Fig. 4.2.1**: *Time averaged wall streamlines on the upper and lower wall for Re=350.*

A confirmation of the three dimensional flow structure for the time averaged flow field is the 2-D Nusselt number distribution on the upper and lower wall shown in fig. 4.2.2 a. The Nusselt number has a nearly symmetric structure in the y-direction with a symmetry line in the middle of the channel width (y=$B_P$/2) as well as the time averaged wall streamlines. Maximum and minimum values of the Nusselt number appear above the rib at the upper wall ($Nu_{min}/Nu_0$=3.4) and direct behind the rib at the lower wall ($Nu_{max}/Nu_0$=0.1). Fig. 4.2.2 b shows the span and time-averaged Nusselt number distribution for the upper and lower wall..

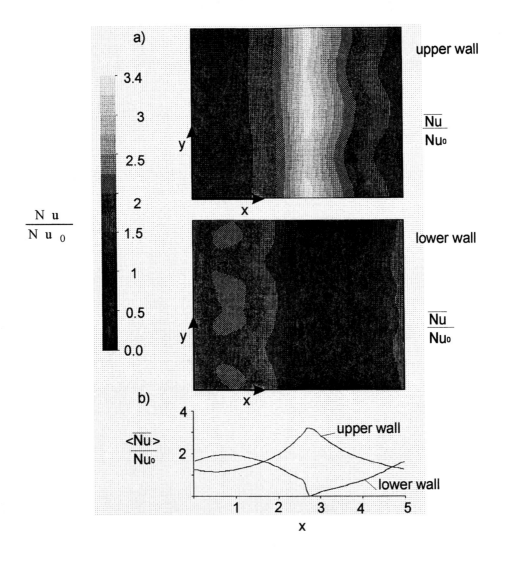

**Fig. 4.2.2**: *Time averaged a) and span averaged b) Nusselt numbers on the upper and lower wall for Re=350.*

In table 4.1 the deviations between the global values for the heat transfer and the flow losses for 2-D and 3-D calculations with the same grid size are listed. The maximum deviation appears for the heat transfer at the front side of the VG (8.6%). The global Nusselt numbers estimated from 2-D and 3-D calculations are identical. The deviation for the flow losses is 5%.

**Table 4.1** *Deviations between two dimensional (2D) and three dimensional (3D) calculations for Nusselt numbers and apparent friction factor, Re = 350*

|      | $Nu/Nu_0$: VG | $Nu/Nu_0$: wall | $Nu/Nu_0$: global | $f/f_o$ |
|------|---------------|-----------------|-------------------|---------|
| 2D   | 1.76          | 1.51            | 1.53              | 5.49    |
| 3D   | 1.76          | 1.51            | 1.53              | 5.78    |
| Δ%   | 0             | 0               | 0                 | 5.0     |

Summarising it can be stated that 3-D effects can be neglected for global heat transfer up to Reynolds number of Re=350.

## 4.3 EXPERIMENTAL RESULTS

Experimental results are presented and compared with numerical results and data from other authors.

### 4.3.1 Flow structure (in time) and flow losses

For a chosen point of the base configuration the time dependent velocity components were compared between experiment and numerics for a Reynolds number of Re=200, see fig. 4.3.1. The number of grid points for the numerical simulation was 400x80. The u-component has less than 1% deviation from the grid independent solution.

The upper part of Fig. 4.3.1 shows the velocity components at the reference point versus time detected via hot wire anemometry. The lower part shows the computed velocity components at the same reference point versus time.

The Strouhal number of the numerical simulation differs by 12% from the experimental value. The mean values of u and v have a deviation of 2% and 66% respectively. The amplitudes have a maximum deviation of 16% in the v-component.

The deviations between experiment and numerics can be explained by the experimental uncertainties in the determination of the Reynolds number, frequency, velocity and positioning of the hot wire probe and the small velocities (especially the v-component). For air and velocities below 0,1 m/s the uncertainties in the hotwire measurements increase rapidly. The present agreement is within the mearsuring accuracy.

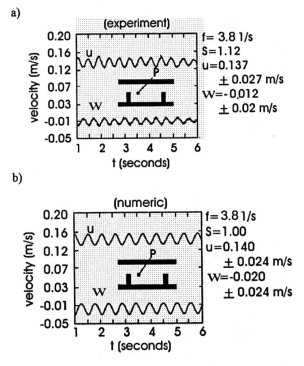

**Fig. 4.3.1** *Comparison of measured (a) and calculated (b) velocity components for the base configuration at Re=200. The reference point is located at the middle of the channel height and $L_P/4$ behind the rib.*

Grosse-Gorgemann [13] evaluated the amplitude of the u-component of the local nonsteady velocity vector as a function of Reynolds number. This amplitude of the two-dimensional velocity vector is defined by :

$$2\,AM_i = \frac{u_i(t)_{max} - u_i(t)_{min}}{[\bar{u}]}.\tag{4.1}$$

For the description of the amplitude with Reynolds number Grosse-Gorgemann used a relation given by Ghaddar et al. [2] (regular Hopf-Bifurkation):

$$AM \approx c_{AM}\,(Re - Re_{crit})^d, \qquad d \approx 0.5.\tag{4.2}$$

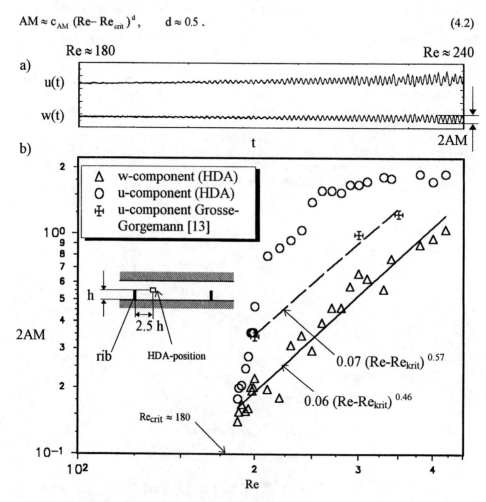

**Fig. 4.3.2** *Amplitude $2 \cdot AM$ of the self-sustained oscillations of the u- and the w-component as a function of time and Reynolds number.*
*a) experiments (ocillatory flow); b) comparison of numerical data and experiment.*

Figure 4.3.2 a) shows experimental results for continuously accelerated flow conditions, and Figure 4.3.2 b) shows experimental and numerical amplitudes versus Reynolds number. The differences between experimental and numerical values for the u-component increase with Reynolds number from about 4% at Re = 200 to 40% at Re = 300. For Re > 320 the amplitudes of the u-component of the hotwire-measurements reached a nearly constant value and the difference to the numerical values got smaller. For Re = 350 the difference between the experimental and the numerical amplitude of the u-component is about 30%.

Grosse-Gorgemann [13] evaluated the critical Reynolds number by extrapolation of the regression line shown in Fig. 4.3.2 to $Re_{crit.}$ = 184. This value is in very good agreement with the critical Reynolds number of the experimental investigations ($Re_{crit.} \approx 180$).

Fig. 4.3.3 shows the standardised apparent friction factor of one period as a function of Reynolds number. The difference between the measurements of Riemann [32] and the correlation of Weber are 5.6%. The friction factors of the numerical simulations of Grosse-Gorgemann [13] and Güntermann [33] are lower than the friction factors of the experimental investigations. The values of Grosse-Gorgemann [13] for Re ≤ 350 are based on two-dimensional simulations while the values for Re > 350 result from three-dimensional simulations. The difference between the 2-D and 3-D simulations of Grosse-Gorgemann [13]

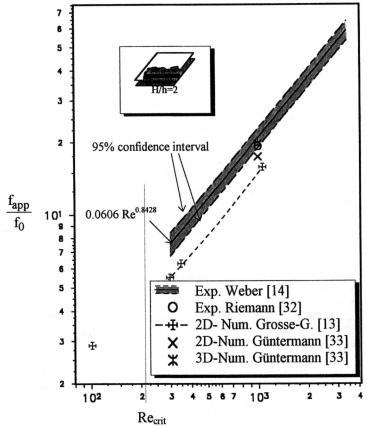

**Fig. 4.3.3** : *Standardised apparent friction factor vs. the Reynolds number for one period; h/H = 0.5, $f_0$ = 24/Re; comparison of experimental and numerical results*

at Re = 350 is less than 6%. Güntermann [33] presents friction factors of a two- and a three-dimensional simulation at Re = $10^3$. The difference between the two-dimensional simulation and the experimental value is 14% while this difference is 3% for the three-dimensional simulation. The calculations of Güntermann [33] were performend on a much coarser grid than Grosse-Gorgemanns [13] calculations. So larger deviations from the experimental results were expected for Güntermanns [33] calculations. The deviations between experimental and numerical results are attributed to the insufficient grid resolution for the numerical calculations.

4.3.2 Heat transfer

Fig. 4.3.4 shows the span averaged Nusselt numbers of one period (h/H = 0.5) for the plane and ribbed wall with fully developed periodic conditions in the Reynolds number range of 500 ≤ Re ≤ 3000 for the plane and the ribbed wall, respectively.
The positions of the local heat transfer maxima and minima within one period are independent of Reynolds number. The local heat transfer maxima at the plane wall are located right before the ribs and can be attributed to the displacement effect of the ribs.
At the ribbed wall the heat transfer has not been measured on the ribs. So the heat transfer is shown to be zero where the ribs are. Between the ribs there are two maxima right in front and behind the ribs and one between the ribs in the area of flow attachment. The maxima in front and behind each rib are due to corner vortices.
Figure 4.3.5 shows the span averaged Nusselt number distribution of the plane and ribbed wall of the numerical and the experimental investigations. It is important to point out two differencies in the conditions. First the Reynolds numbers differ by 10%. Second the thermal boundary conditions for the ribs were different. In the experiment the ribs were adiabatic in the numerics the ribs had constant wall temperature.
The comparison between the numerical and experimental investigations shows considerable deviations in the position of the local heat transfer maximum on the ribbed wall (32 ≤ x/H ≤ 34). Here this maximum of the numerical simulation in the reattachment region is positioned two winglet heights downstream of the maximum of the experimental result. The Nusselt number distributions in Fig. 4.3.4 show that this deviation of the position of the local heat transfer maximum can not be attributed to the different Reynolds numbers of the experimental and numerical investigations. The maxima in front and behind the rib are present in the numerical and experimental results. The deviation between numerical and experimental results can be caused by the low resolution of the numerical simulation (200x40 points), the different boundary conditions at the rib (numeric: $T_w$=const., experiment: $q_w$=0) and probably due to the not sufficient accuracy in time of the numerical algorithm for the here investigated Reynolds number i.e. the nearly continuous frequency spectrum at this Reynolds number, see sec. 4.1, can not be dissolved with the applied algorithm.

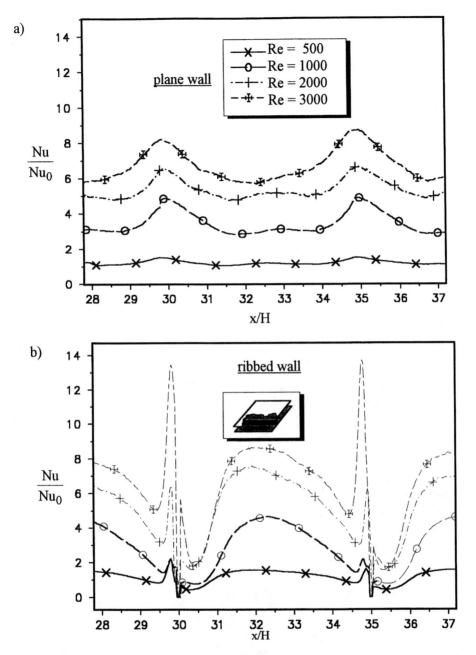

**Fig. 4.3.4** Span averaged Nusselt numbers vs. x/H of one period at the plane and the opposite ribbed wall with hydrodynamically and thermally periodic fully developed flow conditions; $h/H = 0.5$, $Nu_0 = 7.54$, each 40.th value as symbol

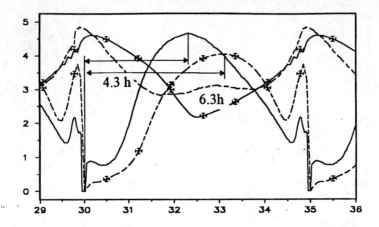

**Fig. 4.3.5** *Spanwise averaged Nusselt numbers of one period at the plane and the opposite ribbed wall with hydrodynamically and thermally periodic fully developed flow conditions: comparison of experimental and numerical results; the experimental investigations at Re = 1000 used adiabatic ribs while the ribs have constant wall temperature in accordance with the walls in the numerical investigations at Re = 1100.*

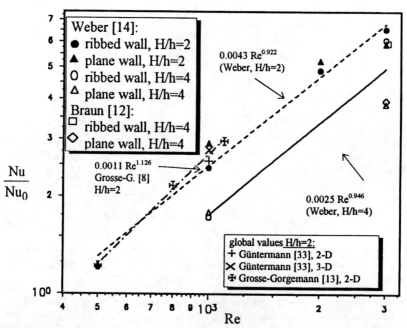

**Fig. 4.3.6** *Mean Nusselt number enhancement versus Reynolds number for comparison of different numerical and experimental investigations*

Deduced from experimental and numerical data Fig. 4.3.6 shows the time and area averaged Nusselt numbers for the ribbed and unribbed wall for rib heights 0,5 H and 0,25H as well as the global Nusselt number for the rib height equal 0.5H.

The correlations for the global Nusselt numbers from experiment ( Weber [14] ) and two dimensional numerical simulations ( Grosse-Gorgemann [13] ) for a rib height equal 0.5H can be directly compared in the Reynolds number range of Re=500 to Re=1100. The deviations are 1.2% for Re=500 and 2% for Re=1000, respectively.

Comparison of the Nusselt numbers estimated from two and three dimensional simulations done by Güntermann [33] and Nusselt numbers deduced from experiments for the Reynolds number Re=1000 gives a deviation of 6% and 1.8%, respectively

For the investigated rib heights 0.25H and 0.5H the experimental and numerical results show a higher Nusselt number for the plane wall for Reynolds number smaller 2000. For Reynolds number Re=3000 the experimental results and the results from a Large Eddy Simulation done by Braun [12] show a reversed behaviour, the Nusselt number for the ribbed wall is higher than for the plane wall.

## 5. CONCLUSIONS

Numerical and experimental investigations of periodically ribbed channels which generate transverse vortices have been carried out. The base configuration had a rib height ratio h/H=0.5, rib thickness ratio $\delta$/H=0.1 and a lateral pitch of Lp/h=10. Reynolds number, rib height and pitch have been varied. The main results are:

• The accuracy of the numerical algorithm increases quadratically with increasing number of grid points for the steady flow and less than quadratically for the unsteady flow.

• In the intersection set of numerical and experimental investigations (Re=200-1000) the agreement of numeric and experiment depends on the resulution of the domain, boundary conditions and investigated Reynolds number. The deviations reach from 2% for the mean velocity at a Reynolds number Re=200 up to 20% for the apparent friction factor at a Reynolds number Re=1000.

• For closed cavities the critical Reynolds number for the beginning of transition can be reduced drastically compared to plane channel value ($Re_{0,crit}$=15392, Hagen-Poiseuille flow) and to open cavities ($Re_{crit}$=4080, Ghaddar [2]). The reduction depends mainly on the rib height ratio h/H.

• For the base configuration the transition process started with a single finite amplitude oscillation at $Re_{crit}$=185. With increasing Reynolds number the amplitudes and number of frequencies increase.

• For steady flow the transverse vortices cause only an increase in the apparent friction factor but no significant increase in Nusselt number. Locally strong changes in friction coefficien and Nusselt number occur.

• When self-sustained large amplitude oscillations are excited, the global Nusselt number increases significantly.

- For fixed Re and rib height the highest amplitude oscillations and maximum global Nusselt numbers result for $(L_p-\delta)/h \approx 8$, in the transition between open and closed cavity flow.

- Up to Reynolds number Re=350 3-D effects can be neglected for the global heat transfer. Further investigations of the influence from 3-D effects on time average flow and temperature fields are in progress.

## NOMENCLATURE

| | | |
|---|---|---|
| a | [m²/s] | thermal diffusivity |
| A | [m²] | heat transfer area |
| $A_f$ | [m²] | area of free flow |
| Am | [m/s] | amplitude of the velocity |
| $\bar{b}$ | [kg/m] | spanaverage mass transfer density |
| b | [-] | weighting factor |
| B | [m] | channel width |
| b(x,y) | [kg/m²] | local mass transfer density |
| $c_f$ | [-] | apparent friction factor |
| $D_{AB}$ | [m²/s] | binary diffusion coefficient for A in a system A-B |
| E | [V] | voltage measured by a heat wire |
| f | [-] | apparent friction factor |
| f | [1/s] | frequency |
| f(E) | [-] | speed calibration function |
| g(θ) | [-] | yaw calibration function |
| H | [m] | channel height |
| k | [-] | weighting factor |
| L | [m] | channel length |
| M | [kg] | mass of a component i in a flow system |
| n | [-] | grid space |
| p | [N/m²] | pressure |
| $\dot{q}$ | [W/m²] | heat flux density |
| R | [-] | reattachment point |
| S | [-] | separation point |
| t | [s] | time |
| T | [K] | temperature |
| u | [m/s] | velocity |
| U | [m/s] | velocity |
| $\dot{V}$ | [m³/s] | volumetric flow rate |
| $x_1, x_2, x_3$ | [m] | cartesian coordinates |
| x,y,z | [m] | cartesian coordinates |

*greek symbols*

| | | |
|---|---|---|
| α | [W/m²K] | heat transfer coefficient |
| ρ | [kg/m³] | fluid density |
| ν | [m²/s] | kinematic viscosity |
| $\beta_{A-B}$ | [m/s] | local mass transfer coefficient |
| $\bar{\beta}$ | [m/s] | mean mass transfer coefficient |

| θ | [°] | fluid flow angle |
| θ | [ ] | dimensionless temperature |
| Γ | [-] | upwind factor |
| δ | [m] | rib thickness |
| τ | [s] | time for one period of oscillation |

*Superscripts*

| – | | mean |

*Subscripts*

| crit | critical values |
| 0 | plane channel values |
| eff | effective |
| n,t,b | normal, tangential and binormal components |
| i,j | carthesian components |
| ∞ | reference condition |
| p | periodical |

*Characteristic nondimensional Numbers*

$$Le = \frac{a}{D} \qquad \text{Lewis number}$$

$$Nu = \frac{\alpha \cdot 2H}{\lambda} \qquad \text{Nusselt number}$$

$$Pr = \frac{\nu}{a} \qquad \text{Prandtl number}$$

$$Re = \frac{\bar{u} \cdot 2H}{\nu} \qquad \text{Reynolds number}$$

$$S = \frac{f \cdot 2H}{\bar{u}} \qquad \text{Strouhal number}$$

$$Sh = \frac{\beta \cdot H}{D} \qquad \text{Sherwood number}$$

# References

[1] Sarohia, V.: „Experimental Investigation of Oscillations in Flows over Shallow Cavities", A/AA Journal, Vol. 15, Nu. 7, pp. 984-991, 1977.

[2] Ghaddar, N.K.; Koczak, K.Z., Mikic, B.B.; Patera, A.T.: „Numerical Investigation of Incompressible Flow in Grooved Channels. Part 1: Stability and Self-Sustained Oscillations", J. of Fluid Mechanics, Vol. 163, pp. 99-127, 1986.

[3] Ghaddar, N.K.; Koczak, K.Z.; Mikic, B.B.; Patera, A.T.: „Numerical Investigation of Incompressible Flow in Grooved Channels. Part 2: Resonance and Oscillatory Heat-Transfer Enhancement, J. of Fluid Mechanics", Vol. 168, pp. 541-567, 1986.

[4] Amon, C.H.; Mikic, B.B.: „Flow Pattern and Heat Transfer Enhancement in Self-Sustained Oscillatory Flows", A/AA 89-0428, 1989.

[5] Amon, C.H.; Mikic, B.B.: „Numerical Prediction of Convective Heat Transfer in Self Sustained Oscillatory Flows.", J. Thermophysics, Vol. 4, No. 2, pp. 239-246, 1990.

[6] Greiner, M.; Ghaddar, N.K.; Mikic, B.B.; Patera, A.T.: „Resonant Convective Heat Transfer in Grooved Channels", Heat Transfer 1986, 8th International Conference Brighton UK, Vol. 6, pp. 2867-2872., 1986.

[7] Greiner, M.: „An Experimental Investigation of Resonant Heat Transfer Enhancement in Grooved Channels", Int. J. Heat Mass Transfer, Vol. 34, No. 6., pp. 1383-1391, 1991.

[8] Tropea, C.; Gackstatter, R.: „The Flow over Two Dimensional Surface Mounted Obstacles at Low Reynolds Numbers", ASME J. Fluid Engeniering, Vol. 107, pp. 489-494, 1985.

[9] Durst, F.; Founti, M.; Obi, S.: „Experimental and Computational Investigation of the Two- Dimensional Channel Flow over two Fences in Tandem", ASME J. Fluid Engeneering, Vol. 110, pp. 48-54, 1988.

[10] Rowley, G. J.; Patankar, S. V.: „Analysis of Laminar Flow and Heat Transfer in Tubes with Internal Circumferential Fins", Int. J. Heat Mass Transfer, Vol. 27, pp. 553- 560, 1984.

[11] Webb, B. W.; Ramadhyani, S.: „Conjugate Heat Transfer in a Channel with Staggered Ribs", Int. J. Heat Mass Transfer, Vol. 28, pp. 1679-1687, 1985.

[12] Braun, H. : „Grobstruktursimulation turbulenter Geschwindigkeits- und Temperaturfelder in Spaltströmungen mit Wirbelerzeugern", Dissertation, Institut für Thermo- und Fluiddynamik, RUB 1996,Cuvillier-Verlag, ISBN 3-89588-765-X.

[13] Grosse-Gorgemann, A.: „Numerische Untersuchung der laminaren oszillierende Strömung und des Wärmeübergangs in Kanälen mit rippenförmigen Einbauten", Dissertation, Fak. f. Maschinenbau, RUB 1995, VDI-Verlag, ISBN 3-18-308719-7.

[14] Weber, D.: „Experimente zu selbsterregt instationären Spaltströmungen mit Wirbelerzeugern und Wärmeübertragung", Dissertation, Institut für Thermo- und Fluiddynamik, RUB 1995, Cuvillier-Verlag, ISBN 3-89588-633-5.

[15] Peric, M.: „A Finite Volume Method for the Prediction of Three-Dimensional Fluid Flow in Complex Ducts", PhD Thesis, Imperial College London, 1985.

[16] Stone, H. L.: „Iterative Solution of Implicit Approximation of Multidimensional Partial Differential Equations", SIAMJ Numerical Analysis 5, pp. 530-558, 1968.

[17] Lau, S.; Schulz, K.; Vasanta Ram, V. I.: „A Computer Operated Traversing Gear for Three Dimensional Flow Surveys in Channels", Exp. in Fluids, Vol. 14 (6), pp. 475-476, 1993.

[18] Jörgensen, F. E.: „Directional Sensitivity of Wire and Figer Film Probes", DISA Inform. 11, pp. 31-37, 1971.

[19] Döbbeling, K.; Lenze, B.; Leuckel, W.: „Computer-Aided Calibration and Measurements with a Quadruple Hot-Wire Probe", Exp. in Fluids, Vol. 8, pp. 257-262, 1990.

[20] Kakac, S.; Shah, R.K.; Aung, W.: „Handbook of Single-Phase Convective Heat Transfer", Wiley & Sons 1987.

[21] Kottke, V.; Blenke, H.; Schmidt, K. G.: „Eine remissionsfotometrische Meßmethode zur Bestimmung örtlicher Stoffübergangskoeffizienten bei Zwangskonvektion in Luft", Wärme- und Stoffübertragung, Vol. 10, pp. 9-21, 1977.

[22] Kottke, V.; Blenke, H.; Schmidt, K. G.: „Messung und Berechnung des örtlichen und mittleren Stoffübergangs an stumpf angeströmten Kreisscheiben bei unterschiedlicher Turbulenz", Wärme- und Stoffübertragung, Vol. 10, pp. 89-105, 1977.

[23] Schulz, K.; Fiebig, M.: „Accurate Inexpensive Local Heat/Mass Transfer Determination by Digital Image Processing Applied to the Ammonia Absorption Method ", Proc. of the 2nd ETS Rome, pp.1079-1088, 1996.

[24] Eckert, E. R. G.:"Analogies to heat transfer processes", Measurements in heat transfer, ed. by E. R.G. Eckert and R. J. Goldstein, pp. 397-423, Hemisphere Publishing, New York 1976.

[25] Mizushima, T.; Nakajima, M.: „Simultaneous Heat and Mass Transfer", Chem. Eng. Japan, Vol. 15, pp. 30-34, 1951.

[26] Zhang, Zheng-Ji: „Einfluß von Deltaflügel-Wirbelerzeugern auf Wärmeübergang und Druckverlust in Spaltströmungen", Dissertation, Institut für Thermo- und Fluiddynamik, RUB 1989.

[27] Grosse-Gorgemann, A.; Weber, D.; Fiebig, M.: „Experimental and Numerical Investigation of Self-Sustained Oscillations in Channels with Periodic Structures", Experimental Thermal and Fluid Science, 1996.

[28] Patz, A.: „Visualisierung oszillierender Querwirbelinnenströmungen in Kanälen mit kleinen charakteristischen Abmessungen und kleinen Geschwindigkeiten mittels Rauchdrahtmethode", Konstruktiver Entwurf, Institut f. Thermo- u. Fluiddynamik, RUB 1994.

[29] Patera, A.T.; Mikic, B.B.: „Exploiting Hydrodynamic Instabilities. Resonant Heat Transfer Enhancement", Int. J. Heat Mass Transfer, No. 8, Vol. 29, pp. 1127-1138, 1986.

[30] Amon, C.H.; Patera, A.T.: „Numerical Calculation of Stable Three-Dimensional Tertiary States in Grooved-Channel Flow", Phys. Fluids, Part A1, Vl. 12, pp. 2005 - 2009, 1989.

[31] Grosse-Gorgemann, A.; Weber, D.: „Wärmeübertragung oszillierender Querwirbelinnenströmungen, Ergebebnisbericht, DFG-Forschungsgruppe Wirbel u. Wärmeübertr., RUB 1992.

[32] Riemann, K.A.: „Wärmeübergang und Druckabfall in Kanälen mit periodischen Wirbelströmungen bei thermischem Anlauf", Dissertation, Institut für Thermo- und Fluiddynamik, Lehrstuhl für Wärme- und Stoffübertragung, RUB 1993.

[33] Güntermann, T.: „Dreidimensionale stationäre und selbsterregt schwingende Strömungs- und Temperaturfelder in Hochleistungs- Wärmeübertragern mit Wirbelerzeugern", Dissertation, Institut für Thermo- und Fluiddynamik, Lehrstuhl für Wärme- und Stoffübertragung, RUB 1992, VDI-Verlag, ISBN 3-18-146019-2.

[34] Webb, R.L.; Eckert, E.R.G.; Goldstein, R.J.: „Heat Transfer and Friction in Tubes with Repeated Rib Roughness"; Int. J. Heat Mass Transfer, No. 14, pp. 601-617, 1971.

[35] Darzin, P.G.; Reid, W.H.: „Hydrodynamic stability" Cambridge University Press., Cambridge, 1984.

# Flow Structure and Heat Transfer of Impinging Jets

N.K. Mitra, H. Laschefski, T. Cziesla
Institut für Thermo- und Fluiddynamik
Ruhr-Universität Bochum
D-44780 Bochum
Germany

## SUMMARY

Direct simulation of laminar and transitional impinging axial and radial jets have been carried out. For the turbulent cases both large eddy simulations with a localized dynamic subgrid model and direct simulations are being performed for axial impinging jets streaming out of a rectangular slot. Both single jets and a jet from a bank of jets have been handled. The impinging jet is characterized by the backflow of the ambient fluid into the computational domain. Computational scheme has been developed to take this backflow into account. For laminar flows effect of free convection and swirl at the nozzle exit as well as the effect of exit angle on the flow and heat transfer have been investigated. Free convection induces instability at a lower Reynolds number. Swirl increases heat transfer for radial jets more strongly than that for axial jets. Exit angle inclination of radial jets also increases heat transfer resulting in a higher average heat transfer for 60° inclination than that for an axial jet. Flow fields of impinging jets especially those of a row of jets show complex interaction of vortices. Flow fields also contain locally laminar and transitional zones. For turbulent jets the large scale vortical structures are overlapped with turbulent eddies. DNS results have been obtained in a Reynolds number range of steady (laminar) to nonsteady periodic to aperiodic flows. Comparison of LES with direct simulation shows good agreement and confirms the accuracy of the used dynamic SGS model and the boundary conditions. Longitudinal vortices produced by the jet interaction and their effect on heat transfer are being analysed from DNS and LES results.

## INTRODUCTION

Jets impinging on a surface can result into large heat or mass transfer at the point of impingement. Hence they find important practical applications in heating, cooling or drying of product surfaces in paper, glass, metal or textile industries. The jets discharge from round or rectangular slots and often a bank of such jets are used in application. These jets can be axial or radial (Figs. 1,2). The transfer coefficient of an axial jet is very large at the impingement point and falls down rapidly away from the impingement point.
A radial jet (Fig. 2) discharges from the side of the feed tube and reattaches on the impinging plate because of the Coanda effect. Here instead of a reattachment point one obtains a reattachment line in form of a closed curve. The transfer coefficient is moderately high on the reattachment line and decreases away from it. Right at the exit of the feed tube the radial jet can discharge parallel to the impingement plate in which case the exit angle of the radial

jet is $\vartheta=0°$. The radial jet can be directed towards the impingement plate surface ($\vartheta>0°$) or away from it ($\vartheta<0°$). For a round feed tube the radius of the reattachment circle depends on geometrical and flow parameters. The main advantage of a radial jet is that the

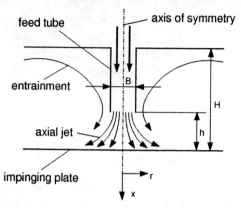

Figure 1: Schematic diagram of an impinging axial jet

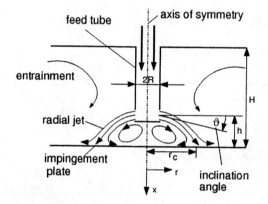

Figure 2: Schematic diagram of an impinging radial jet

moderately high heat or mass transfer can be distributed on a larger area than for an axial jet, the size and the location of this area as well as the flux density can be easily controlled by changing the geometrical parameter (e.g. the distance between the jet and the impinging plate h or the exit angle $\vartheta$) or the flow parameters (e.g. the Reynolds number at the jet discharge). For an axial jet these parameters can change essentially the flux density at the reattachment point.

A large number of experimental investigations of heat and mass transfer by a single or a bank of round or rectangular axial jets have been reported in technical journals. Martin [1] summarized the investigations of the Karlsruhe group. A general summary of such investigations has been given by Viskanta [2].

For laminar and turbulent radial jets impinging on a plate Page et al. [3] made an analytical study of reattachment assuming that the velocity profile of the jet can be described by the boundary layer theory. Their results predict the reattachment radius of turbulent jet reasonably well, but they cannot predict the near or far fields of the flow. Ostowari et al. [4] have presented experimental results of heat transfer on the impinging plate for turbulent radial jets. The flow structure of impinging jets depends on the geometrical parameters such as R, h, L (see Figs. 1,2) and the angle of inclination $\vartheta$ of the jet exit with respect to the impinging plate. The flow structure of a reattaching jet is complex, especially that of a reattaching radial jet since a separation bubble in form of a vortex ring will appear around the axis below the feed tube. The shape and the size of this vortex ring as well as the radius of the reattachment circle depend on the geometry, exit angle and Reynolds number. The effect of the Reynolds number is even more complicated since at a sufficiently high Reynolds number the laminar jet may undergo transition to turbulence before or after hitting the impinging surface. For an axial jet a second peak value of heat transfer may appear downstream of the reattachment point (see Martin [1]). The transition to turbulence after the reattachment on the impinging plate may be the reason. An impinging radial jet may show self-sustained oscillation. Ostowari et al. [4] observed inherent nonsteadiness of the reattachment location of turbulent radial jets. Both axial and radial jets develop as wall jets after impingement. The complete near and far field flow structure of impinging jets can be simulated only by solving full, nonsteady Navier-Stokes and energy equations. The results will show seperated zones and steady or unsteady reattachment. The main difficulty in computing the flow field of an impinging jet comes from the entrainment. The solid walls (e.g. the impinging plate) and the inflow boundary of the nozzle exit are clearly defined for the computational domain. The rest boundaries are free. However, they cannot be handled as outflow boundaries. Through some part of the boundary fluid is sucked into the computational domain, through other part the fluid leaves the computational domain. This behavior of the flow has been observed by Page et al. [3] and Ostowari et al. [4].

Numerical simulations of impinging jets have been reported by Huang et al. [5], Deshpande and Vaishnav [6], Polat et al. [7], Chuang and Wei [8] and Hosseinalipour and Mujumdar [9]. Deshpande and Vaishnav [6] studied impingement of axisymmetric, steady laminar jet. Hosseinalipour and Mujumdar [9] compared different turbulence models for heat transfer on impingement plate and found that the stagnation region is difficult to predict.

A laminar impinging jet can be simulated by solving complete Navier-Stokes and energy equations. Direct time-marching numerical solution (DNS) of nonsteady three dimensional Navier-Stokes equations can be used to simulate steady laminar flows as well as transitional and turbulent flows. The transition should manifest itself as unsteady periodic and aperiodic flow states. The problem of DNS for turbulent flows is that the size of the turbulent eddies that can be directly simulated is restricted by the grid size. The alternative is the large eddy simulation (LES) where eddies larger than grid size are directly simulated and the influence of those smaller than grid size is taken into account by a subgrid stress model (SGS). Both DNS and LES of turbulent jets require large computer storage and time.

For pure turbulent jets (high Reynolds number) one can consider simulation by solving Reynolds averaged Navier-Stokes equations with turbulence models. Because of the strong streamline curvatures a standard k-$\varepsilon$-model will be inadequate to simulate the flow near the stagnation area. Recently Reynolds stress model [10] and nonlinear k-$\varepsilon$-model [11] have been tried. Both these models show promise.

The purpose of our work is to investigate the flow fields and heat transfer of impinging axial and radial single jets and banks of jets by direct Navier-Stokes simulation and large eddy simulation. These tools of investigations are suitable because they are not restricted by the Reynolds number range and the programming is much simpler since the LES code with an activated or inactivated SGS model can be used for all cases. The computational results generate a data base which can be used to test or propose new turbulence models.

In the subsequent sections we present the basic equations, the computional schemes and exemplary results of different cases.

## BASIC EQUATIONS

Computations have been performed for axisymmetric round jets (both axial and radial) and for rectangular jets. LES and DNS have been performed only for rectangular jets. We have assumed the flow medium to be incompressible with constant properties and we have neglected the dissipation terms in the energy equation. The governing equations for the round jets, see Figs. 1 and 2, read in nondimensional form as

Continuity:
$$\frac{\partial(ur)}{\partial x} + \frac{\partial(vr)}{\partial r} = 0 \qquad (1)$$

Axial momentum:
$$\frac{\partial u}{\partial t} + \frac{1}{r}\frac{\partial(u^2 r)}{\partial x} + \frac{1}{r}\frac{\partial(uvr)}{\partial r} = -\frac{\partial p}{\partial x} + \frac{1}{Re}\nabla^2 u + \frac{Gr\,(T-T_{in})}{Re^2\,(T_w-T_{in})} \qquad (2)$$

Radial momentum:
$$\frac{\partial v}{\partial t} + \frac{1}{r}\frac{\partial(uvr)}{\partial x} + \frac{1}{r}\frac{\partial(v^2 r)}{\partial r} - \frac{w^2}{r} = -\frac{\partial p}{\partial r} + \frac{1}{Re}\nabla^2 v \qquad (3)$$

Circumferential momentum (in case of jet with swirl):
$$\frac{\partial w}{\partial t} + \frac{\partial(uw)}{\partial x} + \frac{1}{r}\frac{\partial(rvw)}{\partial r} = \frac{1}{Re}\nabla^2 w \qquad (4)$$

Energy equation:
$$\frac{\partial T}{\partial t} + \frac{\partial(uT)}{\partial x} + \frac{1}{r}\frac{\partial(rvT)}{\partial r} = \frac{1}{Re\,Pr}\nabla^2 T. \qquad (5)$$

The above equations consider a circumferential velocity $w(x,r,t)$ and free convection. The circumferential momentum equation is needed to be solved for the case of jet discharging with a swirl from the nozzle. In the absence of swirl $w$ is set equal to zero. In the absence of free convection the Grashof number $Gr$ is set equal to zero.

The equations have been nondimensionalized by defining the following nondimensional variables axial velocity $u=\tilde{u}/\tilde{u}_{in}$, radial velocity $v=\tilde{v}/\tilde{u}_{in}$, $w=\tilde{w}/\tilde{u}_{in}$, $p=\tilde{p}/\rho\tilde{u}^2_{in}$, $T=\tilde{T}/\tilde{T}_\infty$, $x=\tilde{x}/R$, $r=\tilde{r}/R$, where bold italicized variables represent dimensional values and $\tilde{T}_\infty$ is the ambient temperature. Nondimensional parameters are Reynolds number $Re = \tilde{u}_{in}R/\nu$, Prandtl number Pr and Grashof number $Gr = gR^3\Gamma(\tilde{T}_w-\tilde{T}_m)/\nu^2$ where g is gravitational acceleration, $\Gamma$ volume coefficient of expansion and $\nu$ kinematic viscosity. Note that $\tilde{u}_{in}$ is defined as the average velocity at the nozzle exit, based on the volume flow rate. For rectangular slot jets and rows of jets we have computed both steady laminar and turbulent flows by DNS and LES. Figs. 3 and 4 show the geometries of axial and radial slot jets.

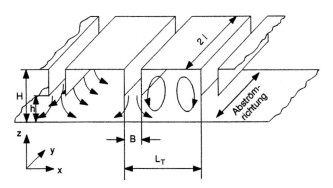

Figure 3:   Schematic of rows of axially impinging slot jets

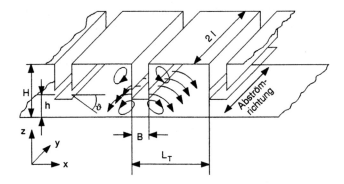

Figure 4:   Schematic of rows of radially impinging jets

The governing equations for the rectangular jets for the large eddy simulation (LES) and direct numerical simulation (DNS) in cartesian coordinates are the following

$$\frac{\partial \overline{u_i}}{\partial x_i} = 0 \qquad (6)$$

$$\frac{\partial \bar{u}_i}{\partial t} + \frac{\partial \bar{u}_i \bar{u}_j}{\partial x_j} = -\frac{\partial \bar{p}}{\partial x_i} + \frac{1}{Re}\frac{\partial^2 \bar{u}_i}{\partial x_j^2} - \frac{\partial \tau_{ij}}{\partial x_j} \qquad (7)$$

$$\frac{\partial \bar{T}}{\partial t} + \frac{\partial (\bar{u}_j \bar{T})}{\partial x_j} = \frac{1}{Re\ Pr}\nabla^2 T - \frac{\partial q_j}{\partial x_j} \qquad (8)$$

with

$$\begin{aligned}\tau_{ij} &= \overline{u_i u_j} - \bar{u}_i\,\bar{u}_j = \overline{u_i' u_j'} \\ q_j &= \overline{u_j T} - \bar{u}_j\,\bar{T} = \overline{u_j' T'}.\end{aligned} \qquad (9)$$

Here $x_i=(x,y,z)$ and $u_i=(u,v,w)$ are cartesian coordinates and velocity components respectively. The quantities with overbar are filtered variables, $u_i'$ are the unresolved subgrid scales [12]. Eqs. (6-8) are used to describe purely laminar jets, for direct Navier-Stokes simulation (DNS) of turbulent jets and for large eddy simulation (LES) of turbulent jets. For the laminar flow and DNS the subgrid stress and heat transfer $\tau_{ij}$ and q are set equal to zero. The velocity components in Eqs. (6-8) have been nondimensionalized by $\breve{u}_{in}$, lengths by 2B and the temperature by $\breve{T}_\infty$, the fluid temperature at the slot exit. The Reynolds number is $Re = \breve{u}_{in} 2B/\nu$.

**Subgrid Scale Closure**

In Eqs. (7 and 8) in the last terms $\tau_{ij}$ and $q_j$ represent the unresolved subgrid scales. Since the small scales tend to be more isotropic than the large scales, they can be modelled by the eddy viscosity model of Smagorinsky [13].

$$\tau_{ij} - \frac{1}{3}\tau_{kk}\delta_{ij} = -2\,\bar{\nu}_t\,\bar{S}_{ij} = -2\,C\,\bar{\Delta}^2\,|\bar{S}|\,\bar{S}_{ij} = -2\,C\,\beta_{ij} \qquad (10\ a)$$

and in an analogous way

$$q_j = -\frac{2}{Pr_t}\bar{\nu}_t\frac{\partial T}{\partial x_j} \qquad (10\ b)$$

where $\nu_{ij}$ and $Pr_t$ are subgrid turbulent viscosity and Prandtl number, $\delta_{ij}$ is the Kronecker delta, $|\bar{S}| = (2\,\bar{S}_{ij}\,\bar{S}_{ij})^{\frac{1}{2}}$

$$\bar{S}_{ij} = \frac{1}{2}\left(\frac{\partial \bar{u}_i}{\partial x_j} + \frac{\partial \bar{u}_j}{\partial x_i}\right) \qquad (11)$$

is the large scale strain rate tensor. In finite difference procedures the length scales, defined by the filter, are given by the grid size by the filter, are given by the grid size

$$\overline{\Delta} = (\overline{\Delta}_1 \overline{\Delta}_2 \overline{\Delta}_3)^{\frac{1}{3}}. \tag{12}$$

The value of C in Eq. (10) depends on the flow type under consideration. For different geometries the value must be determined again. This is what made LES limited to simple geometries like channel flows so far and unapplicable to complex flows of technological importance. Germano et al. [14] and later Lilly [15] suggested a method to calculate the value of C for each time step and grid point dynamically from flow field data. The basic idea of the dynamic subgrid model is the definition of a coarser test filter and to sample the scales in the range between this test filter and the model filter. The background is the assumption, that the scales which are only a little bit larger than the subgrid scales (= test filter scales) behave in a similar manner in comparison to the subgrid scales, so that analogous expressions for their definition can be used. The test filter scales (denoted by a hat or [...]^)

$$T_{ij} = [\overline{u_i u_j}]\hat{} - \hat{\overline{u}}_i \hat{\overline{u}}_j \tag{13}$$

are expressed by using the Smagorinsky closure again

$$T_{ij} - \frac{1}{3} T_{kk} \delta_{ij} = -2 C \hat{\Delta}^2 |\hat{\overline{S}}| \hat{\overline{S}}_{ij} = -2 C \alpha_{ij} \tag{14}$$

and $\hat{\Delta} = (\hat{\Delta}_1 \hat{\Delta}_2 \hat{\Delta}_3)^{\frac{1}{3}}$ ; $\hat{\Delta}_i = 2\Delta_i$ . $\tag{16}$

A relation for the resolved scales of motion between the test and the grid scale can be obtained by substracting (10a) from (14).

$$L_{ij} = T_{ij} + \hat{\overline{\tau}}_{ij} = [\overline{u_i}\,\overline{u_j}]\hat{} - \hat{\overline{u}}_i \hat{\overline{u}}_j = -2 C \alpha_{ij} + 2 [C \beta_{ij}]\hat{}. \tag{17}$$

There are five independent equations (see Eqs. 10, 14 and 17) which cannot be explicitly solved for the model constant C because it appears in a filter operation (last term of Eq. 17). Some authors [14, 16, 17] assumed

$$[C \beta_{ij}]\hat{} = C \hat{\beta}_{ij} \tag{18}$$

so that Eq. (17) can be written as

$$E_{ij} = L_{ij} + 2 C \alpha_{ij} - 2 C \hat{\beta}_{ij} \tag{19}$$

where $E_{ij}$ is the residual. Application of the least squares approach to minimize the residual gives the following equation for C

$$C(y,t) = -\frac{1}{2} \frac{< L_{ij}(\alpha_{ij} - \hat{\beta}_{ij}) >}{< (\alpha_{mn} - \hat{\beta}_{mn})(\alpha_{mn} - \hat{\beta}_{mn}) >} . \quad (20)$$

Here <> stands for averaging on a plane which is necessary for stable solution for C. It is mentioned [14-18] that this averaging represents a great problem of the dynamic model procedure. Since the distribution of C can change rapidly in space and time it is difficult to avoid numerical errors. Besides negative values of C can appear and lead to instabilities. Negative values of C can be interpreted as backscatter [15], i.e. upscale energy transfer, because in that case the turbulent viscosity $v_t$ in (10) is negative. For 3D flow like the rectangular impinging jet there is no homogeneous space direction. This is why we use instead of a plane averaging a local average over the test filter cell, which was also used by Zang [16] for recirculating flows.

Additionally we used a modification of Eq. (18) suggested by Piomelli and Liu [19]. From a mathematical point of view Eqn. (18) has an inconsistency, which results from neglecting that C is a function of space [19,20]. However, the final the goal of the dynamic model procedure, is to get a distribution of C depending on space and time. Piomelli and Liu [19] pointed out that Ghosal et al. [20] developed a consistent procedure without that inconsistency. For calculating C with this method an additional iteration is necessary which leads to a higher computational effort. A simpler approach which is not only mathematically consistent but also needs no iteration procedure is given by Piomelli and Liu [19].

Eq. (17) is written as

$$C \alpha_{ij} = L_{ij} - 2 [ C^* \beta_{ij} ]\hat{} . \quad (21)$$

On the right hand side an estimate of the coefficient C, expressed by $C^*$ and assumed to be known, is used. In that case minimization of the sum of the squares results

$$C(x,y,z,t) = -\frac{1}{2} \frac{(L_{ij} - [ C^* \beta_{ij} ]\hat{})\alpha_{ij}}{\alpha_{mn} \alpha_{mn}} . \quad (22)$$

This is the equation for calculating the model coefficient C. Piomelli and Liu [19] pointed out that there is no significant difference between zeroth- and first-order-approximation for estimating $C^*$. We used zeroth-order-approximation

$$C^* = C^{n-1} . \quad (23)$$

During the calculation numerical errors denoting high peak values of C appeared, although we used the local averaging procedure of Zang [16]. A limit of C after local averaging was necessary. We used

$$0. \leq C \leq 0.04 . \quad (24)$$

(Note: the minimum C=0 is necessary to avoid negative turbulent viscosity; the maximum C=0.04 means for the Smagorinsky coefficient $C_s$=0.2, so that C ranges between the twice of $C_s$=0.1, which was often used [21,22])

## BOUNDARY CONDITIONS

At the feed tube exit either constant or parabolic velocity profile is used. No-slip condition is assumed at the fixed walls. The temperature of impinging surface is $T_w$ and is different from the fluid temperature $T_{in}$ at the exit of the feed tube. The ambient fluid has the temperature $T_\infty$ which is equal to $T_{in}$. For LES and DNS the near-wall area over the impingement plate is computed directly by grid-refining.

A characteristic feature of the impinging jet flow is the inflow of ambient fluid into the computional domain. The amount of the backflow at the exit planes is not apriori known. Consequently the outflow boundary conditions become difficult to prescribe. Constant pressure and vanishing derivates of velocity are used at the outflow. The implementation of these conditions has been adressed in the section of Method of Solution.

## METHOD OF SOLUTION

The basic equations have been numerically solved. Two different codes have been developed and used for the solution. They are named as FIVO and FRACAS. These codes have been developed in cooperation with other projects A4, B1 and B3. FIVO has been used to solve Eqs. (1-5) for laminar steady, periodic and aperiodic flows in both polar and cartesian coordinates. FRACAS has been used to solve Eqn. (6-8) for large eddy simulation and direct simulation of turbulent jets.

FIVO is based on a finite volume technique. It uses SIMPLEC [23] pressure correction. At present this code can use cartesian, polar or body filled coordinates. The discretization is performed on a co-located grid and the momentum interpolation technique has been used. Diffusive and pressure terms have been discretized by central differences and the flux blending of first order upwind and central difference has been used for the convective terms. A spatial second order accuracy can be obtained by the use of deferred correction technique [24]. The time gradients have been discretized by the forward time difference and are of first oder accuracy. Second order temporal and spatial accuracy have been also obtained by using Crank-Nicholson and Adam Bashforth differences [25]. The algebraic equations are solved by a Strongly Implicit Procedure [26]. Details of the computational scheme have been presented in [27] and will not be repeated here.

FIVO is a time-marching scheme meaning that time-dependent solution of nonsteady equations are obtained until a steady, periodic or aperiodic solution is obtained. A periodic solution is recognized when time gradients become constant and do not change in computations from one time step to the next. The solution is declared aperiodic when after computations for a large number of time steps (generally 10000) no steady or periodic solution is obtained.

FRACAS is a fractional step finite difference scheme [28]. Here also all variables are arranged in co-located grids. Adam Bashforth discretization for convective terms and Crank Nicholson discretization for viscous terms guarantee second order spatial and temporal accuracy.

The basic computational philosophy of both FIVO and FRACAS are similar. The implementation of outflow boundary condition is shown below for FRACAS. The time discretization of the Navier-Stokes equations can be written as

$$\frac{\overline{u}_i^{n+1}-\overline{u}_i^n}{\Delta t}+\nabla \overline{p}^{n+1} = -\frac{3}{2}\text{conv}^n+\frac{1}{2}\text{conv}^{n-1}+\frac{1}{2\text{Re}}(\text{diff}^n+\text{diff}^{n+1}) \qquad (25)$$

where the convective and diffusive terms are denoted by "conv" and "diff". The index "n" symbolizes the actual time step, "n-1" the previous and "n+1" the next one. Replacement of the velocity of the following time step $u_i^{n+1}$ by an intermediate velocity $u_i^*$

$$\overline{u}_i^* = \overline{u}_i^{n+1} + \Delta t\, \nabla \overline{p}^{n+1}, \qquad (26)$$

which does not fulfill the continuity equation, yields the following term

$$\frac{\overline{u}_i^*-\overline{u}_i^n}{\Delta t} = -\frac{3}{2}\text{conv}^n+\frac{1}{2}\text{conv}^{n-1}+\frac{1}{2\text{Re}}(\text{diff}^n+\text{diff}^{n+1}). \qquad (27)$$

Application of Eq. (6) to Eq. (27) gives finally the Poisson equation for the pressure field

$$\nabla^2 \overline{p}^{n+1} = \frac{1}{\Delta t}\nabla \overline{u}_i^*. \qquad (28)$$

Integration of the pressure Poisson equation (28) yields

$$\iiint_G (\nabla^2 \overline{p}^{n+1})\, dV = \frac{1}{\Delta t}\iiint_G (\nabla \overline{u}_i^*)\, dV. \qquad (29)$$

Application of the Gauss theorem gives

$$\iint_{\partial G}\left(\frac{\partial \overline{p}^{n+1}}{\partial n}\right) dS = \frac{1}{\Delta t}\iiint_G (\nabla \overline{u}_i^*)\, dV. \qquad (30)$$

It can be shown that the pressure boundary condition "zero first derivative" cannot be used for jet flows [29]. Since the intermediate velocity field $u^*$ is the solution of the momentum equation and does not fulfill the global continuity in contrast to flows without backflow across the exit plane (e.g. channel flows), the term on the left hand side in Eq. (30) does not need to be zero either. Our investigations for laminar impinging jet flows gave good results with constant pressure at the exit plane [30]. We implemented the pressure boundary condition according to Childs and Nixon [31] and Grinstein et al. [32]

$$\overline{p}_{\text{exit plane}} = 0.7\, \overline{p}_{\text{interior}} + 0.3\, \overline{p}_\infty. \qquad (31)$$

Equation (31) denotes that the pressure at the exit plane is composed of the pressure of the last interior cell of the domain and the known ambient pressure, which is equal to zero in its nondimensional form. For the two dimensional single slot jet we compute the 3D flow field of Fig. 3 with periodicity conditions on y-z planes at x=0 and x=5 and exit boundary conditions with inflow on x-z planes at y=0 and y=4. The computational results in x-direction are averaged. Resubstitution of Eq. (26) yields

$$\overline{u}_i^{n+1} = \overline{u}_i^* - \Delta t\, \nabla \overline{p}^{n+1} \qquad (32)$$

to obtain the final velocity field.

From computed temperature field the local Nusselt number on the impingement plate is defined as

$$\text{Nu} = \frac{-\frac{\partial T}{\partial x}\big|_w}{T_w - T_\infty} \quad . \tag{33}$$

Here $T_w$ and $T_\infty$ are nondimensional temperatures of the impingement plate and the ambient fluid respectively. We used $T_w=2$ and $T_\infty=1$. The Nusselt number distribution can be averaged on the surface area A to obtain a mean Nu:

$$\overline{\text{Nu}} = \frac{1}{A} \iint \frac{-\frac{\partial T}{\partial z}\big|_w}{T_w - T_\infty} \, dx \, dy \quad . \tag{34}$$

where A is the nondimensional area of the impingement surface. We notice that Nu directly represents the heat transfer on the impingement plate.

## RESULTS

We have computed flow and heat transfer by different form of axial and radial jets e.g.,
i.) single laminar axisymmetric circular axial and radial impinging jets
ii.) bank of laminar slot (three dimensional) jets
iii.) bank of turbulent slot jets.

The computational results have been presented in different conferences and published in archival journals and conference proceedings [29,33-52]. We present here some exemplary results and summarize the rest with references where they can be looked into. For the laminar flow cases (i) and (ii) the computations have been started at a low Re for which steady flow has been obtained. Then the Re is stepwise increased in order to investigate the route of transition from laminar to turbulent flow. The flowmedium is air with Prandtl number Pr=0.72.

### Single Laminar Axisymmetric Jet

For the case (i), see Figs. 1 and 2 parametric studies on the effect of geometry (h/H) [33,34,41], effect of free convection [35,40], effect of the exit angle for the radial jet [37,45], effect of swirl at the exit of the feed tube [44,49] have been studied.

### Comparison of Axial and Radial Jets

Computations have been performed for Re=100, 250 and 500 with air (Pr=0.71) as flow medium in a grid of 102*202 for H/L=5/10 and 102*302 for H/L=5/15 on an IBM 6000/530 workstation. Steady solutions have been obtained for all Re-numbers.

Figures 5a and 5b show the velocity vectors of axial and radial jets respectively for a nondimensional feed tube exit height of h=1 and Re=250. After impinging the axial jet flow develops like a wall jet with growing cross sections. Ambient fluid is sucked in through upper half of the exit plane. The flow with radial jet shows a seperation bubble with a vortex ring in the middle around the axis of the feed tube. Figures 5c and 5d show the streamlines which depict the vortex structures of the axial and radial jets. The reattachment circle radius $r_c$=3.3R clearly shown by the skin friction coefficient distribution on the impinging plate for the radial jet in Figure 5e. For different exit heights h the flowfield of the axial jet remains qualitatively unchanged. In comparison the flowfield of the radial jet changes tremendously.

With growing feed tube exit height h the reattachment circle becomes larger. For h=2, $r_c$ is 5.3R and for h=2.5, $r_c$ is 6.5. For h>2.5 the flow does not impinge on the lower but on the upper enclosing wall [34].

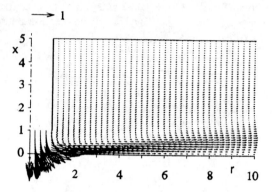

Figure 5a: Velocity vector field of the axial jet for Re = 250, h = 1, H/L = 5/10

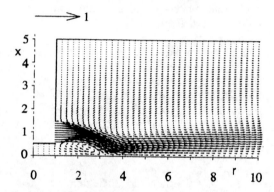

Figure 5b: Velocity vector field of the radial jet for Re = 250, h = 1, H/L = 5/10

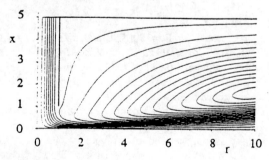

Figure 5c: Streamlines of the axial jet for Re = 250, h = 1, H/L = 5/10

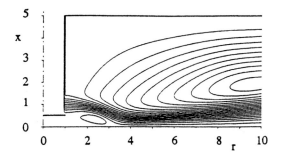

Figure 5d: Streamlines for the radial jet for Re = 250, h = 2, H/L = 5/10

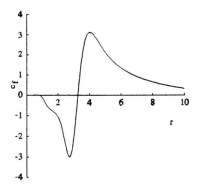

Figure 5e: Skin-friction distribution along the impingin plate for the radial jet, Re = 250; h = 1; H/L = 5/10

For h=3 it reattaches with the same radius $r_c$ on the upper wall as for h=2 on the lower wall. If the feed tube exit is closer to the lower wall an area of low pressure below the jet causes the fluid to reattach on the bottom wall. If the feed tube exit is closer to the upper wall it is the other way around. This is due to the Coanda effect [30]. For h=2.5 the reattachment of the jet is probably due to the asymmetry caused by the dead water zone below the feed tube end and the bottom plate.

In order to analyse the heat transfer, local and average Nusselt numbers are compared. The Nusselt number Nu(r) is defined as in Eq. 33.

The average Nusselt number $Nu_{av}$ is given by:

$$Nu_{av} = \frac{2}{L^2} \int_{r=0}^{r=L} Nu(r) r \, dr. \tag{35}$$

Figure 6 compares the average $Nu_{av}$ for axial and radial jet for Re=250. The dotted line shows the local Nusselt number distribution for the radial jet. From the peak of 18.75 at the stagnation point the local Nusselt number decreases rapidly with increasing r. In the radial case the local Nusselt number first increases, with increasing r up to its moderate maximum of 3.1 at the reattachment point, thereafter and decreases and tends to the value of the axial jet near the exit. The average value $Nu_{av}$ for the radial jet is ~30% smaller than $Nu_{av}$ for the axial jet.

In order to study the effect of Re, a set of computations are performed for Re=100 and an exit height h=2. Once again the flowfield converges to a steady state. The main difference from the solutions for Re=250 is that a recirculation zone appears between the jet and the upper wall at a radial distance of r=6.5 for both axial and radial jets, see Figs. 7a-7d. This secondary vortex causes the jets to seperate from the lower wall and the cold fluid to be sucked in across the exit even near the lower wall, see the skin friction distribution on the lower wall for the axial jet in Fig. 7b. The reattachment circle, as expected for a lower Re, has a smaller radius $r_c$=4.6R than for the comparable case with higher Re (for Re=250, h=2, $r_c$=5.3R). The ratio of the average Nusselt number is nearly one.

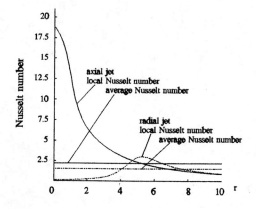

Figure 6: Local and average Nusselt number distribution, Re = 250, h = 2, H/L = 5/10

For Re=500, the secondary vortices as observed for Re=100, vanishes for both radial and axial jets. The flow fields are qualitatively similar to that for Re=250. The reattachment circle for radial jet has a radius of $r_c$=5.9R where the maximum of Nusselt number is observed with a value of 5.4. The ratio of the average Nusselt numbers is $Nu_{av}$(radial)/$Nu_{av}$(axial)=0.65. The previous results suggest that secondary vortices will increase heat transfer on the impinging plate near the exit. In order to investigate if secondary vortices appear for high Re flow (Re>100) in a longer domain further computation are carried out in a geometry with H/L=5/15 for Re=250. Fig. 8a and 8b show the velocity vector fields for the axial and radial jet respectively.

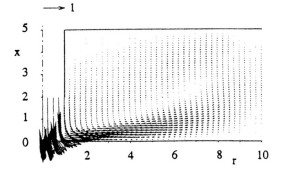

Figure 7a: Velocity vector field of the axial jet for Re = 100, h = 2, H/L = 5/10

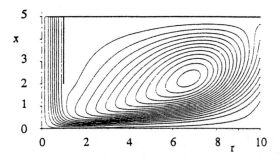

Figure 7b: Streamlines of the axial jet for Re = 100, h = 2, H/L = 5/10

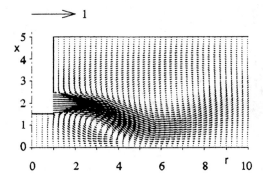

Figure 7c: Velocity vector field of the radial jet for Re = 100, h = 2, H/L = 5/10

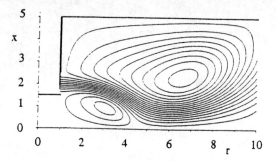

Figure 7d:   Streamlines of the radial jet for Re = 100, h = 2, H/L = 5/10

The secondary vortex above the jets can be clearly seen at a radial distance 10<r<12. In both cases the backflow appears close to the lower wall across the exit and the fluid leaves the computational domain through the upper part of the exit. The average Nusselt numbers for the axial and radial jets and the local Nusselt number distributions for both cases are shown in Fig. 9. Compared to the short channel H/L=5/10 the global heat transfer for the axial and radial jets for H/L=5/15 is nearly the same. The advantage of the radial jet is that for r>4.5 the local Nusselt number is higher than or equal to the Nusselt number of the axial jet. In 90% of the area of the impinging plate the heat transfer with the radial jet is higher than the heat transfer with the axial jet.

The radial jet for Re=250 with a feed tube exit height of h=3 reattaches at the upper adiabatic wall (H-h=1.5). For this configuration no secondary recirculation zone has been observed. For this case with Re=250 and h=3 we have also calculated the flow field in a longer channel H/L=5/15.

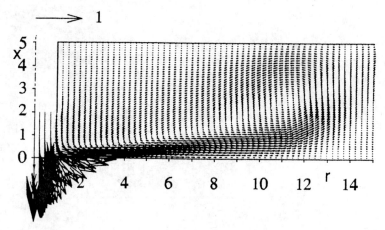

Figure 8a:   Velocity vector field for the axial jet for Re = 250, h = 2, H/L = 5/15

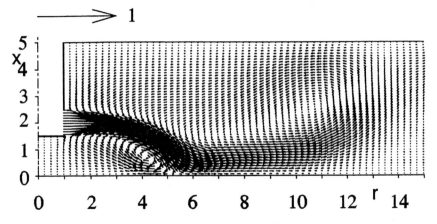

Figure 8b: Velocity vector field of the radial jet for R = 250, h = 2, H/L = 5/15

Surprisingly the flow reattaches on the lower impinging plate and not on the confining top wall. The radius of reattachmet $r_c \sim 8.1$. The heat transfer of this radial jet compared to the axial is nearly the same as shown in Fig. 9.

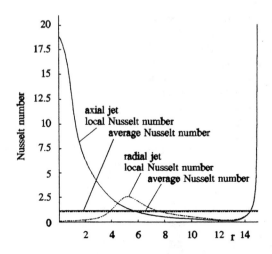

Figure 9: Average and local Nusselt number distribution for Re = 250, h = 2, H/L = 5/15

## Effect of Free Convection

Computations have been performed for Re=200, Gr=10000 and 40000 as input parameters. Air has been assumed as the working fluid; hence Prandtl number of this study is 0.72. The wall temperature $T_w$ is 5 % larger than $T_{in}$. We used h=1.25, H=3. For an axial jet with Re=200 and Gr=0 the flow is steady. The same case with Gr=10000 gives periodic flows. Figures 10a through 10f show the evolution of velocity vectors in a time period $\tau$. Figures 11a through 11f show the corresponding streaklines.

After impinging the plate at the center the free jet without free convection spreads like a wall jet and the ambient fluid is sucked in across the upper part of the boundary. With free convection the ambient fluid moves into the computational domain across the lower part of the boundary. This secondary flow meets the wall jet-like structure of the free jet and the flow lifts off the plate. This seperation line moves from r=1.7 to 2.3. Figure 11 also shows the formation of two vortex rings. Without free convection only one such ring forms near the enclosing upper plate. Figure 12 shows the wall friction coefficient $c_f$ ($c_f = \mu \partial u / \partial r \mid_w / \rho u^2_{av}$) for the axial jet with and without free convection. The negative value of $c_f$ indicates the reverse flow.

Figure 13 compares the Nusselt number distributions on the plate for the axial jet with Gr=0 and 10000. For the parabolic profile at the exit of the feed tube and Gr=0, the Nu-distribution shows, as expected, a peak value of nearly 23 at the reattachment point. Away from this point, Nu drops down fast.

With uniform (top-hat) velocity profile, we obtain for both Gr=0 and 10000, the maximum Nu of 12 at r=0.6. It should be noted that the average velocity at feed tube exit is same for uniform and parabolic profiles. Hence on the axis at the exit the velocity for the parabolic profile case is double that of the uniform profile case. The shift in the location of the peak value for uniform profile is due to the entrainment effect. For Gr=10000, Nu-distributions show similar behaviour as that for Gr=0 up to a distance of r=1.5. Up to this point, the time dependence of Nu is weak. With free convection Nu shows a minimum at r=2 for all time. Thereafter Nu swings and at r=5 a new peak in Nu appears. This is due to the entrainment.

Figures 14 and 15 show $c_f$ and Nu distributions for the radial jet for Re=200 and 0° angle of discharge.

Here again we have computed with both parabolic and uniform velocity profiles at discharge for the case of Gr=0. However, the difference in results for these two initial conditions is not as dramatic as that for the axial jet. The multiple plume structure of the radial jet with free convection manifests itself as multiple separation and reattachment points on the impinging plate. The multiple plume structure (or mushroom like thermals) is also evident in the streaklines and isotherms (not shown here). Without free convection the jet reattaches at r=2.7, see Fig. 14. In order to characterize the unsteadiness, the variation of the radial component of velocity (v) was traced over a considerably large duration of time at a position below the nozzle, near the line of symmetry, in the wake for both axial and radial jets. Figures 16a and 16b show the frequency response of the signal traces for the axial and radial jets respectively. Apparently, four low-frequency oscillations are present in both cases. However, nondimensional frequencies of 0.0415 and 0.0484 correspond to the highest kinetic energies which induce in the flow field for the axial and radial jets respectively. The nondimensional frequency or the Strouhal number is defined as $S = f^* R / u_{in}$ where $f^*$ is the frequency. Presence of other frequencies is unable to impart any influence on the flow field owing to the low level of kinetic energy.

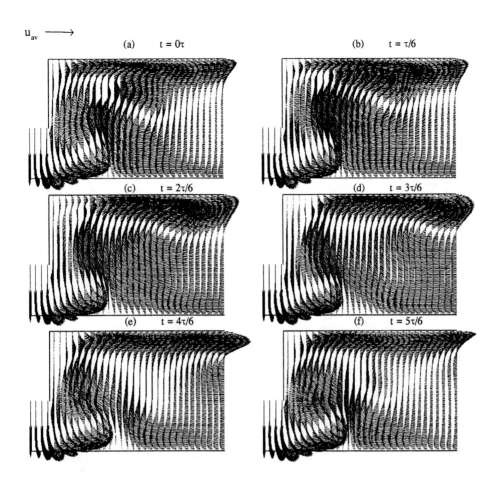

Figures 10a-10f: Velocity vectors for the axially impinging jet; Re=200, Gr=10000, periodic flow, uniform profile at the jet discharge

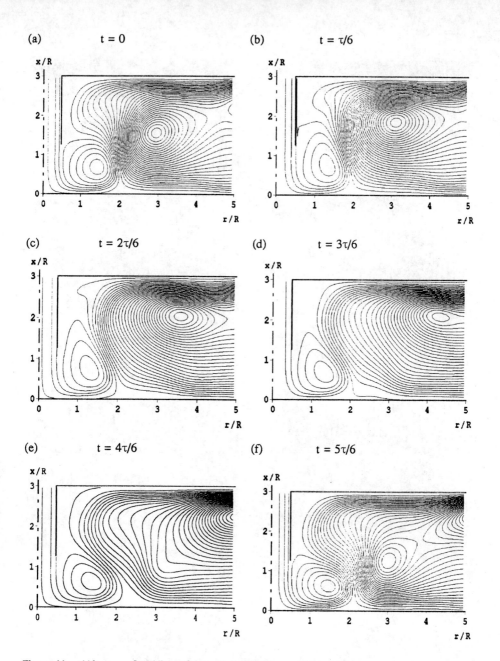

Figures 11a - 11f: Streaklines of the periodic flow for an axial jet over the time period $\tau$; Re = 200, Gr = 10000.

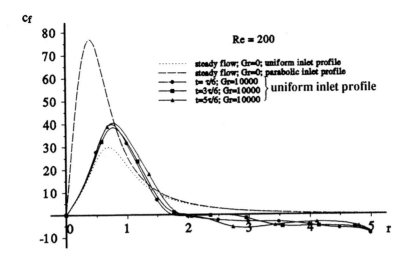

Figure 12: Local $c_f$-distribution of axial free jet over a time period $\tau$

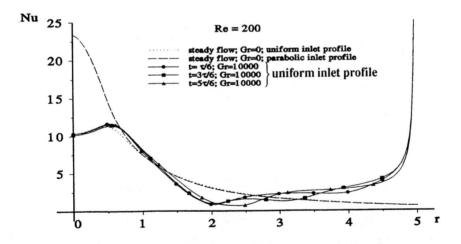

Figure 13: Local Nu-distribution for the axial free jet over a time period $\tau$

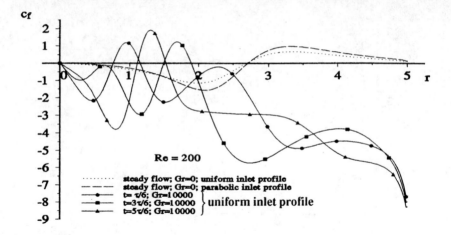

Figure 14: Local $c_f$-distribution for the radial free jet issuing at an angle of 0° on the impingement plate over a time period $\tau$

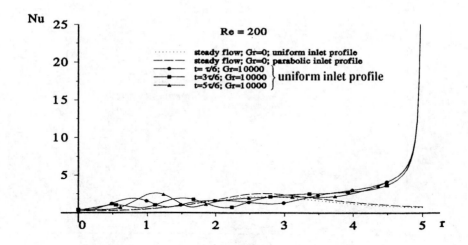

Figure 15: Local Nu-distribution for the radial free jet issuing at an angle of 0° on the impingement plate over a time period $\tau$

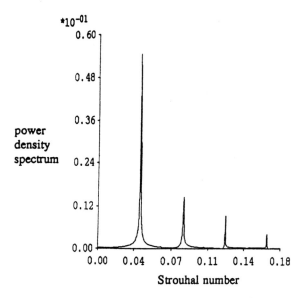

Figure 16a: Fast Fourier analysis for the time-dependent axial (u) velocity component for the axial jet; Re = 200, Gr = 10000

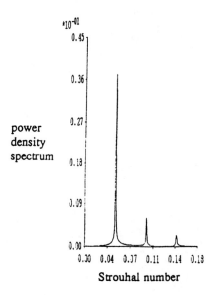

Figure 16b: Fast Fourier analysis for the time-dependent axial (u) velocity component for the radial jet; Re = 200, Gr = 10000, $\vartheta = 0°$

Table 1: Time and area averaged Nusselt number on the impingement plate for Re = 200.

| Grashof | $Nu_{av}$ | | |
|---|---|---|---|
| | axial jet | radial jet with 0° inclination | radial jet with 60° inclination |
| 0 (steady flow) | 2.064 | 1.322 | 2.148 |
| 10000 (periodic flow) | 4.101 | 3.545 | 4.040 |
| 40000 | 4.873 (nonperiodic) | 4.404 (periodic) | 4.563 (nonperiodic) |

Table 2: Comparison Nusselt number and nondimensional force on the impingement surface

| Angle of inclination | Average $Nu_{av}$ | Location of maximum Nu | Reattachment point | nondimensional force |
|---|---|---|---|---|
| Axial jet | 2.18 | 0. | 0. | 4.225 |
| Radial jet -10° | 1.31 | 7.475 | 7.6 | -0.521 |
| 0° | 1.52 | 5.274 | 5.3 | -0.09 |
| 10° | 1.60 | 4.175 | 4.15 | 0.245 |
| 20° | 1.68 | 3.475 | 3.4 | 0.607 |
| 30° | 1.77 | 2.975 | 2.9 | 1.031 |
| 40° | 1.88 | 2.575 | 2.55 | 1.563 |
| 50° | 2.03 | 2.325 | 2.25 | 2.295 |
| 60° | 2.22 | 2.075 | 2.05 | 3.443 |
| 70° | 2.52 | 1.875 | 1.85 | 5.671 |
| 80° | 3.16 | 1.675 | 1.65 | 12.665 |

Without sacrificing much of the objective of a high heat transfer rate, radial jets have been found to have a wider spread in comparison to the axial jets. The radial jet with an exit angle is intermediate between the two limiting configurations: the often studied axial and pure radial jets. For a radial jet with Reynolds and Grashof numbers are 200 and 10000 respectively and exit angle of 60° the reattachment radius does not vary with the time but the flow field oscillates near the upper plate. The situation looks complex because of entrainment, influence of angle of inclination and development of buoyancy driven secondary flows. Isotherms confirm the secondary flow associated with plumes, which first appear as small buds. With increasing time, the buds form into a mushroom shape. The plumes persist maintaining approximately the same shape, but growing slightly. Gradually, the interactions become stronger and the plumes breakup. At this stage, the flow becomes well mixed. However, the flow field here exhibits some unsteady behaviour, which is characterized by a low-frequency, small amplitude oscillation of the plumes and vortices. Hence the flow may be characterized as quasi-steady.

In order to compare the heat transfer with a radial jet, area averaged values of the local Nusselt number distributions on the impingement plate have been computed. Furthermore for periodic flows the Nusselt number has also been time-averaged. Table 1 shows the average Nusselt numbers for Re=200. Without free convection (Gr=0) one obtains steady flow and the average Nusselt number Nu=2.06 for the axial jet. The radial jet with the axis at the discharge parallel to the plate (0° inclination) gives 34% smaller Nu. But when the radial jet axis at the tube exit stays at an angle of 60° inclination with the plate, the average Nu becomes 5% larger than for the axial jet. With free convection Nu can increase by more than 100%. In order to investigate the effect of Gr we computed axial an radial jets for Gr=40000 and 0° and 60° inclination for the radial case. The axial jet flow was found to be nonsteady nonperiodic. The radial jet with 0° inclination has periodic structure with dominant frequency of 0.138. With $\vartheta=60°$ the radial jet is nonsteady nonperiodic. With Gr=40000, average (statistically stationary) Nu for axial jet increases by 140% and is still 6% larger than the corresponding Nusselt number for the radial jet with 60° angle of inclination, see Table 1.

## Effect of Exit Angle of Radial Jet

A large number of computations have been performed with angles of inclination $\vartheta=-10°$ to 80° with increments of 10°. For all these cases the charcteristic Reynolds number is kept fixed at 250. The nondimensional height of the jet axis from the impingement plate is 2 and the height of the confining wall is 5. The nondimensional radius of the impingment plate is chosen to be 10.

With increasing $\vartheta$ the reattachment radius becomes smaller. For $\vartheta=0°$ the nondimensional reattachment radius is 5.2. The corresponding values for $\vartheta=20°$ is 3.3 and $\vartheta=70°$ it is 1.9. The center of the ring vortex also moves towards the center of the plate as $\vartheta$ increases. The vortex core also becomes thinner as $\vartheta$ increases. Otherwise the flow structure remains qualitatively same for all values of $\vartheta$.

For radial jet with $\vartheta=-10°$ the Nu is extremely small at the separated dead water zone in the center of the impingement plate. Nu increases first slowly, then steeply to a peak of 2.3 slightly ahead of the reattachment circle. After that it falls down. This behavior is qualitatively repeated for all angles of inclination. However, the peak value of Nu increases with $\vartheta$

and its location which is always ahead of the reattachment point also moves with increasing $\vartheta$ towards the center. The increase in the peak Nu is not linearly dependent on the angle, but increases with the angle of inclination. It should also be mentioned that the peak Nu for the axial jet is still nearly 20% larger than the highest peak of 16 for the radial jet for $\vartheta=80°$. However, the decrease in the Nu away from the peak for the radial jets are slower than that for that of the axial jet. This suggests that average Nusselt number for the radial jets may show more impressive performance than the local Nu. Table 2 shows that at $\vartheta=60°$ $Nu_{av}$ for the radial jet becomes nearly the same as that for the axial jet. With $\vartheta=80°$ the heat transfer for the radial jet can be 50% larger than for the axial jet. It should be here noted that the average value of the Nusselt number (i.e. the average heat transfer) depends on the size (i.e. the radius) of the impingement plate. The present size r=10 has been chosen arbitrarily. For an even larger impingement surface the radial jet will be more advantageous than the axial jet because of the steep decrease in the Nusselt number for the axial jet.

Table 2 also shows that maximum Nu appears slightly before the reattachment point for $\vartheta=-10°$ and $0°$ and slightly after the reattachment point for $\vartheta>0°$. This behaviour has also been observed in experiments. It is also interesting to note that for $\vartheta\leq0°$ the total pressure force on the impingement plate is negative i.e. impingement surface feels a suction force towards the jet. This has also been observed in experiments and has found application in industries. The pressure force for $80°$ jet is much larger than that for the axial jet because of the larger momentum flux of $80°$ jet than the axial jet.

## Effect of Swirl

Computations have been performed for radial and axial jets in air (Pr=0.72) for various Reynolds numbers which varied between 10 and 1000. At low Reynolds numbers, steady solutions are obtained. However, as the Reynolds number is increased, first periodic, then chaotically unsteady solutions, are observed. No attempt has been made in this work to identify the transition points between the different regimes, as previous work by Huang et al. [5] has examined this question in a similar geometry.

A simplistic solid-body rotational profile has been assumed at the jet exit, such that $w=r\omega$. The swirl number $\bar{S}$ defined as the ratio of the axial flux of the angular momentum and the axial flux of the axial momentum ($\varpi = 2u_{in}S/R$) has been varied between 0 and 1 at Reynolds numbers of 200, 500 and 1000.

Figure 17 illustrates the Nusselt number distribution on the lower wall due to varying swirl numbers. In agreement with the previous work of Huang et al. [5], small swirl number has minor influence on the Nusselt number distribution. As the imposed swirl approaches $\bar{S}=1$, much greater effects were observed. Indeed, the peak Nusselt number at $\bar{S}=1$ was reduced by approximately 34%. Because of the large area considered, the local values at the impingement point have somewhat less influence on the average Nusselt number. Consequently, this large reduction at the impingement point led to only a 3.3% drop in the average Nusselt number. However, since the impingement point pressure played a larger role in the average pressure, the 78% local reduction led to a 48% reduction in the total surface pressure for $\bar{S}=1$.

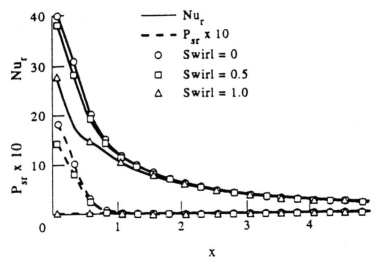

Figure 17: Local Nusselt number and nondimensional surface pressure distribution for an impinging jet at Re = 500

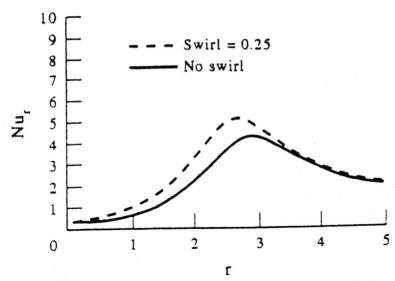

Figure 18: Radial jet local Nusselt numbers at inclination angle $\vartheta = 0°$ for Re = 500

In contrast to the transport coefficient reductions observed with axial jets, significant local and average Nusselt number increases can be obtained with swirling radial jets. The local effect of swirl on the Nusselt number distribution can be observed in Fig. 18, where results are presented for swirl numbers of zero and 0.25. In each of the cases, the value of the average heat transfer enhancement, relative to the no-swirl values, is 12.0%.

Figure 19 shows the variation in average surface pressure and heat transfer coefficient with swirl for various Reynolds numbers. With increasing swirl number, the average Nusselt number increases significantly and the surface suction force is also increased. The percentage increase in heat transfer, relative to the no swirl case, varies with the Reynolds number; the total enhancement at $\bar{S}=1.0$ was 77% for the Re=200 case, 66% at Re=500, and 65% at Re=1000. The nondimensional suction force increased strongly with swirl and Reynolds number. At $\bar{S}=1.0$ and Re=200, 500, and 1000, surface suction relative to the $\bar{S}=0$ case increased by factors of 1.6, 6.6 and 11.9, respectively.

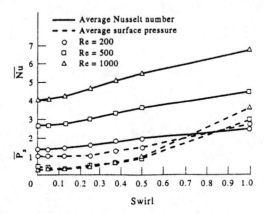

Figure 19: Average Nusselt number and surface pressure for radial jets with different Reynolds numbers as a function of entrance swirl

## Laminar Slot Jets

Figures 3 and 4 show the three dimensional schematic diagrams of banks of axially and radially impinging slot jets. The front views have been shown in accompanying sketches. In Fig. 3 jets issue from rectangular parallel slots of width B. The distance (pitch) between the axes of two neighboring parallel slots is $L_T$, the length of the slot is 2l and the height of the slot exit and the impinging plate is h. The jets are semi-enclosed and the height of the enclosing plate from the impinging plate is H. For the axially discharging jet (Fig. 3) the fluid after hitting the impinging plate moves sidewise only a small distance as it is restrained by the neighboring jets. The fluid moves out in directions normal to the paper (y axis in Fig. 3). In contrast to the axial jet (Fig. 3) the "radial" or knife jet of Fig. 4 shows that the jet fluid discharges from the sides. The height h for this case is the distance between the impinging plate and the midpoint of sidewise slot. The height of the slot is B/2 so that the total flow cross section of the two radial slots is equal to the flow cross section of the axial jet. The computational domain consisting of an element are depicted by dotted lines in Figs. 3 and 4. We investigated the evolution of the flow structure in dependence of the Re [29] especially how the flow structure changes from steady laminar to periodic to aperiodic chaotic flow. Thereby we kept the geometry fixed. The characteristic length is assumed to be 2B, ie. all length scales (h, H, l, $L_T$) have been nondimensionalized with 2B and the Reynolds number has been defined with 2B. Figures 20, 21 show pathlines axial and radial jets for h=1, f=B/$L_T$=0.125 and Re=150. For the axial jet the fluid coming out of the slot hits the plate and then turns in a counterclockwise direction and forms a longitudinal vortex as it comes out of the computational domain.

For the radial jet the fluid divides into two parts. The upper part of the jet turns up wheras the lower part turns down. The divided jets come out in form of a clockwise (lower) and a counterclockwise (upper) vortice. These basic flow structures change with the change in geometrical parameter and Re. We investigated the flow evolution for one geometry (h=1, H=3, l=2.5 and $L_T$=4) and changed Re. For the axial the flow became first periodic and later aperiodic, see Fig. 22 which shows the nondimensional frequency Strouhal number S (S = f*2B/$u_{in}$, where f* is the frequency) against the Reynolds number. For Re<335 the flow was steady.

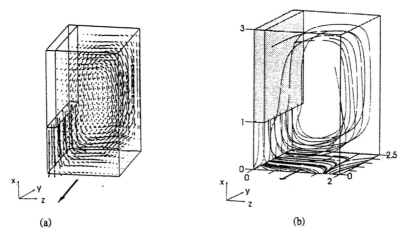

Figure 20: Velocity vectors (a) and Streamlines (b) of an axial jet at Re = 150, f = 0.125 and h = 1

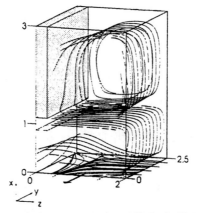

Figure 21: Streamlines of impinging radial jet at Re = 150, f = 0.125, h = 1 and $\vartheta = 0°$

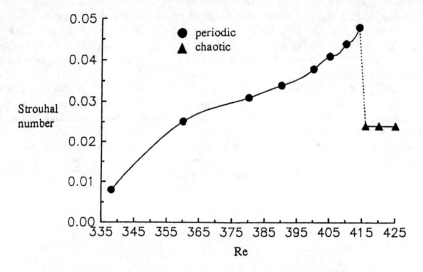

Figure 22: Dependence of the dominant Strouhal number of axial jet on the Reynolds number

For the radial jet we transition from steady to periodic flow at lower Re ($\approx 210$), see Fig. 23. With increasing Re we observed period doubling and eventually chaos but periodic windows.

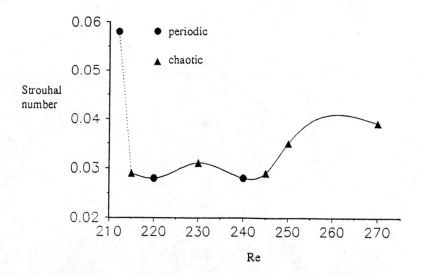

Figure 23: Dependence of the dominant Strouhal number of radial jet on the Reynolds number

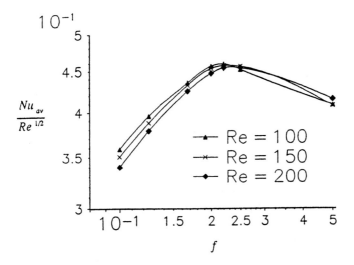

Figure 24a: $Nu_{av}/Re^{1/2}$ vs. relative jet area. $f=B/L_T$ for axial jets for different Re and h/2B=1

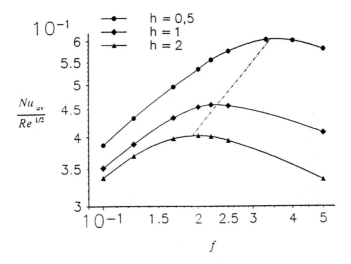

Figure 24b: $Nu_{av}/Re^{1/2}$ vs. relative jet area. $f=B/L_T$ for axial jets for different h and Re=150

The average Nusselt number $Nu_{av}$ in the steady flow range shows a boundary layer type can used to describe the dependence on Re.
As the jet height h decreases, $Nu_{av}$ increases. However, Fig. 24 depicts that for each h, an optimum value of f can be found. It should be recalled that Martin [1] also found such optimum values for nozzle spacing for turbulent jets. Physically the optimum f indicates that

if the jets are too closely spaced (large f), the interaction between the jets will hinder some of the jet fluid in reaching the impingement plate, and this will reduce the heat transfer. If the jets are widely spaced (small f), part of the impingement plate will not be affected by the impingement of jets. Instead, a boundary layer type of flow will bring about poor heat transfer. Another interesting feature is that the division by $Re^{1/2}$ collapses the results of three Re approximately to the same point. Apparently, a boundary layer type scaling can be used to describe the Re dependence of
$Nu_{av}$ in the investigated range of Re. Figure 24 shows that for a given Re and H the optimum relative nozzle area, $f_{op}$ is a function of h. Numerical analysis shows that $Nu_{av}$ is roughly proportional to $f^{1/2}$ up to $f_{op}$. The results of Figure 24 can be given by the following formulas:

$$f_{op} = [-8.5 \ h^2 + 35.7 \ h - 6.5 \ ]^{1/2} \qquad (36)$$

and

$$Nu_{av} = \frac{5}{4} Re^{1/2} \left[ \frac{2}{\frac{f}{f_{op}} + \frac{f_{op}}{f}} \right]^{1/2} f_{op}^{2/3} \qquad (37)$$

for

$$0.25 \leq h \leq 2.5$$

$$0.1 \leq f \leq 0.5$$

$$50 \leq Re \leq 300.$$

If Re is increased over 300, the flow field becomes periodic, and one or more dominant frequencies are obtained from the Fourier analysis of the time series of a velocity signal at a chosen point in the middle of the flow field. Some investigations of the transitional flow structure in dependence of Re have been presented [51]. For the Re very close to 400 the dominant frequencies vanish, and the Fourier analysis of the time series shows a broad spectrum entailing the flow to become unsteady but aperiodic. Figure 25 shows the time averaged Nusselt number $Nu_{av}/Re^{1/2}$ versus Re for f=0.125, h=1 and H=3. For the aperiodic flow, a large number (50,000) of time steps has been considered for the averaging. In the steady flow regime of the Reynolds number, $Nu_{av}/Re^{1/2}$ varies only slightly from 0.39 to 0.41. Sudden increases appear for the periodic and aperiodic flows.
Figure 26a shows the scaled average Nusselt number $Nu_{av}/Re^{0.75}$ versus the relative jet area f for three values for the radial jets. We notice two remarkable features. Unlike for the axial jet, an optimum value for f does not exit for the radial jets and the scaling by $Re^{0.75}$ collapses all computed results to one curve. This scaling is same as that of a wall jet [29]. The monotonic increase in $Nu_{av}$ with f means that the heat transfer will increase continuously through closer spacing of the jets.

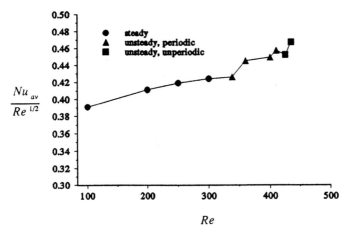

Figure 25: $Nu_{av}/Re^{1/2}$ vs. Re for the axial jets

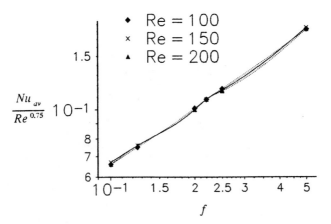

Figure 26a: $Nu_{av}/Re^{0.75}$ vs. $f = B/L_T$ for the radial jets with $\vartheta=0°$ for different Re and h=1

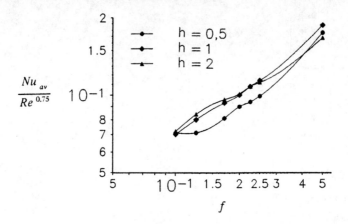

Figure 26b: $Nu_{av}/Re^{0.75}$ vs. $f = B/L_T$ for the radial jets with $\vartheta=0°$ for different h and Re=100

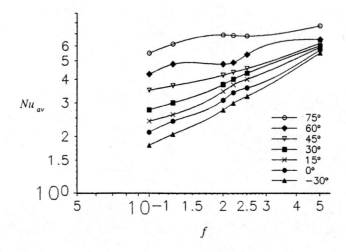

Figure 27: Influence of the exit angle of the radial jets on average Nusselt number at Re=100 and h = 1

Figure 26b shows the influence of the distance of the jet height h on heat transfer. A comparison with Figure 24 shows that $Nu_{av}$ for the radial jets with $\vartheta=0°$ is not as strongly dependent on h as for the axial jets. For the axial jets,
$Nu_{av}$ increases as h decreases. This is not the case for the radial jets of $\vartheta=0$. In general, $Nu_{av}$

is larger for higher h. The crossing of the curves in Figure 26b is dependent on the special flow structure, which is influenced by the different nozzle-plate distances h and relative nozzle areas f.

The influence of the exit angle $\vartheta$ on heat transfer is shown in Figure 27 for various values of f. The curves do not indicate the existence of an optimum f for any $\vartheta$. With increasing $\vartheta$, the $Nu_{av}$ increases because of the increasing axial momentum. The influence of $\vartheta$ on heat transfer becomes weak at smaller jet spacing (larger f). For $\vartheta=-30°$, i.e., for an upwardly inclined radial jet, a suction force is imposed on the impingement surface [46]. In practical applications concerning the paper industry, such suction forces can be utilized to hold the product surface (e.g. paper) to the jet during heating or cooling.

## Turbulent Slot Jets

As have been mentioned before we computed turbulent axial jets mainly by large eddy simulation and some by direct simulations.

For the statistical results 50 samples over 65 nondimensional time-units are used. For the LES with grid refining near the impingement plate instead of wall functions 82x82x54 (= 363096) grids are used. The LES using the logarithmic wall law are done on 82x66x50 (=270600) grids. Besides we made a DNS (analogous numerical method as mentioned above) on 102x82x68 (568752) grids. The geometry of the nozzle with H=3, h=1, $L_T$=4 (here $L_T$ = W) is same as in Fig. 3.

Figure 28 shows the flow structure for Re=600 based on the nozzle exit velocity and double slot width B. Streamlines are plotted in half of the domain for the case of LES with grid refining at the lower wall. In the other half vector velocity plots are shown for yz-planes at x=0 and x=5 (exit planes). The fluid leaving the slot nozzle (1.75≤y≤2.25, x=1) impinges directly on the lower plate, so that the stagnation line (y=2, x=0) appears. Then two wall jets develops in opposite directions. At the periodic sides of the domain fluid seperates and some part reattaches again on the lower plate exactly at the periodic boundaries (y=0 and 4, x=0). So corner vortices develop. Upwash fountains at the periodic boundaries (y=0, y=4) are obtained. Resulting from the geometrical symmetry of the periodic nozzle element helical main flow structures appears, i.e. the movement of the fluid towards the exit planes (z=0, z=5) can be described as a longitudinal vortex. In each quarter of the domain such structures appear. In Figure 28 only two oppositely directed structures are shown (left half of the domain). Since jet flows entrain ambient fluid there is inflow over the two exit planes. Our results demonstrate, that the used dynamic subgrid model together with the exit boundary conditions is able to treat a complex flow with strong streamline curvatures and backflow, because no damping of the helical structures appeared.

Figure 29 shows the time averaged pressure distribution on the impingement surface. In Figure 30 the surface pressure distribution averaged over the z-direction for different calculations (LES with grid refining, LES with wall function, DNS) is shown. Below the nozzle (1.75≤y≤2.25) the peak values caused by the stagnation appear. In the directions towards the periodic sides the pressure decreases because of the wall jet flow acceleration. The minimum values at about y=1.2 and 2.8 characterize the beginning of the seperation zones. Finally there are local maxima at the reattachement zones at y=0 and 4. The pressure decrease from the middle towards the exit planes (z=0, z=5) below the slot results from fluid acceleration directed parallel to the z-direction. This effect has been found by Martin [1] also. Figure 30 shows that the surface pressure distribution obtained from the LES, with or without the wall function, agree very well with DNS.

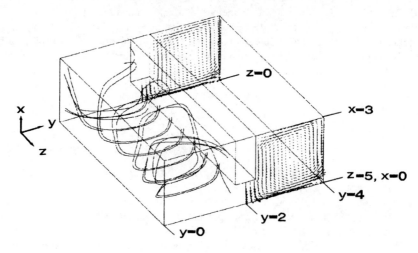

Figure 28:   Streamlines and vectorplots of different yz-planes, time-averaged

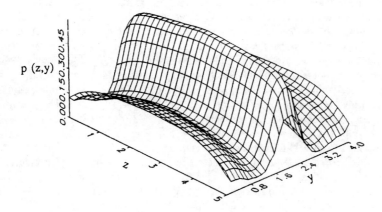

Figure 29:   Nondimensional surface pressure distribution, time-averaged

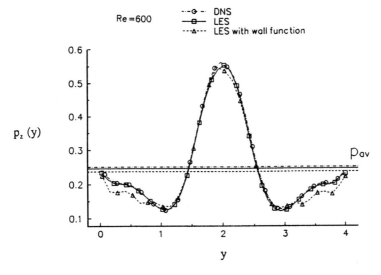

Figure 30: Comparison of nondimensional shear stress distribution, averaged in z-direction

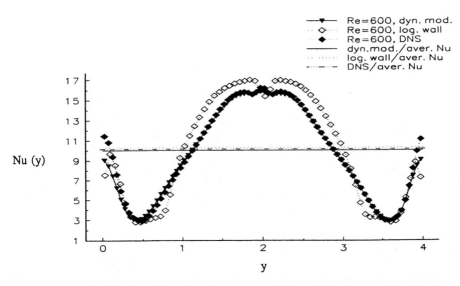

Figure 31: Nusselt number vs. y-direction (averaged in z-direction) for LES with dynamic model with grid refining (without any wall function), logarithmic law of the wall and DNS and averaged Nusselt numbers
($Nu_{av,dyn.\,mod.}$=10.02, $Nu_{av,log.\,wall}$=10.29, $Nu_{av,DNS}$=10.15, Re=600).

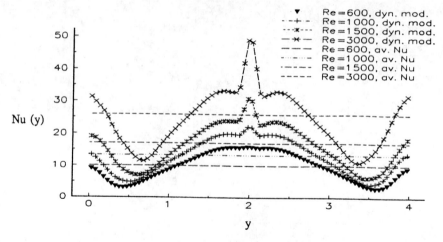

Figure 32: Comparison of Nusselt number distributions averaged in z-direction and averaged Nusselt numbers $Nu_{av}$ for LES (dynamic model) with Re=600, Re=1000, Re=1500 and Re=3000 ($Nu_{av,600}$=10.02, $Nu_{av,1000}$=13.34, $Nu_{av,1500}$=17.06, $Nu_{av,3000}$=25.9).

Figure 31 shows the Nusselt number distribution for Re=600. The curves denoted by triangles are result of LES using a dynamic model. The result of the calculation basing on the logarithmic law of the wall is denoted by empty diamonds. Nusselt numbers are plotted (solid diamonds).

Figure 31 manifests in an excellent manner the quality of the numerical code when using a dynamic model. The curves for the dynamic model and DNS show almost the same run. However, the curve using a logarithmic law of the wall differs in some regions. All three curves fit together for y=0.4, y=0.8, y=2 (stagnation point) and caused by symmetry for y=3.2 and y=3.6. For 0.4<y<0.8 and 3.2<y<3.6 the curve for the logarithmic law runs beneath and for y between 0.8 and 3.2 above the other two runs. Further it has local minimum at the stagnation point in contrast to the other runs who show there a local maximum. Table 1 shows that the area averaged Nu does not differ much from one another. For Re=600 we get $Nu_{av,\,log.\,wall}$=10.29, $Nu_{av,\,dyn.\,mod.}$=10.02 and $Nu_{av,DNS}$=10.15. $Nu_{av}$ increases with Re. The relationship of $Nu_{av}$ and Re can be given as

$$Nu_{av} = 0.202\ Re^{0.6066}. \tag{38}$$

An increase of heat transfer at the edges of the abscissa is observable in Fig. 31. Since we computed a row of impinging jets there is an head-on collision of two adjacent wall jets. Consequently, a thinning of the boundary layer in combination with strong fluctuations occur which lead to an increase of the Nusselt numbers (compare chapter "Arrays of Impinging Jets").

A comparison of Nu for four different Reynolds numbers used for LES with dynamic model is presented in Fig. 32. A 67% increase of the Reynolds number from Re=600 to 1000 leads to a 33% higher mean Nusselt number. In analogous way 28% and 52% higher averaged heat transfer rates can be observed if Re=1000 and 1500 and Re=1500 and 3000 are compared respectively. With increasing Reynolds numbers the stagnation point Nusselt number increases as well[1]. For instance, there is 50% difference in heat transfer between y=2 and y=2.2 for Re=3000. The spacing between the local minima on the left-hand and right-hand side of the plot decreases with increasing Reynolds number. This phenomenon shows that higher Reynolds numbers enhance the thinning of the boundary layer in the collision zone of adjacent jets.

Figures 33a and 33b and Figs. 34a-34c show three-dimensional plots of the Nusselt number distributions on the impingement surface. All figures show a good agreement with the results of Martin [1]. He found that there is a pressure loss for z towards 0 and 5 with a maximum pressure for z=2.5 due to acceleration and friction of the outlet flow. Consequently, the nozzle exit velocity accelerates towards the exit planes (z=0 and z=5) because of the demand of energy conservation. Heat and mass transfer coefficients follow this variation and have maximum values in the edges of the z-direction. Therefor Figs. 33 and 34 have a qualitative contour like a saddle. With increasing Reynolds number the local heat transfer rates increase as well.

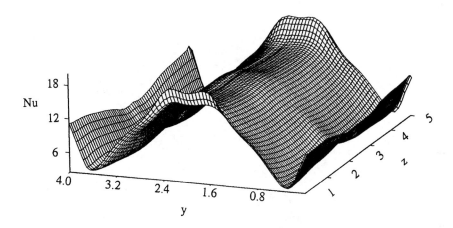

Figure 33a: 3D-plot of the local Nusselt number distribution on the impingement plate for LES with dynamic model (Re=600).

---

[1] The authors assume that the peaks below the nozzle lips at y=-0.5 and +0.5 are caused by fluid acceleration of wall jet flow (see Viskanta [2]).

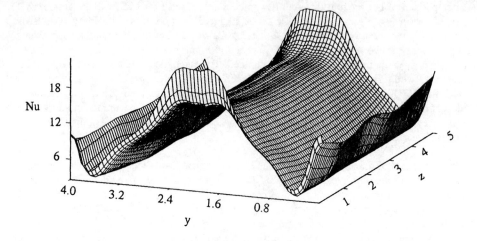

Figure 33b: 3D-plot of the local Nusselt number distribution on the impingement surface for DNS with Re=600.

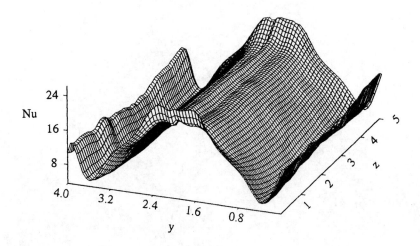

Figure 34a: 3D-plot of the local Nusselt number distribution on the impingement plate for LES with dynamic model (Re=1000).

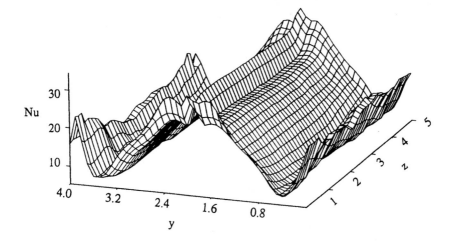

Figure 34b: 3D-plot of the local Nusselt number distribution on the impingement plate for LES with dynamic model (Re=1500).

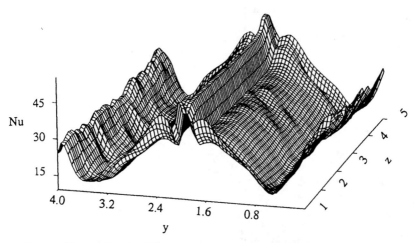

Figure 34c: 3D-plot of the local Nusselt number distribution on the impingement surface for LES with dynamic model (Re=3000).

## CONCLUDING REMARKS

Direct numerical simulation of flow and temperature fields of impinging jets in laminar and transition regime and direct and large eddy simulation of turbulent flow regime have been performed. Both axially and radially discharging round and rectangular (slot) jets have been considered. Furthermore, the effects of free convection, initial swirl, exit angle for radial jets on heat transfer have also been investigated. Following conclusions can be drawn from the investigations:

### Single impinging jet from round nozzle (laminar):
For axisymmetric single impinging radial jets (round nozzles) with inclination angles larger than $60°$ higher heat flux than for axial jets can be obtained.
For an axial or radial jet impinging vertically downward a warmer plate the consideration of mixed convection shows higher heat transfer than forced convection alone. For this case the vectoring of a radial jet ($60°$) did not show larger heat transfer than that for the axial jet.
Swirl at the nozzle exit profile decreases the heat flux for axial jets. In contrast the swirl leads to a transfer rate increase for the radial jets.

### Array of impinging jets from rectangular slot nozzles:
For arrays of laminar impinging jets, radial jets with inclination angles larger than $60°$ gives higher heat flux than the axial jets.
The axial jet flow becomes turbulent at Re of 435 or higher. The correponding Re for radial jets is 275.
For arrays of turbulent impinging axial jets for Re=600 the LES without any wall function and with the direct calculation of near wall flow gives better agreement with the DNS than the LES with the wallfunction.

## ACKNOWLEDGEMENT

This work has been financially supported by the DFG.
We thank Prof. Gautam Biswas of IIT, Kanpur, India for his help with insightful discussion at every step of this work.

# REFERENCES

[1] MARTIN, H.: "Heat and Mass Transfer between Impinging Gas Jets and Solid Surfaces", Advances of Heat Transfer, 13 (1977) pp. 1-60.

[2] VISKANTA, R.: "Heat Transfer to Impinging Isothermal Gas Flame Jets", Experimental Thermal and Fluid Science, 6 (1993) pp. 111-134.

[3] PAGE, R.H., HADDEN, L.L., OSTOWARI, C.: "Theory for Radial Jet Reattachment Flow", AIAA J., 27 (1989) pp. 1500-1505.

[4] OSTOWARI, C., PAIKERT, B., PAGE, R.H.: "Heat Transfer Measurement of Radial Jet Reattachment on a Flat Plate", Proc. National Fluid Dynamics Conf., Cinn. OH., July 1988.

[5] HUANG, B., DOUGLAS, W.J.M., MUJUMDAR, A.,S.: "Heat Transfer under a Laminar Swirling Impinging Jet - A Numerical Study", Proc. of the 6th Int. Heat Transfer Conf., 5 (1978), Hemisphere, pp. 311-316.

[6] DESHPANDE, M.D., VAISHNAV, R.N.: "Submerged Laminar Jet Impingement on a Plane", JFM, 114 (1982) pp. 213-236.

[7] POLAT, S., HUANG; B., MUJUMDAR; A.S., DOUGLAS, W.J.: "Numerical Fluid Mechanics and Heat Transfer", Annual Review of Numerical Fluid Mechanics and Heat Transfer, edited by C.L. TIEN, T.C. CHAWLA, 2 (1988), Hemisphere, pp. 157-197.

[8] CHUANG, S.H., WIE, C.-Y.: "Computations for a Jet Impinging Obliquely on a Flat Surface", Int. J. Numerical Methods in Fluids, 12 (1991) pp. 637-653.

[9] HOSSEINALIPOUR, S.M., MUJUMDAR, A.S.: "Comparative Evaluation of Different Turbulence Models for Confined Impinging and Opposing Jet Flows", Num. Heat Transfer, A, 28 (1995) pp. 647-666.

[10] LESCHZINER, M.A., INCE, N.Z.: "Computational Modeling of Three Dimensional Impinging Jets with and without Cross Flow Using Second Moment Closure", UMIST Rep. TFD/94/06, (Aug. 1994), Manchester, U.K..

[11] SUGA, K.: "Development and Application of a Nonlinear Eddy Viscosity Model Sensitized to Stress and Strain Invariants", Ph.D. Thesis, UMIST, (Dec. 1995), Manchester, U.K..

[12] FERGIZER, J.H.: "Large Eddy Simulation of Turbulent Flows", UMIST, (Dec. 1995) Manchester, U.K..

[13] SMAGORINSKY, J.S.: "General Circulation Experiments with Primitive Equations. I. The Basic Experiment", Monthly Weather Report, 91 (1963) pp. 99-164.

[14] GERMANO, M., PIOMELLI, U., MOIN, P., CABOT, W.: "A Dynamic Subgrid-Scale Eddy Viscosity Model", Phys. Fluids A, 3 (1991) pp. 1760-1765

[15] LILLY, D.K.: "A Proposed Modification of the Germano Subgrid-Scale Closure Method", Phys. Fluids A, 4 (1992) pp. 633-635.

[16] ZANG, Y., STREET, R.L., KOSEFF, J.R.: "A Dynamic Mixed Subgrid-Scale Model and its Application to Turbulent Recirculating Flows", Phys. Fluids A, 5 (1993) pp. 3186-3196.

[17] YANG, K.S., FERZIGER, J.H.: "Large-Eddy Simulation of Turbulent Obstacle Flow using a Dynamic Subgrid-Scale Model", AIAA J., 31 (1993) pp. 1406-1413.

[18] HOFFMANN, G., BENOCCI, C.: "Numerical Simulation of Spatially-Developing Planar Jets", Proc. AGARD, (Apr. 1994), Chania, Greece.

[19] PIOMELLI, U., LIU, J.: "Large-Eddy Simulation of Rotating Channel Flows using a Localized Dynamic Model", Proc. AGARD, (Apr. 1994), Chania, Greece, pp. 1-9

[20] GHOSAL, S., LUND, T.S., MOIN, P.: CTR Annual Research Briefs, Center for Turbulence Research, Stanford University, 3 (1993).

[21] BRAUN, H., FIEBIG, M., MITRA, N.K.: "Large-Eddy Simulation of Separated Flow in a Ribbed Duct", Proc. AGARD, (Apr. 1994), Chania, Greece.

[22] CZIESLA, T., BRAUN, H., BISWAS, G., MITRA, N.K.: "Large-Eddy Simulation in a Channel with Exit Boundary Conditions", ICASE (NASA Langley) Report 198304, No. 96-18, March 1996.

[23] VAN DOORMAL, J.P., RAITHBY, G.D.: "Enhancement of SIMPLE Method for Predicting Imcompressible Fluid Flows", Num. Heat Transfer, 7 (1984) pp. 147-163.

[24] KHOSLA, P.K., RUBIN, S.G.: "A Diagonally Dominant Second-Order Accurate Implicit Scheme", Computer & Fluids, 2 (1974).

[25] HAHNE, H.W.: "Numerische Untersuchung von Geschwindigkeits- und Temperaturfeldern in Spaltstroemungen mit Einbauten", Dissertation, RUB, 1997.

[26] STONE, H.L.: "Iterative Solution of Implicit Approximations of Multidimensional Partial Differential Equations", SIAM J. Numerical Analysis, 5 (1968) pp. 530-558.

[27] CZIESLA, T.: "Programmdokumentation des finiten Volumenverfahrens zur numerischen Simulation ungeführter und geführter Freistrahlströmungen mit Prallplatte, Studienarbeit, Januar 1994.

[28] KIM, J., MOIN, P.: "Application of a Fractional-Step Method to Incompressible Navier-Stokes Equations", J. Computer Physics, 59 (1985) pp. 308-323.

[29] LASCHEFSKI, H.: "Numerische Untersuchung der dreidimensionalen Strömungsstruktur und des Wärmeübergangs bei ungeführten und geführten laminaren Freistrahlen mit Prallplatte", Dissertation, Fortschrittsberichte, VDI, Reihe 19, Nr. 72, VDI-Verlag, Düsseldorf (1994).

[30] LASCHEFSKI, H., BRAESS, D., HANEKE, H., MITRA, N.K.: "Numerical Investigations of Radial Jet Reattachment Flows", Int. J. Numerical Methods in Fluids, 18 (1994) pp. 629-646.

[31] CHILDS, R.E., NIXON, D.:"Unsteady Three-Dimensional Simulations of a VTOL Upwash Fountain", AIAA-Paper, 86-0212 (1986).

[32] GRINSTEIN, F.F., ORAN, E.S., BORIS, J.P.:"Numerical Simulation of Axisymmetric Jets", AIAA-Journal 25, 92 (1987).

[33] LASCHEFSKI, H., HOLL, A., GROSSE-GORGEMANN, A., MITRA, N.K., PAGE, R.H.: "Comparison of Heat Transfer on a Flat Plate by Impinging Axial and Radial Jets", Proc. ENCIT 92, 4th Brasilian Thermal Science Meeting, (Dec. 1992), Rio de Janeiro, Brasil, pp. 131-136.

[34] LASCHEFSKI, H., HOLL, A., GROSSE-GORGEMANN, A., MITRA, N.K., PAGE, R.H.: "Flow Structure and Heat Transfer of Radial and Axial Jet Reattachment on a Flat Plate", Fundamentals of Forced Convection Heat Transfer, HTD-Vol. 210, (ASME), edited by M. A. Ebadian, P. H. Oosthuizen, (1992) pp. 123-131.

[35] POTTHAST, F., LASCHEFSKI, H., MITRA, N.K., FIEBIG, M., BISWAS, G.: "Influence of Free Convection on Flow Structure and Heat Transfer of Impinging Radial and Axial Jets", in Enhanced Cooling Techniques for Electronics Applications, HTD-Vol. 263, (ASME), edited by V. Garimella, M. Greiner, M.M. Yovanovich, V. W. Antonettie, (1993) pp. 19-32.

[36] LASCHEFSKI, H., CZIESLA, T., MITRA, N.K., PAGE, R.H.: "Numerical Simulation of Flow and Heat Transfer of Laminar Impinging Jets", Eurotherm 31, Proc. of the Seminar, (May 24-26 1993), Bochum, Germany.

[37] CZIESLA, T., LASCHEFSKI, H., MITRA, N.K.: "Numerical Investigation of the Influence of the Inclination Angle on Impinging Radial Jet Heat Transfer", Heat Transfer '94, Proc. of the Seminar, (August 22-24 1994), Southampton, UK, pp. 83-90.

[38] WINKELSTRÄTER, M., LASCHEFSKI, H., HEIDRICH,G., MITRA, N.K.: "Numerical Simulations of Flow and Heat Transfer of Rectangular Impinging Jets in Rows", Heat Transfer '94, Proc. of the Seminar, (August 22-24 1994), Southampton, UK.

[39] LASCHEFSKI, H., CZIESLA,T., MITRA, N.K.: "Flow and Heat Transfer of Laminar Impinging Jets", HTD-301, presented at ASME-Winter-Annual-Meeting Chicago, USA, November 6-11, 1994.

[40] POTTHAST, F., LASCHEFSKI, H., BISWAS, G., MITRA, N.K.: "Numerical Investigation of Flow Structure and Mixed Convection Heat Transfer of Impinging Radial and Axial Jets", Num. Heat Transfer, A, 26, 2, (1994) pp. 123-140.

[41] LASCHEFSKI, H., BRAESS, D., HANEKE, H., MITRA, N.K.: "Numerical Investigations of Radial Jet Reattachment Flows", Int. J. Numerical Methods in Fluids, 18 (1994) pp. 629-646.

[42] WINKELSTRÄTER, M., SOEST, C., LASCHEFSKI, H., MITRA, N.K., FIEBIG,M.: "Numerical Investigations of Flow and Heat Transfer of Laminar 2D and 3D Impinging Jets", in Proc. 10. Heat Transfer Conf. Brighton, UK, 1994, edited by G. F. Hewitt, 3, pp. 119-124.

[43] WINKELSTRÄTER, M., LASCHEFSKI, H., CZIESLA, T., MITRA, N.K.: "Transition to Chaos in Impinging Three Dimensional Axial and Radial Jets", Proc. 14 Int. Conf. on Numerical Methods in Fluid Dynamics, (July 11-15 1994), Bangalore, India, Springer, pp. 368-371.

[44] OWSENEK, B.L., CZIESLA, T., SCHOPHAUS, U., BISWAS, G.: "Influence of Swirl on Heat Transfer of Impinging Axial and Radial Jets", Second ISHMT-ASME Heat and Mass Transfer Conference, (Dec. 28-30 1995), Surathkal/Mangalore, India.

[45] CZIESLA, T., LASCHEFSKI, H., MITRA, N.K.: "Influence of the Exit Angle on Radial Jet Reattachment and Heat Transfer", Journal of Thermophysics and Heat Transfer, 9, 1, (Jan.-Mar. 1995) pp. 169-174.

[46] CZIESLA, T., BRAUN, H., BISWAS, G., MITRA, N.K.: "Large Eddy Simulation in a Channel with Exit Boundary Conditions", ICASE (NASA Langley) Report 198304, No. 96-18, März 1996.

[47] CZIESLA, T., MITRA, N.K.: "Large Eddy Simulation with Dynamic Subgrid Stress Model of a Rectangular Impinging Slot Jet", in Proc. 15th Int. Conference on Numerical Methods in Fluid Dynamics, (24.-28. June 1996), Monterey, USA.

[48] CZIESLA, T., MITRA, N.K.: "Large Eddy Simulation of Rectangular Impinging Jets", in Proc. Second ERCOFTAC Workshop on Direct and Large Eddy Simulation, (16.-19. Sep. 1996), Grenoble, France.

[49] OWSENEK, B.L., CZIESLA, T., BISWAS, G., MITRA, N.K.: "Numerical Investigation of Heat Transfer in Impinging Axial and Radial Jets with Superimposed Swirl", Int. J. Heat and Mass Transfer, 40, 1 (1997) pp. 141-147.

[50] LASCHEFSKI, H., CZIESLA, T., BISWAS, G., MITRA, N.K.: "Numerical Investigation of Heat Transfer by Rows of Rectangular Impinging Jets", appears in Numerical Heat Transfer, A, 30, (July 1996), pp. 87-101.

[51] LASCHEFSKI, H., CZIESLA, T., MITRA, N.K.: "Evolution of Flow Structure in Impinging Three Dimensional Axial and Radial Jets", appears in Int. J. Numerical Methods in Fluids.

[52] CZIESLA, T., TANDOGAN, E., MITRA, N.K.: "Large Eddy Simulation of Heat Transfer from Impinging Slot Jets", appears in Numerical Heat Transfer.

# HEAT TRANSFER AND FLOW LOSSES IN STEADY AND SELF-OSCILLATING CHANNEL FLOWS WITH RECTANGULAR VORTEX GENERATORS (RVG)

H. Neumann, H.-W. Hahne, U. Müller, and M. Fiebig
Lehrstuhl für Wärme- und Stoffübertragung,
Institut für Thermo- und Fluiddynamik
Ruhr-Universität Bochum, 44801 Bochum, Germany

## SUMMARY

The flow field, temperature field, heat transfer and flow loss enhancement in channel flow with periodically mounted longitudinal and transverse rectangular vortex generators (RVGs) are studied experimentally and numerically in the laminar to transitional Reynolds number range. Comparison of experiments and numerics have been carried out. A base RVG configuration with a height ratio of $e/H=0.5$, a longitudinal pitch ratio of $Lp/e=10$, a length ratio of $l/e=4$, a thickness ratio of $\delta/e=0.1$, and an angle of attack of $\beta=45°$ is used. The lateral pitch ratio of one period consisting of two diverging winglets in the flow direction is equal to $Bp/e=8$ with a distance ratio between the leading edges of the winglets of $s/e=(1-\sin \beta)\cdot l/(2\cdot e)$. In addition to the Reynolds number the angle of attack, longitudinal pitch ratio, and height ratio are varied in the ranges $0°\leq\beta\leq90°$, $5\leq Lp/e\leq15$ and $0.25\leq e/H\leq0.5$, respectively. It is shown that the critical Reynolds number of plane channel flow can be reduced by more than an order of magnitude from $Re_{crit,\,0}=3470$ to $Re_{crit}\approx200$ for $\beta=90°$ and $e/H=0.5$. Smaller angles of attack and height ratios increase the critical Reynolds number and reduce the turbulence level. Furthermore, the transition Reynolds number range is stretched by the use of RVGs. Relative to plane channel flow the functional dependence of heat transfer and pressure loss on Reynolds number in the range $200\leq Re\leq3000$ can be expressed for the experiments by one power law. Maximum heat transfer is obtained at $\beta\approx45°$ while the pressure loss increases monotonically with angle of attack. This increase of the pressure loss with angle of attack is attributed to the dominant form drag of the RVGs. Like the heat transfer the frictional resistance reaches a maximum value at $\beta\approx45°$.

## INTRODUCTION

Compact heat exchangers are characterized by high heat duties per unit volume and high heat transfer coefficients. This implies small hydraulic diameters, low Reynolds numbers and fins. Vortex generators (VGs) can be used as fins or to modify fins [1]. Mounted VGs increase the heat transfer surface as fins and simultaneously enhance the convective heat transfer on the primary surface. Two kinds of VGs can be distinguished: Transverse vortex generators (TVGs) and longitudinal vortex generators (LVGs). TVGs generate vortex structures with their axes mainly transverse to the primary flow direction, while LVGs generate vortex systems with vortex axes mainly along the primary flow direction, see Fiebig [1-5]. Local heat transfer enhancement in steady channel flow with TVGs on one channel wall occurs in the regions of flow reattachment and acceleration. The global heat transfer coefficient remains unchanged at the plane channel value [3, 5-8]. LVGs in contrast cause already a global heat transfer enhancement in steady channel flows by the corkscrew motion of the fluid between the core and the channel walls [3-5,8]. Another mechanism of heat transfer enhancement are self-induced velocity fluctuations (transition, turbulence). Using VGs the transition Reynolds number is decreased. Self sustained fluctuations, transition to turbulence start already at a lower Reynolds number compared to plane channel flow. The flow fluctuations in a channel with TVGs start with a small amplitude oscillation of the vortex. This oscillation causes an increased mixing and heat transfer by the same mechanism as turbulence increases mixing and heat transfer.

The turbulent transport quantities were investigated by Vasanta Ram and Lau [10] for the base configuration at $\beta=30°$ and $2\cdot10^4 \leq Re \leq 4\cdot10^4$ with a four-component hot-wire anemometer, see also Fiebig [4]. The vortex structure, local heat transfer and flow losses in turbulent flow $5\cdot10^3 \leq Re \leq 14\cdot10^3$ were studied numerically and experimentally by Neumann, Braun, and Fiebig [11] for different angles of attack (90°, 45°). In addition the height ratio of the RVGs was reduced to e/H=0.05 to limit the RVGs to the viscous sublayer. The transition region was especially studied for $\beta=90°$ by Hahne, Weber, and Fiebig [12].

Here the transition from laminar to turbulent flow and its dependence on RVG angle of attack $\beta$, height e, and longitudinal pitch Lp (Fig. 1) is investigated. The influence of self-sustained oscillations on heat transfer and flow losses by TVGs and LVGs is of particular technical interest.

For these fundamental investigations of laminar and transitional channel flows the geometric base configuration, shown in Fig.1 has been chosen. It is characterized by lateral and longitudinal periodicity and its property that it transforms from a LVG into a TVG when $\beta$ is increased. Because of its rectangular form the vortex generator is called a RVG. For angle of attack $\beta=0°$ the VGs represent inline fins of half the channel height (e/H=0.5) with a lateral and longitudinal pitch ratio of Bp/e=8 and Lp/e=10, respectively, a length ratio of l/e=4, and a thickness ratio of $\delta$/e=0.1. For $\beta=90°$ the VGs represent compact transverse ribs.

Both experimental and numerical methods are used for the investigations. Especially the experimental methods had to be advanced. Self-sustained oscillations of the flow are experimentally investigated by hot-wire-anemometry (base configuration with e/H={0.5, 0.25} and $\beta$={45°, 90°}). The apparent friction factors are determined from the static pressure drop. A surface covering high resolution mass transfer technique is used to determine quantitatively the complex two-dimensional heat transfer distribution. For this technique the analogy between heat and mass transfer is used to determine local Nusselt numbers from the Sherwood numbers. Furthermore the experimental results serve as a reference case for the numerical simulations of the channel flow with RVGs. A 3D unsteady finite volume code has been developed by our group to determine numerically in detail the struc-

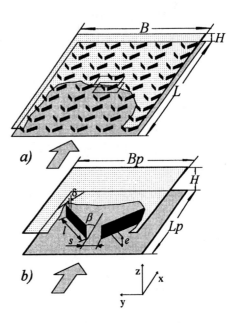

*Fig. 1:* Vortex generator base configuration with: e/H = 0.5; $\delta$/e = 0.1; Lp/e =10; Bp/e = 8; l/e = 4; s/e = (1-sin $\beta$)·l/(2·e); $0° \leq \beta \leq 90°$

ture of the flow and temperature fields. The numerical results allow a quantitative description of the flow to analyse the physical mechanisms which causes the heat transfer and flow loss enhancements.

## EXPERIMENTAL METHODS

Hot-wire measurements

For the investigation in the transition region measurements with a two-component hot-wire anemometer are performed in an open wind tunnel. The test section has a cross sectional area of height $H = 20$mm and width $B = 22.5 \cdot H$. The overall length of the test section is $150 \cdot H$.

The hot-wire equipment consists of a DISA type 55P61, 5-µ, tungsten-wire, X-wire probe, and a PSI model 6100 constant temperature anemometer. The data were sampled via a fast Keithley-Metrabyte DAS-20 analog/digital converter with a Keithley-Metrabyte simultaneous-sample-and-hold SSH-4 module. The data were directly stored on a PC hard disc at a sample frequency of 2048 Hz for each channel. The reference velocity was monitored via an orifice meter that was connected to an MKS Baratron 100 mbar pressure gauge. The signal was also stored on the PC hard disc.

The hot-wire probe is positioned by a computer-controlled traversing mechanism that is described in detail in [13]. Its design principle is based on three circular discs mounted eccentrically within each other. The discs are free to rotate in the $x$-$y$ plane which is the plane of the wall, see Fig. 1. The probe can be positioned in the $x$-$y$ plane within a circular area of 340mm and in any desired $z$ position. A rotation of the probe around the z-axis is also possible.

Because of the small flow velocities of $0.1$m/s $< \bar{u} < 1$m/s, an *in-situ* calibration of the probe is performed before and after each run. For calibration, only the ribbed wall is exchanged for a plane channel wall, so the hot-wire equipment is kept turned on and stays at the same position during the calibration and the actual measurement. Fully developed laminar flow is established then, since the length of the test section upstream of the probe is larger than $200 \cdot H$.

Jörgensen's directional sensitivity equation [14] describes the cooling of each wire i of the probe

$$U_{\text{eff},i}^2 = U_{n,i}^2 + k_{t,i} U_{t,i}^2 + k_{b,i} U_{b,i}^2 \quad (i=1,2), \tag{1}$$

where $U_{\text{eff}}$ is the effective cooling velocity and $U_n$, $U_t$, $U_b$ are the components of the velocity vector U, normal, tangential and binormal to the wire i. If one assumes that the flow vector is two-dimensional, and that the instantaneous velocity field is uniform over the sensing volume of the probe, the binormal velocity component $U_b$ can be neglected. Introducing the angle $\theta$ between flow direction and probe axis, Eq. (1) can be expressed as

$$f_i(E_i) = U_{\text{eff},i}^2 = U^2 \, g_i(\theta) \quad i=1,2, \tag{2}$$

$$g_i(\theta) = \left(\cos^2\theta + k_i^2 \sin^2\theta\right). \tag{3}$$

$g_i(\theta)$ and $f_i(E_i)$ are the yaw and the speed function, respectively. These functions must be determined by calibration. The calibration procedure is performed as follows: The probe is positioned in the plane channel as shown in Fig. 2 (a). Its wires are positioned in the x-z plane.

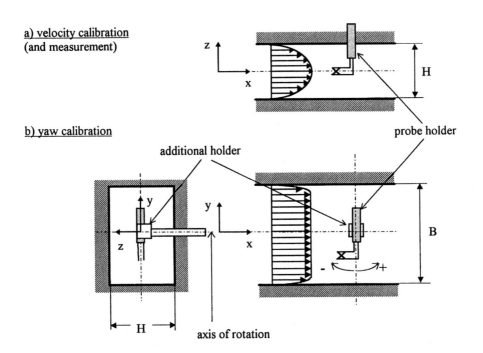

**Fig. 2:** Test section and probe arrangement during calibration and measurement. a) sideview on velocity calibration, b) side and top view on yaw calibration; from [20]

The probe is rotated around the y-axis first, while the flow velocity U is kept constant. A step motor controlled rotation of the probe in the interval of -40° < θ < 40° is performed in a single run at constant angular velocity within approximately 25 seconds. From this we get the dependency of the effective cooling velocity on the flow angle θ as

$$\frac{U_{\text{eff},i}^2}{U^2} = g_i(\theta). \tag{4}$$

Following Döbbeling et al. [15], it is assumed that $g_i(\theta)$ is not a function of U. This assumption is reasonable for laminar flow and, especially when the Reynolds number interval that is investigated is small, and the calibration is performed within this interval. In order to prove the validity of this assumption, the yaw angle calibration was repeated for several U. The probe is rotated into the x-y plane next, see Fig. 2 (b). It is worth mentioning that this rotation does not require the disconnection of the probe cables. For θ = 0° we defined $U_{\text{eff}} = U$ (Eqs. (2) and (3)), i.e. $g_i(\theta) = 1$ (i=1,2). Hence we can perform the speed calibration, simplifying Eq. (2) to

$$f_i(E_i) = U_{\text{eff},i}^2 = U^2 \qquad i=1,2. \tag{5}$$

The flow speed U of the fully developed laminar flow in the plane channel is varied continuously during the speed calibration. Mass flow rate and the two anemometer offset voltages are recorded simultaneously with the A/D-converter and stored at the hard disk of the PC with a frequency up to 12500 Hz. The entire Reynolds number interval is covered by the calibration procedure within several minutes. It was shown at an early stage of the investigation, that the continuous variation of U, e.g. decrease of U from 1 m/s to 0.1m/s within 3 minutes, leads to the same speed function as a fit based on several steady state calibration points, each with a constant U. Approximately 20000 samples are recorded during the speed calibration.

The speed calibration functions $f_i(E_i)$ for the two wires (Eq. (4)) are determined by use of a 4th order polynomial best fit approximation to the recorded data. The functions $f_i(E_i)$ are used to calculate the effective velocities $U_{eff,i} = f_i(E_i)$ from the $E_i$ that were sampled in the yaw calibration. Following Eq. (3), the yaw calibration functions $g_i(\theta)$ are also determined by a 4th order polynomial best fit approximations.

After calibration the plane channel wall is exchanged for a ribbed wall and the measurements are performed. Up to 170000 samples are recorded at typical sampling frequencies between 100 and 1000 Hz. U and $\theta$ are calculated from the sampled $E_i$ by solving

$$\sum_{i=1}^{2}\left(\frac{f_i(E_i)}{U^2} - g_i(\theta)\right) = 0 \qquad i=1, 2. \tag{6}$$

Eq. (6) is solved iteratively by a conventional Newton-Raphson algorithm. The computation time on a PC is less than 15 minutes for 170000 samples.

The frequencies of the fluctuations are read from the power density spectra of the flow velocity. Those were calculated by transforming the calculated velocity components via a discrete Fast-Fourier-Transform- (FFT-) algorithm.

Pressure drop measurements

For the investigation of the flow losses and the heat transfer the open wind tunnel as shown in Fig. 3 is used. The flow losses are taken from the measured static pressure drop. For this purpose three pressure taps are positioned laterally at the plane wall opposite of each of the first ten ribs. Three pressure taps at each cross section increase the statistical precision of the results. The 30 pressure tap's positions are shown in Fig.3.

Following Kakac et al. [16] the apparent friction factor $f_{app}$ is defined from the pressure drop as:

$$f_{app} = \frac{2\Delta p}{\rho \bar{u}_m^2} \frac{A_f}{A}. \tag{7}$$

For the case of periodically fully developed flow the apparent friction factor is the same for each period. A constant apparent friction factor for successive periods in the flow direction is a necessary condition for periodically fully developed flow.

Heat/Mass transfer measurements

The analogy between mass and heat transfer is used to determine the local heat transfer coefficients. The optical *Ammonia-Absorption Method* (AAM) is used that yields the local surface mass transfer distribution for constant wall concentration. The colour distribution of a filter paper caused by the chemical reaction is analysed photometrically. Because of the fast chemical reaction at the surface the ammonia concentration on the wall is equal to zero. This corresponds to the thermal boundary condition of constant wall temperature. The evaluation of local Nusselt numbers using the analogy between mass and heat transfer and using the periodically constant mass transfer coefficient is described in detail by Weber [20].

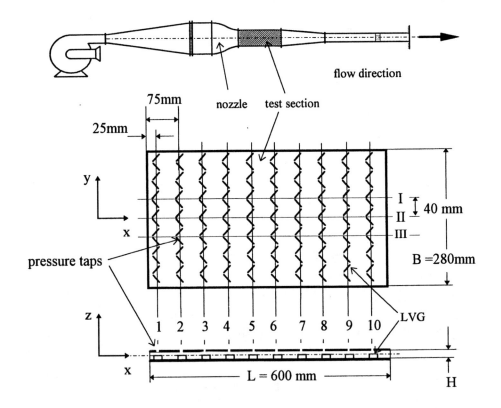

**Fig. 3:** Wind tunnel and test section for the pressure drop and the mass transfer measurements; the x-y-positions of the pressure taps are marked with the intersection of the lines I-1 to III-10; from [20]

For the mass transfer experiments a small amount of reaction gas ($NH_3$) is added to the air stream in a short gas pulse. The ammonia is partially absorbed by the filter paper, which is soaked with a reactive solution of $H_2O$, $MnCl_2$ and $H_2O_2$. A colour reaction takes place with the ammonia. This colour reaction is a well known ammonia detection method and can could be described as follows

$$NH_3 + H_2O \Leftrightarrow NH_4OH \Leftrightarrow NH_4^+ + OH^-$$
$$2NH_4OH + MnCl_2 \Rightarrow 2NH_4Cl + Mn(OH)_2 \quad (8)$$
$$Mn(OH)_2 + H_2O_2 \Rightarrow MnO_2 + 2H_2O.$$

The locally transferred $NH_3$ mass becomes visible as a colour density distribution. This colour intensity is digitally determined by a commercial flat bed scanner (HP scanner IIc).

The mass transfer coefficient is defined as:

$$\beta_{NH_3-air} = \frac{\dot{m}_{NH_3}}{\left(\rho_{NH_3B} - \rho_{NH_3wall}\right)}. \quad (9)$$

Here $(\rho_{NH_3,B} - \rho_{NH_3,wall})$ is the difference between bulk and wall ammonia density. Because of the chemical reaction at the wall the density of ammonia at the wall $\rho_{NH_3,wall}$ is equal to zero. The absorbed ammonia during the experimental time $t_m$ can be determined by the following integration of Eq. (9):

$$\int_{t=0}^{t=t_m} \dot{m}_{NH_3} dt = \int_{t=0}^{t=t_m} \beta_{NH_3-air} \rho_{NH_3,B} dt. \tag{10}$$

The density of the ammonia in the volume flux is equal to the ratio of the added mass flow of ammonia and the volume flux:

$$\rho_{NH_3,B} = \frac{\dot{M}_{NH_3}}{\dot{V}} \tag{11}$$

With Eq. (10) and (11) the locally absorbed mass density of ammonia is equal to:

$$b = \frac{M_{NH_3\, local}}{A} = \beta_{NH_3-air} \int_{t=0}^{t=t_m} \rho_{NH_3,B} dt = \beta_{NH_3-air} \frac{M_{NH_3}}{\dot{V}}. \tag{12}$$

In Eq. (12) $M_{NH_3\, local}$ is the locally absorbed ammonia and $M_{NH_3}$ is the total mass of ammonia at the investigated cross section during the measurement. The mass of reacted ammonia per unit area b corresponds to the colour intensity determined by a flat bed scanner. The quantitative determination of the mass transfer coefficient as a function of the colour intensity requires a thorough calibration with a surface of known mass transfer distribution. The calibration surface choosen is a circular disc in cross flow. For known $\beta_{NH_3-air}$ the calibration establishes for measured $M_{NH_3}$ and $\dot{V}$ the relation between the colour intensity and the mass transfer density b.

With the knowledge of the local mass transfer density it is possible to determine the local mass transfer coeffiecient by the equation

$$\beta_{NH_3-air}(x,y) = b(x,y) \frac{\dot{V}}{M_{NH_3}(x)}. \tag{13}$$

Here $M_{NH_3}(x)$ describes the mass of ammonia in the air flow at the cross section x during the measuring time. The so determined mass transfer coefficient implies the assumption that the bulk density of ammonia in the air flow is constant over the span
$\rho_{NH_3\infty}(x,y) = \rho_{NH_3\infty}(x) \neq f(y)$.

With the assumption of a power law dependence on Prandtl and Schmidt number the analogy between heat and mass transfer yields the following relation between the local Nusselt number and the local Sherwood number, see Eckert [21]

$$\frac{Nu}{Sh} = \left(\frac{Pr}{Sc}\right)^n = \left(\frac{D_{NH_3-air}}{a_{air}}\right)^n = Le^{-n}. \tag{14}$$

In Eq. (14) D is the ammonia-air diffusion coefficient, a the thermal diffusivity, and n the exponent of the Lewis-number. The diffusion coefficient and its dependence on pressure and temperature are well known quantities, but n depends on the flow situation. For Pr between 0.5 and 2.5 Mitzushima and Nakajima [22] show that the value of n=0.5 is a good choice. For the used ammonia-air mixtures ($0.9 \leq Le \leq 1.1$), the variation of n between 0.37 (turbulent boundary layer) and 0.6 (laminar boundary layer) is not critical, the uncertainty is below 2%. In addition the determination of the local mass transfer rate depends on the measured grey scale value because of the non linear calibration curve (s. Zhang [23]), but for spanwise averaged Nusselt distributions Zhang [23] shows that the statistical error is not more than 4%.

Schulz [17], Kottke et al. [18], and Schulz and Fiebig [19] determine the amount of ammonia in Eq. (12) with the known injected amount of ammonia. For a low volume flux as in the present investigations the ammonia losses in the upstream parts of the test section cannot be neglected. For periodically developed flow, the periodically constant mass transfer coefficient is used by Weber [20] to determine the amount of ammonia remaining at one cross section. With the periodic boundary condition:

$$\beta(x, y) = \beta(x + n \cdot L_P, y) \tag{15}$$

The change of the mass of ammonia in the air flow $\Delta M_{NH_3}(x; x + \Delta x)$ between two cross sections x and x+$\Delta$x is determined by the mass balance.

$$M_{NH_3}(x) - M_{NH_3}(x + \Delta x) = \int_{X=x}^{X=x+\Delta x} \left( \int_y b(X, y) dy \right) dX \equiv \Delta M_{NH_3}(x; x + \Delta x). \tag{16}$$

With the local mass transfer coefficient (Eq. (13)) for points (x,y) and (x + n·L$_P$,y) and Eqs. (15) and (16) it follows for the local mass transfer coefficient in the periodically developed section.

$$\beta(x, y) = \left( b(x, y) - b(x + n \cdot L_P, y) \right) \frac{\dot{V}}{\Delta M_{NH_3}(x; x + n \cdot L_P)}. \tag{17}$$

With Eq. (17) the determination of the local Sherwood number distribution is independent of the injected amount of ammonia. However the practicability of Eq. (17) is restricted by the noise level of the measured mass transfer density. Instead of using Eq. (17) alone for the determination of the Sherwood number one can equate Eqs. (13) and (17) to determine the amount of ammonia at one cross section. So the amount of ammonia at one cross section x = A is determined with the span averaged mass transfer densities in the periodically ribbed section.

$$M_{NH_3}(A) = \Delta M_{NH_3}(A; A + \Delta x;) + \frac{\overline{b}(A + \Delta x) \Delta M_{NH_3}(A + \Delta x; A + \Delta x + n \cdot L_P)}{\overline{b}(A + \Delta x) - \overline{b}(A + \Delta x + n \cdot L_P)} \tag{18}$$

with: $\overline{b}(x) = \int_y b(x, y) dy$.

$\Delta M_{NH_3}$ describes the absorbed ammonia between two cross sections determined by Eq. (16). With Eq. (18) and the here realized resolution it is possible to get nearly 1000 values for the determination of the amount of ammonia at one cross section A. With the knowledge of $M_{NH_3}(A)$ and Eq. (16) the local mass transfer coefficient is:

$$\beta(x, y) = \frac{b(x, y) \dot{V}}{M_{NH_3}(A) - \left( \int_{X=A}^{X=x} \overline{b}(x) dX \right)}. \tag{19}$$

The Nusselt number follows from Eq. (14).

## NUMERICAL METHODS

Governing equations and boundary conditions

The flow and temperature fields are governed by the continuity, nonsteady three dimensional Navier-Stokes, and energy equations. The fluid properties are assumed to be constant and the dissipation term in the energy equation is neglected. The equations then read in nondimensional cartesian tensor notation:

Continuity equation

$$\frac{\partial u_i}{\partial x_i} = 0 \tag{20}$$

Momentum equation
$$\frac{\partial u_i}{\partial t} + \frac{\partial u_i u_j}{\partial x_j} = -\frac{\partial p}{\partial x_i} + \frac{2}{Re}\frac{\partial^2 u_i}{\partial x_j^2} \quad (21)$$

Energy equation
$$\frac{\partial T}{\partial t} + \frac{\partial T u_i}{\partial x_i} = \frac{2}{Re\,Pr}\frac{\partial^2 T}{\partial x_i^2}. \quad (22)$$

Here $x_1$ corresponds to the x-axis, $x_2$ to the y-axis, and $x_3$ to the z-axis, see Fig.1.

The velocity components are made nondimensional by the average main stream velocity $\bar{u}_m$, all lengths have been normalized by the channel height. Only the Reynolds number is based on $2 \cdot H$ (Re = $\bar{u}_m \cdot 2 \cdot H/\nu$). The pressure is normalized with $\rho \cdot \bar{u}_m^2$ where $\rho$ is the density. The Prandtl number Pr equals to Pr = $\nu/a$. In order to realize periodic boundary conditions for pressure, the pressure gradient is divided into a periodic and a constant part. The constant part represents the pressure gradient per period. It is determined with the constant mass flux condition. Periodic boundary conditions for the temperature field can be achieved by normalizing the temperature with the bulk temperature $\theta=(T-T_W)/(T_B-T_W)$ as proposed in [26].

The used hydrodynamic and temperature boundary conditions for the computational domain are:

<u>Solid walls</u>:
no slip: $u_i = 0$ (23)
constant wall temperature: $T_w$ = const (24)

<u>Side boundaries</u>:

symmetry: $u_2|_{x_2=0} = u_2|_{x_2=Bp} = 0$; $\Psi = \{u_i, p_{per}, \theta\}$; $\left.\frac{\partial \Psi}{\partial x_2}\right|_{x_2=0} = \left.\frac{\partial \Psi}{\partial x_2}\right|_{x_2=Bp} = 0$ (25)

or

periodicity: $\Psi = \{u_i, p_{per}, \theta\}$; $\Psi|_{x_2=0} = \Psi|_{x_2=Bp}$; $\left.\frac{\partial \Psi}{\partial x_2}\right|_{x_2=0} = \left.\frac{\partial \Psi}{\partial x_2}\right|_{x_2=Bp}$ (26)

<u>Inlet and outlet</u>:

periodicity: $\Psi = \{u_i, p_{per}, \theta\}$; $\Psi|_{x_1=0} = \Psi|_{x_1=Lp}$; $\left.\frac{\partial \Psi}{\partial x_1}\right|_{x_1=0} = \left.\frac{\partial \Psi}{\partial x_1}\right|_{x_1=Lp}$. (27)

Discretization

A finite-volume method is used to discretize Eqs. (20)-(22) on co-located grids. For the time dependent terms a backward Euler scheme with first order accuracy is used, see Grosse-Gorgemann [9]. The viscous and the pressure terms are discretized with central differences with second order accuracy. The non-linear convective terms are determined through the "deferred-correction approach", which was first suggested by Khosla and Rubin, see [9]. According to this technique, the convective flux is split into an implicit part, expressed through the upwind differencing scheme (UDS) and an explicit part containing the differences between the central scheme and the UDS approximations. When a converged solution is obtained the discretization becomes tantamount to a second order accurate scheme.

## Numerical method

The outcome of the discretization is a system of three non-linear algebraic equations for the three velocities and two linear algebraic equations for the pressure and temperature. To handle this problem an iterative solution method is used which suppresses the nonlinearity by lagging, i.e. one factor of each quadratic term is taken from the previous iteration step.

For solving the linearized and linear algebraic equation systems the SIP-algorithm [25] is used.

To start the computation cycle, estimated fields for velocity, pressure and temperature are necessary. From these fields, corrected pressure and velocity fields are interactively calculated with the modified SIMPLE-C algorithm [24]. From the initial temperature field and the afore mentioned corrected velocity field, a temperature distribution for the first cycle is calculated from the energy equation. The forgoing corrected pressure and velocity fields are used to solve the Eq. (21) to evaluate velocities for the next time step. In summary the solution proceeds in three steps. In the first step, the momentum equations are solved with the help of converged velocity and pressure fields of the previous time step. In the second step, a pressure velocity correction equation is solved following the modified SIMPLE-C algorithm which determines the correct pressure and a divergence-free velocity field. In the third step, the solution of the energy equation is solved.

## Validation

The code has been validated with the spectral element code Nekton [26] and hot wire measurements by Grosse-Gorgemann et al. [27]. For 2D and 3D unsteady flow grid refinement studies were performed to estimate the grid dependence of the results.

The estimation of the numerical error due to the finite differences of the computational domain is carried out with a extrapolation of the results (local values: maximum amplitudes and mean velocity, global values: apparent friction factor and Nusselt number) for a grid size equal to zero (grid independent solution). For the extrapolation the Romberg extrapolation scheme is used, see for example [36]. To apply a second order extrapolation scheme three solutions for three grid sizes ($\Delta X_i$) are necessary. The used grids with the required main memory and computing time are listed in table 1 for the base configuration with $\beta = 45°$ and $Re = 350$.

*Table 1:* Grid resolution and required memory and computing time

| grid | nx | ny | nz | $\Delta X_i$ | memory | time | computer |
|---|---|---|---|---|---|---|---|
| 1 | 77 | 62 | 18 | 0.065 | 47 MB | 0.25 d | IBM 580H |
| 2 | 102 | 82 | 22 | 0.05 | 90 MB | 15 d | IBM 530H |
| 3 | 167 | 142 | 37 | 0.029 | 490 MB | 25 d | IBM 580H |

Figure 4 shows the dependence of the relevant values for the flow as a function of the standardized grid size ($\Delta X_i/\Delta X_{max}$). The maximum amplitude Am of the axial velocity at the reference point is nearly independent of the grid size. The mean velocity at the reference point and the apparent friction factor have a oscillatory behavior with decreasing grid size. The averaged Nusselt number for the vortex generators, the plates and the global Nusselt number are shown in Fig. 5. The Nusselt numbers for the vortex generators decrease continuously with decreasing grid size whereas the Nusselt number for the plates and the global Nusselt number show an oscillating behavior like the values for mean velocity at the reference point and the apparent friction factor.

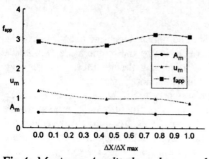

**Fig.4:** Maximum Amplitude and mean velocity at a reference point as well as the apparent friction factor as a function of the standardized grid size $(\Delta X/\Delta X_{max})$

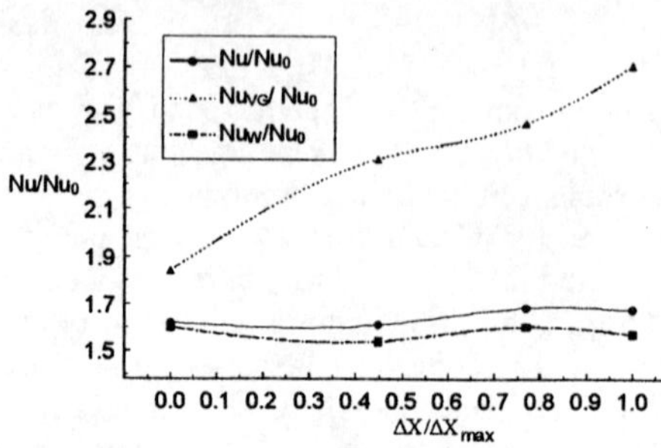

**Fig. 5:** Nusselt numbers for the vortex generators and plates and the global Nusselt number as function of the standardized grid size $(\Delta X/\Delta X_{max})$

The error distribution as a function of the grid size is shown in table 2. The error is defined by:

$$\Delta g = \frac{g_i - g_{\Delta x \to 0}}{g_{\Delta x \to 0}} \quad [\%] \tag{27}$$

with $g \in (Nu, f_{app}, A_m, u_m)$.

Most numerical calculations are carried out with the standardized grid size of 0.769, which has a derivation in the global Nusselt number of 3.88% and in the apparent friction factor of 11.63%, see table 2.

**Table 2:** Percentage deviations of the solutions with the different grids from Tab. 1 to the grid independent solution

| grid →<br>deviations ↓ | 1 | 2 | 3 |
|---|---|---|---|
| $\Delta X / \Delta X_{max}$ | 100 | 76.9 | 48.8 |
| $\Delta u_m$ | -33.41 | -21.01 | -22.38 |
| $\Delta Am$ | -10.44 | -8.89 | -3.09 |
| $\Delta f$ | 8.26 | 11.63 | -6.15 |
| $\overline{\Delta Nu}$ | 3.39 | 3.88 | -0.52 |
| $\overline{\Delta Nu}_{sw}$ | -5.78 | 3.14 | -3.13 |
| $\overline{\Delta Nu}_{rw}$ | 2.11 | 4.53 | -4.18 |
| $\overline{\Delta Nu}_w$ | -1.81 | 0.312 | -3.66 |
| $\overline{\Delta Nu}_{VG}$ | 47.43 | 34.02 | 25.72 |

# RESULTS AND DISCUSSION

Flow structure

At an angle of attack of $\beta = 0°$ the secondary flow components are very small. Figure 6 shows the primary flow of the base configuration with $\beta = 0°$ at $z/H = 0.25$. Above and below the two winglets in Fig. 6 the flow is accelerated and between the winglets the flow is decelerated. These differences in the velocity distribution can be attributed to the asymmetric lateral position of the distances between the winglets i.e. the separation of the winglets with regard to the lateral symmetry is only the outerspacing.

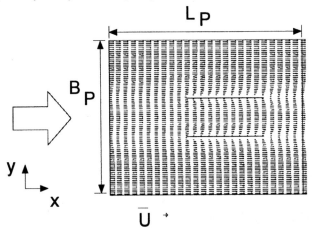

*Fig. 6: Primary flow at: $z/H=0.25$, $\beta=0°$, $Re = 350$ ($e/H=0.5$, $Lp/e=10$, $Bp/e=8$, $l/e=8$, $s/e=4$); from [9]*

Figures 7 and 8 show the secondary flow of the base configuration with angles of attack of $\beta = 15°$ and $\beta = 45°$ respectively.

At an angle of $\beta = 15°$ a steady longitudinal vortex is generated along the side edge of each rectangular winglet. The secondary flow components in Fig. 7 show that the longitudinal vortices generated are nearly circular. The area influenced by this longitudinal vortices across the primary flow extends over the projected width of the VGs. The vortex axis are essentially parallel to the primary flow direction and are situated at $z/H \approx 0.5$, $y/H \approx 1.4$ and $y/H \approx 2.6$ respectively. The axial flow is accelerated to a maximal value of $u_{max} / \bar{u}_m = 2.0$.

At an angle of $\beta = 45°$ the flow is unsteady periodic at $Re = 350$. Compared to the flow at $\beta = 15°$ the shape of the vortices at $\beta = 45°$ is more ellipsoidal. The counterrotating vortices are stretched over the whole cross-section of the periodic element. The axis of the counterrotating vortices are not parallel to the primary flow direction but have a lateral spatial oscillation in each periodic element. The position of each vortex axis is a spiral in the flow direction. Its direction of rotation is the same as the vortex. The time dependent position of each vortex axis at any cross section rotates on a ellipsoidal path against the direction of rotation of the vortex.

More details concerning the flow structure at different angles of attack are given by Grosse-Gorgemann [9].

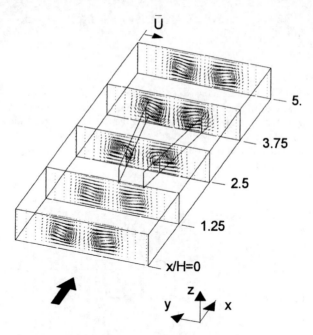

**Fig. 7:** Secondary flow at $x/H = \{0, 1.25, 2.5, 3.75, 5\}$, $\beta = 15°$, $Re = 350$ ($e/H = 0.5$, $Lp/e = 10$, $Bp/e = 8$, $l/e = 4$, $\delta/e = 0.1$, $s/e = 2.965$); from [9]

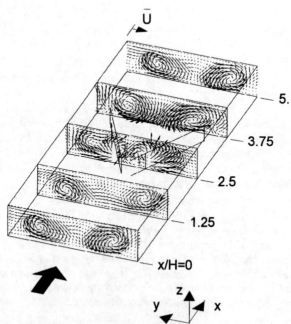

**Fig. 8:** Momentum plot of the secondary flow at $x/H = \{0, 1.25, 2.5, 3.75, 5\}$, $\beta = 45°$, $Re = 350$, ($e/H = 0.5$, $Lp/e = 10$, $Bp/e = 8$, $l/e = 4$, $\delta/e = 0.1$, $s/e = 1.172$); from [9]

## Transition

For the investigation of transition the hot-wire is positioned in the middle of the channel ($z/H = 0.5$) and 3·e downstream of the VG. With continuously accelerated flow the time dependent variations of the u- and w-component are shown in Fig. 9. In plane channel flow the transition occurs in the Reynolds-number range of $3470 \leq Re \leq 3520$. Due to the transition the centerline velocity decreases from $u_c \approx 1.5\bar{u}$ to $u_c \approx 1.2\bar{u}$. This drastic drop in $u_c$ and the frequency spectrum indicate that the flow is turbulent.

Figure 9 b shows the beginning of transition for the channel with TVGs (base configuration, Fig. 1 with $\beta = 90°$). The measurements are made downstream of the eighth VG-row. The continuously small acceleration was to $\partial \bar{u}/\partial t = 5.5 \cdot 10^{-4}$ m/s². Up to $Re \approx 160$ the values of the variation of the velocities was $|u'/\bar{u}|_{noise} = |w'/\bar{w}|_{noise} \leq 0.03$. This was in the noise range of the used instruments. For Reynolds numbers $Re > 160$ the amplitudes Am of the velocity fluctuations increased noticeable. At $Re \approx 250$ the mean Am-values of the u-component reached 90% of the mean velocity and the Am-values of the w-component 45%. At Reynolds number $Re \approx 160$ the transition process started with small self-sustained oscillations and turned into very large unstructured fluctuations at $Re \approx 350$. In contrast to the transition in plane channel flow (Fig. 9 a) the oscillations of the velocity field were coupled with the Reynolds number in a wide Re range.

Figure 9 c shows the appropriate results for the LVG base configuration of Fig. 1 with $\beta = 45°$. For this case the acceleration was about twice that of the 90° TVG case, $\partial \bar{u}/\partial t \approx 12 \cdot 10^{-4}$ m/s². Transition started at a higher Re ($\approx 240$) and was not been completed at $Re \approx 380$. The transition Re range for the TVGs was about twice the plane channel value. It is note worthy hat the power density spectrum shows already at the critical Reynolds number of $Re \approx 240$ more than one dominant frequency. The TVG base configuration (Fig. 1 with $\beta = 90°$) begins with one dominant frequency and a quasi sine oscillation.

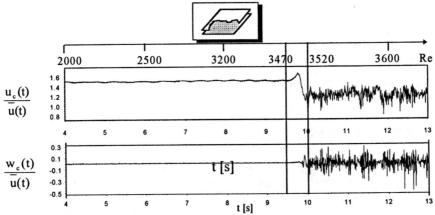

*a) plane channel*

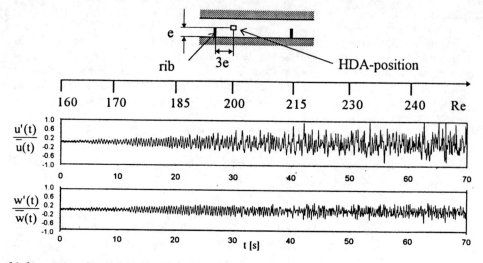

b) base configuration with $\beta = 90°$

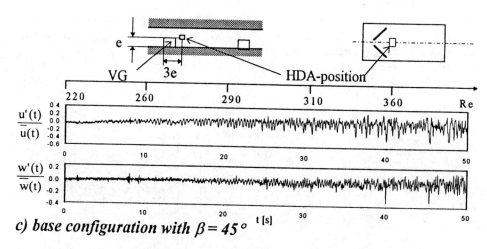

c) base configuration with $\beta = 45°$

*Fig. 9:* Transition of the continuous accelerated flow: u- and w-component at z/H = 0.5 of a) plane channel; b) base configuration with $\beta = 90°$; c) base configuration with $\beta = 45°$; The position of the hot-wire was downstream of the eighth VG-row.; from [20]

Figure 10 shows the RMS-value of the experimentally investigated TVG and LVG configurations versus the Reynolds number. The standardized Root-Mean-Square- (RMS-)Value is defined by Eq. (28).

$$\text{RMS} = \frac{\sqrt{\overline{u'^2} + \overline{w'^2}}}{\overline{u}_m} . \tag{28}$$

Here $\overline{u}_m$ describes the spatial (index m) and temporal ($\overline{\phantom{u}}$) mean velocity in the plane channel. In laminar steady flow the noise generated a mean RMS-value of RMS ≈ 0.02 for the TVG and RMS ≈ 0.03 for the LVG, respectively. The lower critical Reynolds number, the beginning of transition is defined when the RMS-values increase noticeable with Reynolds-number. In Fig. 10 the transition depends on the sign and amount of acceleration of the mean velocity (volume flow). The lower critical Reynolds number (begin of transition) can be deduced from the increase in RMS values, about 160 for the TVG base configuration with e/H=0.5 and 240 for the corresponding LVG configuration (β=45°, e/H=0.5). For e/H=0.25 transition starts for the TVG base configuration at Re≈520 and for the corresponding LVG configuration at about 640. The upper critical Reynolds number, where transition is completed and the turbulence is fully developed is more difficult to determine. It may be defined as the Re value when the RMS values become independent of Re. This value has only been reached for the e/H=0.5 TVG base configuration (Re≈750). For the other configuration transition is not yet completed. It may

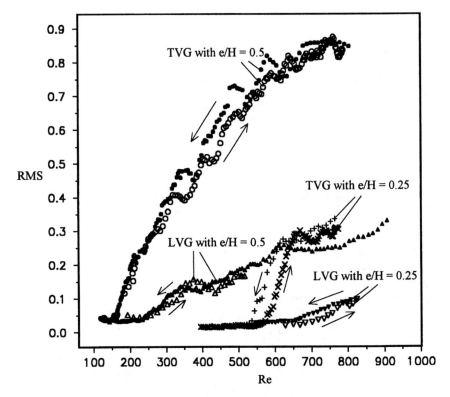

*Fig. 10: Standardized Root-Mean-Square- (RMS) values of the TVG and LVG base configurations with e/H = 0.5 and e/H = 0.25 vs. the Reynolds-number. The increasing and the decreasing of Re is distinguished by the arrows (↑) and (↓); from [20]*

also be expected that the RMS value decreases again for very large velocities or Re.

The transition was influenced by the different acceleration conditions. An expected result was that the lower critical Reynolds numbers of the TVG configurations are smaller than those for the comparable LVG configurations. Furthermore the RMS-values of the TVG configurations were higher than for the corresponding LVG configurations.

The numerical results confirm that the unsteadiness of the flow increases with angle of attack β. Table 3 gives the time behaviour of the flow at different angles of attack at Re = 350.

*Tab. 3: Time behaviour of the flow at different angles of attack at Re = 350; from [9]*

| β | 0°, 15° | 30°, 45°, 60° | 75°, 90° |
|---|---|---|---|
| time behaviour of the results | steady | unsteady periodic | unsteady non-periodic |

The increasing unsteadiness with angle of attack for constant Re manifests itself also in the increasing number of detected frequencies. Figure 11 shows the frequency spectrum of the velocity components u, v, w determined by a Fast Fourier Transformation (FFT) [28]. The dimensionless frequency is given by the Strouhal number.

$$S_{2H} = \frac{f \, 2H}{\bar{u}_m} \qquad (29)$$

The points P(x, y, z) were selected for their high amplitudes (Am). For each angle of attack the point with maximum amplitudes is situated in the area of the trailing edge of the winglet. For a pair of delta-winglets embedded in a boundary layer Torii and Yanagihara [29] also found the maximum amplitudes in this *"up-wash"* area. For the angles of attack of β = {30°, 45°, 60°} Table 4 shows the maximum amplitudes and the Strouhal numbers.

Both Fig. 11 and Table 4 show that identical Strouhal numbers are dominant at β = 30° and β = 45°. At an angle of β = 60° the dominant Strouhal number is decreased to S = 0.195. Furthermore with increasing angle of attack the dominant Strouhal numbers decrease and the spectrum extends to higher Strouhal numbers.

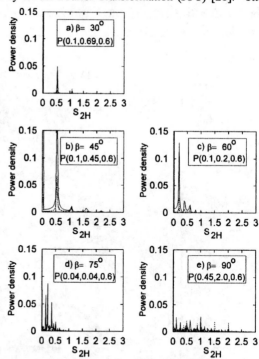

Fig. 11: Strouhal numbers of the velocity components u, v, w at the point P(x,y,z) at Re = 350 for a) β = 30°, b) β = 45°, c) β = 60°, d) β = 75° and e) β = 90°; from [9]

**Tab. 4:** *Strouhal-numbers and amplitudes of each velocity component for different angles of attack at Re = 350; from [9]*

| β   | S     | $\bar{u}$ | Am(u) | $\bar{v}$ | Am(v) | $\bar{w}$ | Am(w) | P(x, y, z)       |
|-----|-------|-------|-------|-------|-------|-------|-------|------------------|
| 30° | 0.537 | 0.943 | 0.107 | 0.073 | 0.056 | 0.199 | 0.048 | (0.1, 0.69, 0.6) |
| 45° | 0.537 | 0.902 | 0.471 | 0.084 | 0.138 | 0.107 | 0.136 | (0.1, 0.45, 0.6) |
| 60° | 0.195 | 0.786 | 0.494 | 0.055 | 0.333 | 0.072 | 0.171 | (0.1, 0.2, 0.6)  |

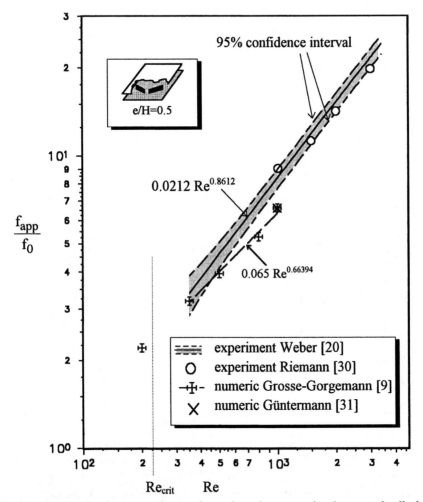

**Fig. 12** *Standardized apparent friction factor $f_{app}$ of one period with a periodically fully developed flow vs. Reynolds number. Comparison of different experimental and numerical values;*
*$e/H = 0.5$, $Lp/e = 10$, $Bp/e = 8$, $l/e = 4$, $\beta = 45°$; $f_0 = 24/Re$; from [20]*

## Pressure Losses

For the base configuration with $\beta = 45°$ and $\beta = 90°$ apparent friction factors are obtained experimentally by Weber [20] and Riemann [30] and numerically by Grosse-Gorgemann [9] and Güntermann [31].

All available results concerning the base configuration with $\beta = 45°$ are shown in Fig. 12. In a Reynolds number range of $1000 \leq Re \leq 3000$ the experimentally determined apparent friction factors of Weber [20] and Riemann [30] agree very well. The values of Riemann [30] are within the 95% confidence interval of the values determined by Weber [20]. The measurements were taken in different wind tunnels and the TVGs had different dimensions. The maximum deviation of the apparent friction factors determined by Riemann to the correlation

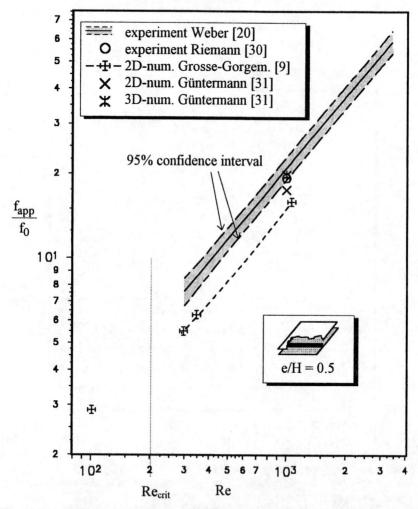

*Fig. 13: Standardized apparent friction factor $f_{app}$ of one period with a periodically fully developed flow vs. Reynolds-number. Comparison of different experimental and numerical values;*
*$e/H = 0.5$, $Lp/e = 10$, $Bp/e = 8$, $l/e = 4$, $\beta = 90°$; $f_0 = 24/Re$ ; from [20]*

$$\frac{f_{app}}{f_0} = 0.0212 \, \text{Re}^{0.8612} \qquad (e/H=0.5) \qquad 350 \leq \text{Re} \leq 3000 \qquad (30)$$

is equal to 8%. For a Reynolds number range of $200 \leq \text{Re} \leq 1000$ the numerically determined friction factors versus the Reynolds number can be described by the following correlation

$$\frac{f_{app}}{f_0} = 0.065 \, \text{Re}^{0.66394}. \qquad (e/H=0.5) \qquad 200 \leq \text{Re} \leq 1000. \qquad (31)$$

The deviation between the experimental (Eq. 30) and numerical results (Eq. 31) increases from 3.1% at Re = 350 to 18.7% at Re = 1000. For Re = 350 and Re = 500 the numerically determined values are within the 95% confidence interval of the values determined by Weber [20]. The deviation between the numerically determined apparent friction factor of Grosse-Gorgemann [9] and Güntermann [31] at Re = 1000 is less than 0.5%.

Figure 13 shows the equivalent results of the apparent friction factor versus the Reynolds number for the base configuration with $\beta = 90°$. The comparison of the results of Weber [20] between Fig. 12 and Fig. 13 shows that the TVG-configuration produces higher flow losses, so the ratio of the apparent friction factor is $f_{app, \beta=90°}/f_{app, \beta=45°}=2.9$ for $350 \leq \text{Re} \leq 3000$. However the results of Grosse-Gorgemann [9] in Fig. 12 and Fig. 13 show a variable ratio of $f_{app, \beta=90°}/f_{app, \beta=45°} = 1.8$ at Re = 200 to 2.4 at Re = 1000.

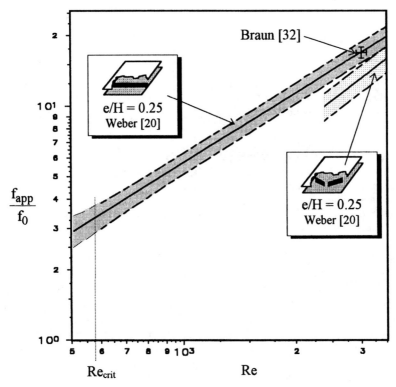

*Fig. 14: Standardized apparent friction factor $f_{app}$ of one period with a periodically fully developed flow vs. Reynolds-number. Comparison of different experimental and numerical values;*
*e/H = 0.25, Lp/e = 10, Bp/e = 8, l/e = 4, $\beta = \{45°, 90°\}$; $f_0 = 24/Re$; from [20]*

The apparent friction factor at $Re = 10^3$, for the TVG configuration experimentally determined by Riemann [30], agrees well with the result of Weber [20], Fig. 13.

The numerically determined values of Grosse-Gorgemann [9] and Güntermann [31] are lower than the experimental values for both configurations (e/H = 0.5, $\beta$ = {45°, 90°}).

For the TVG configuration in Fig. 13 the deviation between the experimental and 2D numerical apparent friction factors of Weber [20] and Grosse-Gorgemann [9] is nearly constant for $350 \leq Re \leq 1100$, but relatively high with 27%. This indicates an insufficient grid resolution in the calculations. The 3D values of Güntermann [31] differs by less than 3% from the experimental values of Weber [20].

The experimentally determined apparent friction factors for the configurations with $\beta$ = {45°,90°} and e/H = 0.25 are presented in Fig. 14. For comparison a Large-Eddy-Simulation (LES) value of Braun [32] is also shown for Re = 3000. The numerical procedure of the LES is described in detail by Braun [32] and by Neumann et al. [11] in this volume. For the TVG configuration ($\beta$ = 90°) with e/H = 0.25 the deviation between the experimental results of Weber [20] and the numerical value of Braun [32] is less than 1.2%.

The functional dependence of the apparent friction factor on Reynolds number from a steady laminar flow well into the transition region allows the following observations.

- For the TVG configuration ($\beta$ = 90°) with e/H = 0.25 the transition starts at $Re \approx 640$. The experimental results indicate that the functional dependence of the apparent friction factor ratio $f/f_0$ on Reynolds number has a power law dependence with n = 1.102 for $640 \leq Re \leq 3000$.

- For the TVG configuration ($\beta$ = 90°) with e/H = 0.5 the numerically determined apparent friction factor ratio $f/f_0$ has a power law dependance with n = 0.798 in the range $100 \leq Re \leq 350$. So there is no change of the gradient in the range of the critical Reynolds number.

- The extrapolation of the experimental values for the TVGs for $350 \leq Re \leq 3000$ (e/H = 0.5, $\beta$ = 90°)) to steady flow conditions at Re = 100 yields an apparent friction factor ratio $f/f_0$ which differs from the numerically determined value of Grosse-Gorgemann [9] by 2.2%. The small deviation of the extrapolated value of the experiments and the numerically determined values indicate that there is little change in the dependence of n on Reynolds number in this range.

- The extrapolation of the experimental values for the TVGs for $350 \leq Re \leq 3000$ (e/H = 0.5, $\beta$ = 90°)) to turbulent flow with $Re = 40 \cdot 10^3$ yields an apparent friction factor ratio $f/f_0$ which differs from the value of Riemann [30] by 8%. The extrapolation of the experimental values for the LVGs (Eq. (30) for $350 \leq Re \leq 3000$ (e/H = 0.5, $\beta$ = 45°)) to turbulent flow with $Re = 50 \cdot 10^3$ yields an apparent friction factor ratio $f/f_0$ which differs from the value of Riemann [30] by 7%. The small difference between the extrapolated and the measured value indicate that there is little change in the gradient $df_{app}/dRe$ between transition and turbulent flow.

- The deviations between numerical and experimental results are mainly due to insufficient grid resolution.

The dominance of the form drag compared to friction drag results in a weaker dependence of the apparent friction factor on Reynolds number. Figure 15 shows the numerically determined apparent friction factor enhancement $f/f_0$ versus angle of attack $\beta$ for Re = 350. The dimensionless geometrical parameters correspond to the base configuration in Fig. 1. The apparent friction factor enhancement $f/f_0$ increases continually with increasing angle of attack. This can be attributed to the continually increasing form drag with angle of attack. The frictional drag reaches a maximum value at an angle of attack of 45°.

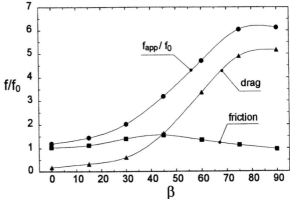

*Fig. 15:* Numerically determined apparent friction factor ratio versus angle of attack $\beta$ for Re = 350 e/H = 0.5;, Lp/e = 10, Bp/e = 8, l/e = 4, $s/e = (1-\sin\beta) \cdot l/(2 \cdot e)$ ; from [9]

Table 5 also shows the normalized form drag and friction drag for different angles of attack.

As can be seen in Fig. 15 and Table 5 the friction drag at $\beta = 0°$ and $\beta = 90°$ are nearly equal to the friction drag of the plane channel flow. A maximum value of the friction drag of $c_f/f_0 = 1.56$ is reached at $\beta = 45°$.

*Tab. 5:* Normalized form drag and friction drag at different angles of attack, Re = 350; from [9]

| $\beta$ | form drag $\dfrac{c_w A_{VG}}{f_0 A_w}$ | frictional drag $\dfrac{c_f}{f_0}$ | $\dfrac{f_{app}}{f_0}$ |
|---|---|---|---|
| 0° | 0.168 | 1.032 | 1.200 |
| 15° | 0.320 | 1.130 | 1.450 |
| 30° | 0.613 | 1.410 | 2.023 |
| 45° | 1.630 | 1.560 | 3.190 |
| 60° | 3.361 | 1.350 | 4.711 |
| 75° | 4.900 | 1.140 | 6.040 |
| 90° | 5.161 | 0.963 | 6.124 |

Heat Transfer Enhancement

If the analogy between momentum and heat transfer should be valid the dependence of friction drag on angle of attack should also hold for the heat transfer.

Figure 16 shows the numerically determined Nusselt number enhancement versus angle of attack for the geometry of Fig. 15. Two different side boundary conditions were chosen, symmetry and periodicity. For an unsteady flow the periodicity condition at the sides allows locally mass transfer to the laterally adjacent periods. With symmetry condition at the sides lateral flow oscillations may be suppressed. For angles of attack of $30° \leq \beta \leq 60°$ the time behaviour of the flow is steady with symmetry side condition and periodic with periodicity side condition. The different side boundary conditions allow to estimate the influence of the oscillations on heat transfer for the same angle of attack at this Reynolds number.

The maximum value of Nu/Nu$_0$ = 1.84 was reached with periodic side boundary condition as the maximum value of the friction drag at β ≈ 45°. On the other hand the maximum heat transfer with symmetry boundary condition was reached at β ≈ 60°. More detailed information is given in Table 6.

For β = 45° the Nusselt number on the ribbed wall is about 34% higher for periodic flow conditions (periodicity side condition) than for steady flow conditions (symmetry side condition). The smooth wall values are about 9% lower. On the average the Nusselt number for the periodic flow condition (periodicity side condition) is about 10% enhanced over the steady flow condition (symmetry side condition).

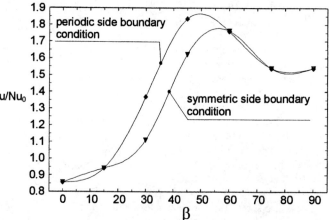

**Fig. 16:** Numerically determined Nusselt-number enhancement versus angle of attack β for Re = 350
e/H = 0.5;, Lp/e = 10, Bp/e = 8, l/e = 4,
s/e = (1-sinβ)·l/(2·e) ; from [9]

**Tab. 6:** Mean Nusselt-number enhancement on the smooth wall (without winglets), on the ribbed wall (with winglets), on both walls without winglet surfaces and on both walls with winglet surfaces for different side boundaries and angles of attack; Re = 350; from [9]

| β | Nu$_{sw}$/Nu$_0$ | Nu$_{rw}$/Nu$_0$ | Nu$_w$/Nu$_0$ | Nu/Nu$_0$ | time behaviour | side boundaries |
|---|---|---|---|---|---|---|
| 0° | 0.93 | 0.86 | 0.90 | 0.86 | steady | sym. or per. |
| 15° | 1.22 | 0.74 | 0.98 | 0.94 | steady | sym. or per. |
| 30° | 1.42 | 0.90 | 1.16 | 1.11 | steady | symmetry |
|  | 1.69 | 1.03 | 1.36 | 1.37 | periodic | periodicity |
| 45° | 1.79 | 1.31 | 1.56 | 1.62 | steady | symmetry |
|  | 1.64 | 1.77 | 1.71 | 1.84 | periodic | periodicity |
| 60° | 1.34 | 1.77 | 1.55 | 1.76 | steady | symmetry |
|  | 1.33 | 1.77 | 1.55 | 1.76 | periodic | periodicity |
| 75° | 1.39 | 1.32 | 1.36 | 1.54 | non periodic | sym. or per. |
| 90° | 1.85 | 1.19 | 1.52 | 1.54 | non periodic | sym. or per. |

Figure 17 shows Nusselt number and apparent friction factor enhancements for varying longitudinal pitch and Reynolds number. For all Reynolds numbers a maximum heat transfer enhancement resulted for longitudinal pitch of Lp/e ≈ 7.5. In the same Lp/e range the apparent friction factor enhancement f/f$_0$ decreases with increasing longitudinal pitch due to the decreasing importance of the VG drag.

The maximum of the heat transfer enhancement at Lp/e ≈ 7.5 can be attributed to the superposition of three effects:

- The area of reversed flow behind each VG is nearly equal for all investigated longitudinal pitches. So the percentage of the size of the heat transfer surface influenced by the dominant longitudinal vortex decreases with decreasing longitudinal pitch.

- Between two inline VGs the velocity and so the heat transfer decreases with increasing longitudinal pitch.

- The average vorticity $\omega_x$ of the flow increases with decreasing longitudinal pitch. Thus increases the mixing of the flow and the heat transfer.

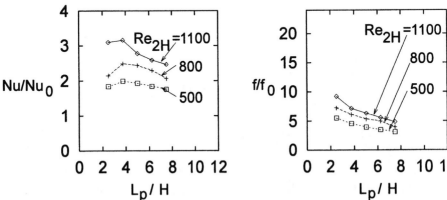

*Fig. 17: Heat Transfer and flow loss enhancement vs. the dimensionsless pitch for different Reynolds numbers*
*e/H = 0.5;, Lp/e = 10, Bp/e = 8, l/e = 4, s/e = 1.172 for β = 45°; from [9]*

Figure 18 shows the span averaged Nusselt numbers of the TVG base configuration in the entrance region of the channel at Re = $10^3$. As shown in Fig. 3 the test section is connected to a nozzle so the oncoming flow in the test section at x/H = 0 is laminar and nearly homogenous. Each period in Fig. 18 a) and b) consists of 197 data values. For comparison a heat transfer correlation of Kakac et al. [16] for a simultaneously hydrodynamically and thermally developing flow in a flat channel with an initially laminar and homogenous flow is presented in Fig. 18 a) and b). After a channel length of x/H ≈ 20 for Re = $10^3$ the correlation of Kakac et al. [16] reaches the Nusselt number of plane Poiseuille flow ($Nu_0$ = 7.54) within 2%.

For the TVG base configuration with e/H = 0.5 at Re = $10^3$ the heat transfer is periodic after the third vortex generator row (x/H > 15), see Fig. 18 a) and b). For higher Reynolds numbers Re > $10^3$ the thermal periodicity is reached at least after the fourth vortex generator row. The mean Nusselt number per period increases for the first periods. For Re = $10^3$ the mean Nusselt number reaches a maximum value for the second period. It is about 6% higher than the value of the first period (5 ≤ x/H ≤10) for the smooth wall and about 60% higher for the ribbed wall. The mean Nusselt number of the following period (10 ≤ x/H ≤15) is 1.5% higher at the smooth channel wall and 11% higher at the ribbed channel wall than the Nusselt numbers of the following periods.

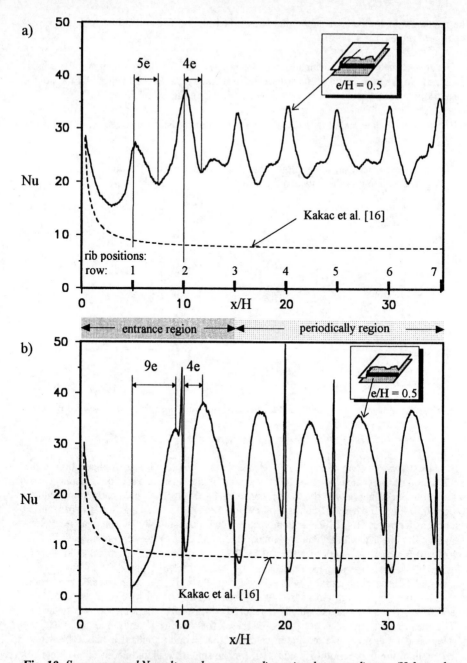

*Fig. 18:* Span averaged Nusselt numbers versus dimensionsless coordinate x/H from the entrance to the periodic region;
 a) smooth wall and b) opposite ribbed wall with TVG base configuration;
 $Re = 10^3$, initially nearly homogeneous laminar flow; for comparison a correlation for hydrodynamic and thermal entrance flow by Kakac [16] is shown; from [20]

The increase in mean Nusselt-numbers in the first two periods is accompanied by a shifting of the local maximum, see Fig. 18 a) and b).

A similar shift can be detected for the LVG base configuration with e/H=0.5 and $\beta = 45°$ at $Re = 10^3$. Figure 19 shows the span averaged and local heat transfer distribution on the smooth and ribbed wall in the entrance region. The shift of the local heat transfer maximum for the first two periods can be seen in Fig. 19 a). The increase in mean Nusselt number for the first periods becomes more distinct at higher Reynolds numbers. Figure 20 shows the span averaged Nusselt numbers from the entrance region up to the periodic region for $Re = 3 \cdot 10^3$. The maximum Nusselt number on the ribbed wall is reached between the second and third winglet row and on the smooth wall between the first and second winglet row.

On the ribbed side the complex local heat transfer distribution $Nu = f(x, y)$ with the LVG base configuration (Fig. 19 b) can be characterized as follows.

- Due to the upstream effects of the LVGs the flow becomes three-dimensional with a resulting two-dimensional heat transfer on the wall in front of the first winglet row.

- A horseshoe vortex forms around the winglets and enhances the heat transfer directly in front of each winglet. This heat transfer enhancement can be detected only in front of the first winglet row. In the following periods the dominant longitudinal vortices influence the horseshoe vortices and the heat transfer effects of the horse shoe vortices cannot be detected.

- The dominant counterrotating longitudinal vortices generate a large area of high heat transfer in the down-wash region, grey area in Fig. 19 b). The minimum heat transfer within one period can be detected in the up-wash regions between the winglets of two laterally neighbouring periods. These areas of minimum heat transfer are marked in light grey.

On the smooth wall the complex heat transfer distribution ($Nu = f(x, y)$) can be characterized as follows:

- In the first part of the test section ($0 \leq x/H \leq 3.5$) the heat transfer enhancement is higher than on the ribbed wall for all investigated Reynolds-numbers. The flow acceleration near the smooth wall by the VGs in the entrance region is the mayor reason.

- For the first winglet row the areas of high heat transfer are a footprint of the VGs at the opposite ribbed wall. This can be attributed to the flow displacement by the VGs. Further downstream the superposition of this flow displacement and the vortices causes a more homogenous heat transfer distribution.

- Areas of high heat transfer on the smooth wall lie opposite to areas of low heat transfer on the ribbed wall.

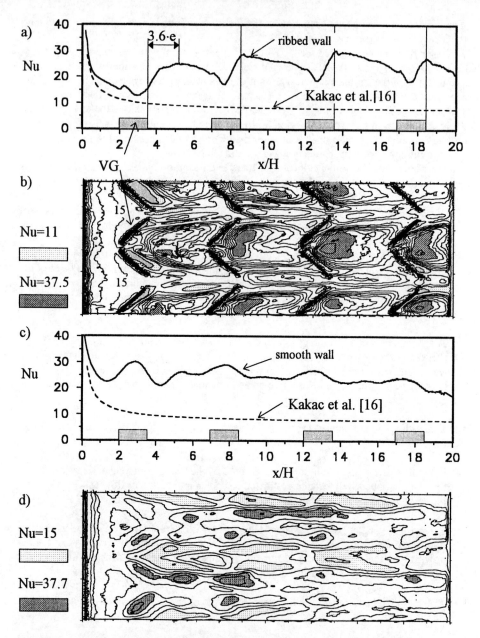

*Fig. 19:* Spanwise averaged (a), c)) and local (b), d)) Nusselt number distribution on the ribbed (a), b)) and the smooth wall (c), d)) versus the dimensionless coordinate x/H for Re = $10^3$; LVG base configuration with e/H = 0.5; $\beta$ = 45°; from [20]

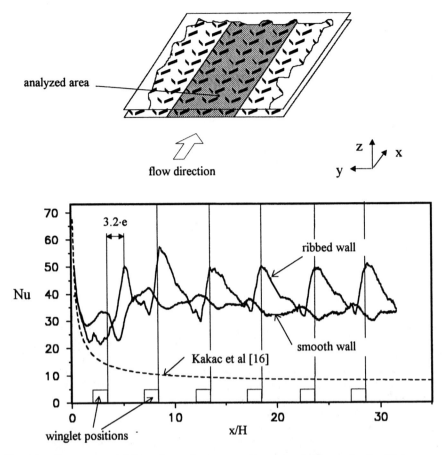

*Fig. 20:* Span averaged Nusselt numbers versus dimensionsless coordinate x/H from the entrance region up to the periodically region for Re = $3 \cdot 10^3$; LVG base configuration with e/H = 0.5 and $\beta$ = 45°; from [20]

Weber [20] attributed the laterally unsymmetrical heat transfer distribution to small disturbances in the entrance flow or to the manufacturing tolerances of the configuration, respectively.

The agreement of the positions of heat transfer minimum and maximum on the ribbed and smooth wall can be better seen for the span averaged values, for example in Fig. 20.

As for TVGs the heat transfer on the LVG ribbed wall is periodically fully developed downstream of the third VG row and on the smooth wall downstream of the fourth VG row.

For the periodically fully developed flow the global Nusselt numbers of both walls for the experimentally investigated configurations are presented in Table 7.

One of the aims of this work is a comparison of experimental and numerical results. While the comparison of the flow losses between the experimental and numerical results is restricted to the global values per period in the periodically fully developed flow region, the span averaged Nusselt numbers can be compared. A good agreement of the local or span averaged heat transfer distribution in the periodic region between experimental and numerical values is an essential criterion for their correctness. Inhomogenous inlet conditions have little influence in the periodic region. Deviation in geometry between numeric and experiment will influence local values more than span averaged values.

*Tab. 7: Global Nusselt numbers of both walls for the experimentally investigated configurations for different Reynolds numbers; $Lp/e = 10$, $Bp/e = 8$, $l/e = 4$, $s/e = (1-\sin \beta) \cdot l/(2 \cdot e)$; from [20]*

| configuration | Re | $Nu/Nu_0$ ribbed wall | $Nu/Nu_0$ smooth wall | $Nu/Nu_0$ global |
|---|---|---|---|---|
| TVG, $e/H = 0.5$, $\beta = 90°$ | 500 | 1.21 | 1.23 | 1.22 |
|  | 1000 | 2.44 | 2.91 | 2.68 |
|  | 2000 | 4.88 | 5.22 | 5.05 |
|  | 3000 | 6.55 | 5.88 | 6.21 |
| LVG, $e/H = 0.5$, $\beta = 45°$ | 1000 | 2.88 | 2.63 | 2.76 |
|  | 2000 | 3.33 | 3.14 | 3.24 |
|  | 3000 | 4.28 | 3.97 | 4.13 |
| TVG, $e/H = 0.25$, $\beta = 90°$ | 1000 | 1.72 | 1.78 | 1.75 |
|  | 3000 | 6.06 | 3.84 | 4.95 |
| LVG, $e/H = 0.25$, $\beta = 45°$ | 1000 | 2.33 | 1.95 | 2.14 |

One mayor difference between experiments and numerics is that inaccuracies of the numerical results increase with Reynolds number while inaccuracies of experimental results decrease with increasing Reynolds number. With increasing Reynolds number the small scale fluctuations and consequently the need of high grid resolution for describing these fluctuations increases. Dang and Morchoisene [33] find that the ratio of large to small scale fluctuations, and hence the necessary grid resolution increases as $Re^{9/4}$. Because of the limited memory resources for the present simulations they were confined to Reynolds numbers $Re \leq 1100$. For decreasing Reynolds numbers the pressure differences between pressure taps fall and the inaccuracy increases, respectively. Also smaller Reynolds numbers reduce the gradient of the ammonia concentration in the periodic flow region. So the mass transfer rate at the entrance region may exceed the calibration limit and may fall below the calibration limit in the periodically region. A reduction of the absolute size of the channel and the configuration enhance the accuracy of the results at lower Reynolds numbers. The disadvantage of this step consists in the increasing inaccuracies of the manufacturing of the configuration, the decreased area for analyzing the local heat transfer, and the decreased velocities for the HDA. As a compromise a Reynolds number of $Re=10^3$ is chosen for the comparison between experimental and numerical determined heat transfer values. At this Reynolds number for both, numerics and experiments a reasonable accuracy can be expected.

For $25 \leq x/H \leq 40$ the experimental data, shown in Fig. 21, give an average Nusselt number increase for one periodic element of $Nu/Nu_0 = 2.88$ for the ribbed wall and $Nu/Nu_0 = 2.63$ for the smooth wall. The numerical data give a value of $Nu/Nu_0 = 2.84$ for the ribbed wall and $Nu/Nu_0 = 2.62$ for the smooth wall. This is an excellent over all agreement. The deviations of local numerical heat transfer data from the experimental data are significantly higher than the deviations of the global values, and rise up to 30%. This cannot be explained by experimental uncertainties due to noise and positioning errors of the VG, only. The numerical Nusselt number distributions on the ribbed and smooth wall show much sharper gradients in both the spanwise and the streamwise direction than the experimentally determined distributions. This is mirrored by the span averaged numerical data of Fig. 21. Maxima and minima of the span averaged Nusselt numbers are over- and underestimated, respectively, up to 20% by the numerical data. The variations of the experimental data for the four different VG-rows, plotted in Fig. 21, are less than 5%.

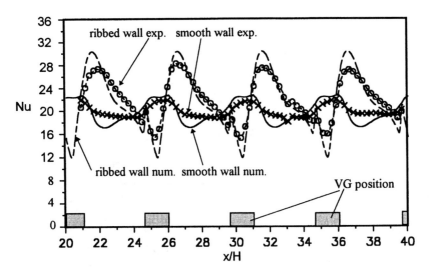

*Fig. 21:* Comparison of experimentally and numerically determined span and time averaged Nusselt number distributions for periodically fully developed flow, LVG base configuration with $e/H = 0.5$, $\beta=45°$, $Re =10^3$. Experimental data are span averaged over 5 periodic elements; from [20]

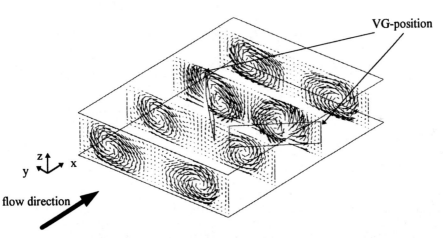

*Fig. 22:* Time averaged flow field and vortex structure at $Re = 10^3$; $Lp/e = 10$, $Bp/e = 8$, $l/e =4$, $s/e = 1.172$ for $\beta = 45°$; from [20]

From comparison of local time dependent numerically determined velocity vectors with hot-wire measurements [34] at Re=1000 it is known that the direct numerical simulation (DNS) reduces the amplitude and frequency of the unsteadiness and hence the Reynolds shear stress. The reason is the grid resolution, which is limited by memory size and speed of currently available workstations. This leads to an underestimation of the flow losses, see Fig. 12. But it cannot explain the deviations of the numerical and experimental Nusselt number data on the ribbed wall and smooth wall, plotted in Fig. 21.

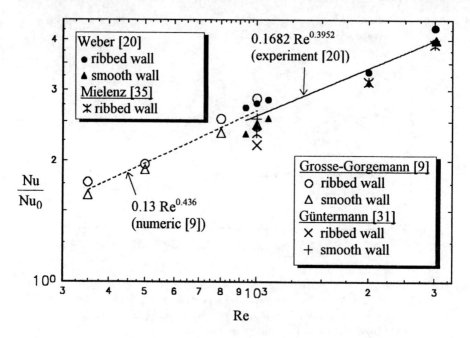

***Fig. 23:*** *Standardized Nusselt number of one period with periodically fully developed flow versus the Reynolds number; Comparison of different experimental and numerical values;*
*e/H = 0.5, Lp/e = 10, Bp/e = 8, l/e = 4, s/e = 1.172 for $\beta$ = 45°; from [20]*

At the ribbed wall, at $x/H \approx 25$, 30, etc., the numerical data (Fig. 21) show a distinct Nusselt number minimum that is due to a recirculation zone behind each RVG. Downstream, in the region of $x/H \approx 26.5$, 31.5, etc., the numerical data indicate fluid flow toward the wall, 'downwash zones' of the longitudinal vortices, generated by the RVG. These are regions of absolute maxima of the Nusselt number. The experimental data of Fig. 21 suggest smaller recirculation zones and give smaller absolute Nusselt number maxima. These deviations may be caused by additional vortex shedding from the front and rear edge of the VG which are suppressed by the DNS for the used grid resolution of 102·82·22 cells, see Fig. 22.

The deviations between numerical and experimental data on the top wall without VGs cannot be clearly explained. It was observed that the Nusselt number distribution for one periodic element on the smooth wall was strongly influenced by the distribution of its spanwise adjacent elements. For example, the Nusselt maxima on the smooth wall of all seven spanwise elements of one streamwise VG row shifted sidewards, whereas the distribution on the ribbed wall was strongly symmetrical. Generally speaking, the lateral boundary conditions do influence the Nusselt number distributions on the smooth wall much more than on the ribbed wall.

In all the comparison between the experimental and numerical determined results show excellent agreement for the global values and a satisfactory agreement for local and span averaged Nusselt numbers. The comparison between experimental and numerical results for global heat transfer enhancement is shown for all reported values in Fig. 23.

Figure 23 also shows experimental results of Mielenz [35]. These results are based on a newly developed unsteady infrared thermography method for periodically fully developed flows. Mielenz' [35] results are restricted to the ribbed wall. The deviation of Mielenz' [35] results from Weber's [20] for the ribbed wall amount to nearly 14% for Re = $10^3$, and 5% for Re = $2\cdot10^3$ and only to 2% for Re = $3\cdot10^3$. Estimated accuracy by Mielenz [35] for the Nusselt number is 20%.

The numerical results of Güntermann [31] are restricted to Re = $10^3$ and the deviation to the experimental results of Weber [20] is 3.4% for the ribbed and 23.9% for the smooth wall.

## CONCLUDING REMARKS

Flow structure and temperature fields in transitional channel flow with longitudinal and transverse vortices were investigated experimentally and numerically. For the basic investigations periodically arranged slender rectangular plates called rectangular winglets or rectangular vortex generators (RVG) were attached to one channel wall. The attachment to the channel wall was vertically and along a side edge of the RVG. This chosen vortex generator configuration is characterized by the possibility to generate mainly longitudinal vortices or transverse vortices by varying the angle of attack. At Reynolds numbers below Re = 1100 the investigations were carried out numerically. Experimental investigations extended the Reynolds number range up to $3\cdot10^3$. The numerical investigations were performed with a finite volume code for direct numerical simulation of the three-dimensional unsteady Navier Stokes equations, continuity and energy equation for constant property Newtonian fluids. Experimentally the flow losses were determined from static pressure measurements. The beginning of transition, the lower critical Reynolds number for the onset of self sustained oscillations was determined by two-component hot-wire anemometry. The analogy between heat and mass transfer was used to determine local heat transfer coefficients for constant wall temperature by a mass transfer technique. This Ammonia-absorption method was further developed for analyzing periodically fully developed conditions.

The main improvements of the experimental techniques were:

- For calibration of the HDA probe at very small velocities ($\approx$ 0.1 m/s) a dedicated windtunnel was designed which allowed an *in-situ* calibration. Here the HDA probe was positioned in a laminar hydrodynamically fully developed plane channel flow with the well known velocity profile. The simultaneous recording of the HDA signal and the volume flow allowed a speed calibration by continuous accelerating or decelerating flow, respectively.

- The periodically constant mass transfer coefficient was used to determine the injected amount of ammonia. This means ammonia losses upstream of the test section did not affect the determination of the mass transfer coefficients.

From the results the following conclusions could be drawn:

- The complex local heat transfer distribution on the ribbed side of the RVG configuration with $\beta$ = 45° is mainly influenced by three different vortex structures: (1) the dominant longitudinal vortices from the side edges of the RVGs, (2) horseshoe-like vortices, which form in front of the RVGs, and (3) the separation at the trailing edges of the RVGs.

- The constancy of the apparent friction factor for successive periods in the flow direction showed that the flow is periodically fully developed after 20 channel heights independent of the height of the vortex generators (0.5H or 0.25H). Periodic vortex generators cause shorter entrance lengths than plane channel flows for high Re.

- Compared with plane channel flow the use of vortex generators causes a reduction of the lower critical Reynolds-number, the beginning of transition and a stretching of the Re range in which transition occurs. With increasing angle of attack and height of the RVGs from e/H = 0.25 to e/H = 0.5 the critical Reynolds number decreases. The amplitudes of the fluctuations and the number of frequencies increases with Re. For example, the TVG configuration ($\beta = 90°$) with e/H = 0.5 reduces the lower critical Reynolds number by more than an order of magnitude from $Re_{crit, 0} = 3470$ for a plane channel flow to $Re_{crit} \approx 200$.

- For the base configuration and Reynolds number of Re = 350 the choice of the side boundary conditions, symmetry or periodicity, had a significant influence on the time behavior of the flow and consequently on the heat transfer.
  For angles of attack of $\beta = \{30°, 45°, 60°\}$ the flow remained steady for lateral symmetry conditions. For lateral periodicity conditions the flow became unsteady with a resulting higher heat transfer.
  For $\beta = \{0°, 15°\}$ the flow was steady and for $\beta = \{75°, 90°\}$ unsteady and non periodic with the same heat transfer for both lateral boundary conditions.

- Variations of angle of attack have shown that the RVG configuration with $\beta \approx 45°$ (LVG) causes a higher heat transfer and lower pressure loss than the corresponding RVG configuration with $\beta = 90°$ (TVG). The continuously increasing pressure loss with angle of attack stems from the continuously increasing and dominant form drag while the frictional drag reaches a maximum value for $\beta \approx 45°$. The latter holds also for the heat transfer, which implies a limited validity of the analogy between momentum and heat transfer.

- Variation of the longitudinal pitch shows a maximum heat transfer at Lp/e $\approx$ 7.5 while the pressure loss decreases continuously with increasing longitudinal pitch for Lp/e $\geq$ 3.75.

- Comparison of the experimental and the numerical results at Re = $10^3$ shows an excellent agreement for the global Nusselt numbers while the local and spanwise averaged Nusselt number deviate about 30% and 20%, respectively. This deviations of the Nusselt numbers can be explained with the limited grid resolution which reduces the amplitude of the unsteady flow motion and the Reynolds shear stress. Consequently numerical simulation underestimated the pressure loss.
  Extrapolation of the experimental fit of the apparent friction factor versus the Reynolds number from the unsteady flow region to the steady flow region at Re = 200 for the LVG configuration ($\beta = 45°$) and Re = 100 for the TVG configuration ($\beta = 90°$) respectively yields a value which differs just about 8.2% and 2.2% respectively from the numerical determined apparent friction factor.

## NOMENCLATURE

| | | |
|---|---|---|
| A | [m] | heat transfer area |
| $A_f$ | [m] | cross section |
| Am | [m/s] | velocity amplitude |
| a | [m²/s] | thermal diffusivity |
| B | [m] | width |
| Bp | [m] | lateral periodic pitch |
| b | [g/m²] | mass of reacted ammonia per unit area |
| D | [m²/s] | diffusion coefficient |
| E | [V] | voltage |
| e | [m] | vortex generator height |
| f | [ - ] | speed function |
| f | [Hz] | frequency |
| g | [ - ] | yaw function |
| H | [m] | channel height |
| k | [ - ] | weighting factor |
| L | [m] | length |
| Lp | [m] | longitudinal periodic pitch |
| l | [m] | vortex generator length |
| $M_{NH3}$ | [kg] | mass of ammonia |
| n | [ - ] | exponent of Lewis number |
| n | [ - ] | number of periods |
| nx,ny,nz | | number of volumes in direction x,y,z |
| p | [Pa] | pressure |
| s | [m] | distance between the leading edges of the vortex generators |
| T | [K] | temperature |
| $U_i$ | [m/s] | velocity components at the hot wire |
| u | [m/s] | velocity component in flow direction |
| $\dot{V}$ | [m³/s] | volume flow |
| v | [m/s] | velocity component transverse to flow direction (y) |
| w | [m/s] | velocity component transverse to flow direction (z) |
| ΔX | [ - ] | dimensionless grid size |
| x,y,z | [m] | cartesian coordinates |

*greek symbols*

| | | |
|---|---|---|
| α | [W/m²K] | heat transfer coefficient |
| β | [°] | angle of attack |
| β | [m/s] | mass transfer coefficient |
| δ | [m] | vortex generator thickness |
| λ | [W/(mK)] | heat conductivity |
| ρ | [kg/m³] | density |
| ν | [m²/s] | kinematic viscosity |
| Δ | [ - ] | difference |
| θ | [°] | angle between flow direction and hot wire |
| θ | [ - ] | dimensionless temperature |

*Subscripts*

| | |
|---|---|
| B | bulk |
| b | binormal to the hot wire |
| c | centerline |
| crit | critical value |
| eff | effective |
| i | counter |
| m | spatial averaged |
| max | maximum |
| $NH_3$ | ammonia |
| n | normal to the hot wire |
| n | per periodic part |
| rw | on the ribbed wall |
| sw | on the smooth wall |
| t | tangential to the hot wire |
| VG | on the vortex generator |
| w | on the walls |
| ¯ | temporal averaged |
| ' | temporal variable |
| 0 | value of the plane channel |
| ∞ | reference |

*dimensionless parameters*

$$f_{app} = \frac{2\Delta p}{\rho \bar{u}_m^2} \frac{A_f}{A} \qquad \text{apparent friction factor}$$

$$Le = \frac{Sc}{Pr} \qquad \text{Lewis number}$$

$$Nu_{2H} = \frac{\alpha \, 2H}{\lambda} \qquad \text{Nusselt number}$$

$$Pr = \frac{\nu}{a} \qquad \text{Prandtl number}$$

$$Re = \frac{\bar{u}_m \, 2H}{\nu} \qquad \text{Reynolds number}$$

$$RMS = \frac{\sqrt{\overline{u'^2} + \overline{w'^2}}}{\bar{u}_m} \qquad \text{standardized root-mean-square value}$$

$$S_{2H} = \frac{f \, 2H}{\bar{u}_m} \qquad \text{Strouhal number}$$

$$Sc = \frac{\nu}{D} \qquad \text{Schmidt number}$$

$$Sh_{2H} = \frac{\beta \, 2H}{D} \qquad \text{Sherwood number}$$

## REFERENCES

[1] Fiebig, M.: "Vortex Generators for Compact Heat Exchangers", J. of Enhanced Heat Transfer, Vol. 2, Nos. 1-2, pp. 43-61, 1995.

[2] Fiebig, M.: "Embedded vortices in internal flow: heat transfer and pressure loss enhancement", Int. J. Heat and Fluid Flow 16, pp. 376-388, 1995.

[3] Fiebig, M.: "Vortices: Tools to Influence Heat Transfer - Recent Developments", Proc. of the 2nd ETS, Rome, Vol. 1, pp. 41-56, 1996.

[4] Fiebig, M.: "Vortices as Tools to Influence Heat Transfer; Overview of the Results of the DFG-Research Group 'Vortices and Heat Transfer'";

[5] Fiebig, M.: „Vortices and Heat Transfer", ZAMM Z. angew. Math. Mech. 77, Vol.1, pp 3-18, 1997.

[6] Herman, C. V. and Mayinger, F.: "Interferometric Study of the Heat Transfer in a Grooved Geometry", Lehrstuhl für Thermodynamik, TU München, Exp. Heat Transfer, Fluid Mechanics and Thermodynamics, Keffer, J.F.; Shah, R.K.; Ganic, E.N. (Eds.), Elsevier Sci. Pub., pp. 522-529, 1991.

[7] Rowley, G. J. and Patankar, S. V.: "Analysis of Laminar Flow and Heat Transfer in Tubes with Internal Circumferential Fins", Int. J. Heat Mass Transfer, vol. 27, pp. 553-560, 1984.

[8] Webb, B. W. and Ramadhyani, S.: "Conjugate Heat Transfer in a Channel with Staggered Ribs", Int. J. Heat Mass Transfer, vol. 28, pp. 1679-1687, 1985.

[9] Grosse-Gorgemann, A.: "Numerische Untersuchung der laminaren oszillierenden Strömung und des Wärmeüberganges in Kanälen mit rippenförmigen Einbauten", VDI-Fortschrittsberichte, Reihe 19: Wärmetechnik/Kältetechnik, Nr. 87, ISBN 3-18-308719-7, 1996.

[10] Vasanta Ram, V. and Lau, S.: " Experimental Investigation of Momentum and Heat Transport in the Turbulent Channel with Embedded Longitudinal Vortices", Overview of the Results of the DFG-Research Group 'Vortices and Heat Transfer'"; B2.

[11] Neumann, H., Braun, H. and Fiebig, M.: "Vortex Structure, Heat Transfer and Flow Losses in Turbulent Channel Flow with Periodic Longitudinal Vortex Generators", Overview of the Results of the DFG-Research Group 'Vortices and Heat Transfer'"; B3.

[12]  Hahne, H.-W., Weber, D. and Fiebig, M.: "Flow and Heat Transfer in Ribbed Channels with Self-sustained Oscillating Transverse Vortices", Overview of the Results of the DFG-Research Group 'Vortices and Heat Transfer'"; A4.

[13]  Lau, S., Schulz, V., and Vasanta Ram, V. I.: "A Computer Operated Traversing Gear for Three Dimensional Flow Surveys in Channels", Exp. in Fluids, vol. 14 (6), pp. 475-476, 1993.

[14]  Jörgensen, F. E.: "Directional Sensitivity of wire and figer film probes", DISA Inform. 11, pp. 31-37, 1971.

[15]  Döbbeling, K., Lenze, B. and Leuckel, W.: "Computer-Aided Calibration and Measurements with a Quadruple Hot-Wire Probe", Exp. in Fluids, vol. 8, pp. 257-262, 1990.

[16]  Kakac, S., Shah, R. K. and Aung, W.: "Handbook of Single-Phase Convective Heat Transfer", Wiley & Sons 1987.

[17]  Schulz, K.: "Entwicklung und Erprobung optischer Verfahren zur Messung des lokalen konvektiven Wärmeübergangs mittels digitaler Bilderfassung", Dissertation, Ruhr-Universität Bochum, Cuvillier Verlag, 1997, in press.

[18]  Kottke, V., Blenke, H. and Schmidt, K. G.: "Eine remissionsfotometrische Meßmethode zur Bestimmung örtlicher Stoffübergangskoeffizienten bei Zwangskonvektion in Luft", Wärme- und Stoffübertragung, vol. 10, pp. 9-21, 1977.

[19]  Schulz, K. and Fiebig, M.: " Accurate Inexpensive Local Heat/Mass Transfer Determination by Digital Image Processing Applied to the Ammonia Absorption Method", Proc. of the 2nd ETS, Rome, pp. 1079-1088, 1996.

[20]  Weber, D.: "Experimente zu selbsterregt instationären Spaltströmungen mit Wirbelerzeugern und Wärmeübertragung", Dissertation, Ruhr-Universität Bochum, Institut für Thermo- und Fluiddynamik, Cuvillier Verlag, 1995.

[21]  Eckert, E. R. G.: "Analogies to heat transfer processes", Measurements in heat transfer, ed. by E. R. G. Eckert and R. J. Goldstein, pp. 397-423, Hemisphere Publishing New York, 1976.

[22]  Mizushima, T. and Nakajima, M.: "Simultaneous heat and mass transfer", Chem. Eng. Japan, Vol. 15, pp. 30-34, 1951.

[23]  Zhang, Zheng-Ji: "Einfluß von Deltaflügel-Wirbelerzeugern auf Wärmeübergang und Druckverlust in Spaltströmungen", Dissertation, Ruhr-Universität Bochum, Institut für Thermo- und Fluiddynamik, 1989.

[24]  Peric, M.: "A Finite Volume Method for the Prediction of Three-Dimensional Fluid Flow in Complex Ducts", Phd Thesis Imperial College London, 1985.

[25]  Stone, H. L.: "Iterative Solution of Implicit Approximation of Multidimensional Partial Differential Equations", SIAM J., Numerical Analysis 5, pp. 530-558, 1968.

[26] Patera, A. T.: "Spectral Element Method for Fluid Dynamics, Seminar Flow in a Channel Expansion", Journal of Computational Physics, Vol. 54, pp. 468, 1984.

[27] Grosse-Gorgemann, A.; Weber, D.; Fiebig, M.: "Numerical and Experimental Investigation of Self-Sustained Oscillations in Channels with Periodic Structures", Proceedings of the 31st Eurotherm Conference, Bochum, pp. 42-50, 1993.

[28] Brigham, E. Oran: "FFT Schnelle Fourier-Transformation", 4. Auflage, R. Oldenbourg Verlag München Wien 1989.

[29] Torii, K. and Yanagihara, J. I.: "The Effects of Longitudinal Vortices on Heat Transfer of Laminar Boundary Layers"; JSME International Journal, Series II, vol. 32, pp. 395-402, 1995.

[30] Riemann, K. A.: "Wärmeübergang und Druckabfall in Kanälen mit periodischen Wirbelströmungen bei thermischem Anlauf"; Dissertation, Ruhr-Universität Bochum, Institut für Thermo- und Fluiddynamik, 1992.

[31] Güntermann, T.: "Dreidimensionale stationäre und selbsterregt-schwingende Strömungs- und Temperaturfelder in Hochleistungs-Wärmeübertragern mit Wirbelerzeugern"; VDI-Fortschrittsberichte, Reihe 19: Wärmetechnik/Kältetechnik, Nr. 60, ISBN 3-18-146019-2, 1992.

[32] Braun, H.: "Grobstruktursimulation turbulenter Geschwindigkeits- und Temperaturfelder in Spaltströmungen mit Wirbelerzeugern"; Dissertation, Ruhr-Universität Bochum, Institut für Thermo- und Fluiddynamik, Cuvillier Verlag, 1996.

[33] Dang, K. and Morchoisene, Y.: "Numerical Methods for Direct Simulation of Turbulent Shear Flows"; Turbulent Shear Flows: Lecture Series 1989-03, Karman Institute for Fluid Dynamics, 1989.

[34] Grosse-Gorgemann, A., Weber, D. and Fiebig, M.: "Experimental and Numerical Investigations of Self-Sustained Oscillations in Channel with Periodic Structures"; to be published in Experimental and Thermal Fluid Science.

[35] Mielenz, O.: "Bestimmung lokaler Wärmeübergangskoeffizienten in Spaltströmungen"; Diplomarbeit Nr. 94-14, Ruhr-Universität Bochum, Institut für Thermo- und Fluiddynamik, Lehrstuhl für Wärme- und Stoffübertragung, 1994.

[36] Bronstein, Semendjajew: "Taschenbuch der Mathematik"; Harri Deutsch Verlag, Thun und Frankfurt/main, 23. Auflage 1987, S. 766 ff, ISBN 3-87144-492-8.

# MEASUREMENT AND ANALYSIS OF THE TURBULENT FLOW QUANTITIES IN A CHANNEL WITH EMBEDDED LONGITUDINAL VORTICES

S. Lau, V. Vasanta Ram
Institut für Thermo- und Fluiddynamik, Ruhr-Universität Bochum
D-44780 Bochum, Germany

## SUMMARY

The subject of study in the present paper is the channel flow in which longitudinal vortices are generated by winglet arrangements typically employed in heat exchangers for heat transfer enhancement. The study covers flow fields of both velocity and temperature. Detailed measurements with quadruple hot-wire probes have been conducted of both the mean and the fluctuations in these quantities in an air flow, heated where necessary. Flow visualisation studies in water flow in the same geometry supplement the probe measurements in air flow. The experimental data have been reduced to establish a data base of the Reynolds averaged flow quantities that is suited for comparison with computational results with turbulence models. The data base contains distributions of the mean velocity and temperature, and of the correlations between the fluctuations in velocity and temperature. The measurements have also been analysed to examine if coherent structures relatable to types known to be present in other simpler turbulent shear flows are also detectable in this flow. The methods of examination employed for this purpose are forms of variable-interval time averaging (VITA) methods and wavelets. While there are hints from this study that the flow could contain structures, the analysis of measured data conducted does not lead to a conclusive evidence on the existence or nature of the structures.

## INTRODUCTION

An important design objective generally pursued in laying out industrial heat exchangers is realising high heat transfer rates at a low overall pressure loss in the flow system. In heat exchangers of the type commonly met with, both heat transfer and pressure loss are dominated by convection. Physical consideration of convective heat transfer in some representative flows hint at the possibilty of achieving enhanced heat transfer rates through the introduction of unsteadiness and vortices in the flow, see Fiebig [1]. While it is easily surmised that vortices exert an influence on heat transfer, flow studies also suggest that the overall pressure losses would be mostly lower when the axes of the vortices are oriented parallel to the stream than, when, say, they are spanwise, see Liu and Lee [2]. The potential of longitudinal vortices for deployment in heat exchangers is therefore higher than of vortices of other orientation. In view of the promise they hold, designers of heat exchangers are seeking for ways to generate longitudinal vortices of the desired size and strength at the proper locations in the flow. Besides the size and strength of the vortices, flow turbulence also affects heat exchanger performance. Flow turbulence and longitudinal vortices exert mutual influence on each other, and it is in this context that the present research work has to be viewed. It is a detailed study of the momentum and heat transfer in the turbulent flow in a channel passage in which longitudinal vortices are generated by winglets.

Engineering means of generating longitudinal vortices in a flow are numerous when viewed in terms of their geometrical properties, and it is convenient to classify the means broadly according to the primary physical principle of vortex generation in the flow. We may distinguish between two phyical principles. In the first, it is the shaping of the walls themselves. Vortex generation herein is explicable through a Goertler-type instability in the boundary layer flow on a concave wall. Physical reasoning behind this instability suggests that it is the streamline curvature going along with the concave shape that gives rise to longitudinal vortices if the flow parameters lie in a certain range. In the second, the longitudinal vortex is generated by introduction of a winglet-like body in the flow. Generally, an array of winglets is arranged on the walls of the heat exchanger. In this class, two physical mechanisms may be identified that cause generation of longitudinal vortices. The first is traceable to Helmholtz´s laws of vortex motion that calls for a turning of vortex lines into the streamwise direction when a flow with spanwise vorticity in the oncoming stream encounters an obstacle that is finite in the spanwise direction. An example of longitudinal vortex forming primarily through such a mechanism may be seen in figs. 92 and 93 of Van Dyke [3]. The second is the rolling up of the separated shear layers along the winglet edges into longitudinal vortices. This is a phenomenon observed when a wing or a winglet is in an irrotational oncoming stream too, hence not dependent on vorticity being present in the oncoming flow. An example for a vortex forming from this mechanism is in fig. 90 of Van Dyke [3]. The generation of longitudinal vortices through winglets is an attractive proposal from an engineering point of view, since it offers the advantage of a broader range for selection of size and shapes for winglets combined with relative ease of fabrication. Common shapes for winglets are the delta and the rectangle. The location, size and strength of the longitudinal vortices thus generated depend upon geometrical details as well as on the flow parameters. An understanding of the flow and turbulence in typical heat exchanger passages affected by these devices is indispenable for their engineering design. Acquiring this understanding defines the scope of our present work.

Within the past decade, extensive studies of the flow and the turbulence in the boundary layer affected by longitudinal vortices have been conducted by Cutler and Bradshaw [4], [5], Eibeck and Eaton [6], Mehta and Bradshaw [7], Shabaka et al. [8] and Shizawa and Eaton [9]. Numerical studies of the turbulent flow and heat transfer in a channel with both delta and rectangular winglets are by Zhu [10]. More recently, direct numerical simulations of the flow and heat transfer in this configuration have been carried out by Braun [11]. An outstanding feature of the flow in a typical heat exchanger passage with winglet generated longitudinal vortices is that there are regions of helically wound streamlines surrounded by regions in which the streamlines are almost straight. Helical winding is equivalent to the streamlines being curved in both the longitudinal and transverse planes. In contrast, the streamline curvature in the surrounding region is uniplanar and relatively mild. Since streamline curvature is a major factor affecting turbulent motion in the flow in question, turbulence structures in neighbouring regions differ widely from each other. The flow therefore belongs to a class that has come to be known as *complex turbulent flows* see Kline et al. [12]. They stand in contrast to *simpler* turbulent flows, examples for which are boundary layers with mild pressure gradients, flow in passages with mild curvature and so on. As a survey through this literature shows, there are also other flow conditions under wich complex turbulent flows may result. Examples are: separation and reattachment, and rapid changes in wall roughness. The complexity of the turbulent motion in these flows is of such nature that hitherto the flows have not been found amenable to treatment through turbulence models in Reynolds averaged equations. It is conjectured that the relative success of turbulence models in simpler flows is ascribable to the closure hypotheses in the Reynolds averaged equations capturing the essential physical mechanisms of the turbulent motion. In contrast, the hypotheses apparently fail to capture the physical mechanisms in complex turbulent flows.

An approach that has been considered to hold promise for the treatment of complex turbulent flows is the so-called *large-eddy simulation*, see Bradshaw [13]. *Large-eddy simulation* divides the turbulent motion into different scales and attemps to obtain details of the larger scale motion through solving equations that are only partially averaged, i.e. averaged over the smaller scales only. These equations, as against the *direct simulation*, contain a turbulence model known as the *subgrid-scale model* that accounts for the effect of the smaller scales on the larger scale motion. The hypotheses on the interaction between motion at the different scales contained in the subgrid-scale model, although postulated to be universal, does need to be verified in the complex turbulent flow of present interest, and for this purpose detailed measurements and analysis of the fluctuating motion in this flow are a necessitiy. The analysis has to separate the motion from a point of view of scales consistent with the theory, so that the *subgrid-scales stresses* may be evaluated therefrom. The task of analysing the fluctuating motion to separate out the *large-scale motion* from the rest and quantifying the scales belongs to the fast growing discipline of *structure identification* in turbulent flows.

The present paper is a report summarising the authors´ work from 1991 up to the present on the turbulent flow and heat transfer in a channel in which longitudinal vortices are generated by rectangular winglets. Following a concise presentation of the theoretical background for interpretating experimental data, the experimental set up and the measurements of the instantaneous velocity and temperature in this flow through multi-wire probes are described. This is followed by an account of the flow visualisation studies supplementing the probe measurements of the flow field.

The probe measurements of the instantaneous velocity and temperature were evaluated to establish a data base of the fields of the mean velocity vector and temperature, Reynolds stresses and turbulent heat flux in this flow. Since these are the physical quantities entering a Reynolds-averaged description of the flow, this data base is suited for purposes of verifying the validity of a turbulence model of interest for the class of complex turbulent flows with embedded longitudinal vortices. A selection of results from this data base is presented in the section under the heading "Long time averages".

The velocity data are also further analysed from the point of view of structure identification. Methods of VITA (Variable Interval Time Averaging) and wavelet analysis have been employed to examine if, in the flow in question, there are structures of the large-scale fluctuating motion identifiable to those observed in simpler flows. An account of this data analysis form the subject of the section under this heading.

# THEORETICAL BACKGROUND

In order that the data base from the experiments serve its scientific purpose, it is necessary to view the measurements in a theoretical context. The theoretical framework for data analysis should be firmly anchored in the equations of motion prior to any averaging. For, over the entire Reynolds number range in consideration here, laminar, transitional or turbulent, the flow is governed by the laws of motion expressed through these equations. The set of measured data, after making due provision for uncertainties and other limitations on accuracy, may be regarded as equivalent to obtaining the solution of the equations of motion realised by nature under the boundary conditions prescribed by the experiment. The equations of motion prior to any averaging may be written in standard notation as follows, see for example Tennekes and Lumley [14] and Gersten and Herwig [15]:

$$\frac{\partial u_i}{\partial x_i} = 0$$

$$\rho\left(\frac{\partial u_i}{\partial t} + u_j \frac{\partial u_i}{\partial x_j}\right) = -\frac{\partial p}{\partial x_i} + \mu \frac{\partial^2 u_i}{\partial x_j \partial x_j} \qquad (1)$$

$$\rho c_p\left(\frac{\partial T}{\partial t} + u_j \frac{\partial T}{\partial x_j}\right) = \lambda \frac{\partial^2 T}{\partial x_j \partial x_j} \quad .$$

The turbulent flow realised in the experiment is a flow with a complex vortical state with details spread over a large range of scales both in length and time. Obtaining a physically meaningful solution for the flow in a given geometry at the Reynolds numbers of interest is a task generally beyond means available today, save at moderate Reynolds numbers through direct numerical simulation (DNS). But, a resolution of details over the entire hierarchy of scales is seldom required in engineering applications. Classically, engineers have sought to arrive at the quantities of interest, which are only suitably averaged quantities, by resorting to averaging the equations themselves and treating the averaged equations. However, there are limitations to this approach. An alternative of relatively recent origin is the large-eddy simulation.

## Data analysis of quantities in Reynolds averaged equations

Engineers have traditionally approached the above problem through subjecting the equations of motion to the procedure of *Reynolds-averaging*. In this procedure, a physical quantity like the velocity $u_i(t, x_j)$ is regarded as a sum of its Reynolds-average $\bar{u}_i(x_j)$ and its fluctuation $u_i'(t, x_j)$, i.e.

$$u_i(t, x_j) = \bar{u}_i(x_j) + u_i'(t, x_j) \ .$$

The *Reynolds-averaged* equations are then:

$$\frac{\partial \bar{u}_i}{\partial x_i} = 0$$

and

$$\rho \bar{u}_j \frac{\partial \bar{u}_i}{\partial x_j} = -\frac{\partial \bar{p}}{\partial x_i} + \mu \frac{\partial^2 \bar{u}_i}{\partial x_j \partial x_j} - \rho \frac{\partial \overline{u'_i u'_j}}{\partial x_j} \qquad (2)$$

$$\rho c_p \bar{u}_j \frac{\partial \bar{T}}{\partial x_j} = \lambda \frac{\partial^2 \bar{T}}{\partial x_j \partial x_j} - \rho c_p \frac{\partial \overline{u'_j T'}}{\partial x_j}.$$

The fundamental problem encountered in this approach is well known. Briefly, it is the *closure problem* of turbulence. Since these *Reynolds-averaged* equations only provide a constraint to be obeyed by the average and the fluctuations of the physical quantity (more precisely, correlations of the fluctuations) solving these to obtain the average quantitiy is not possible even in principle without invoking a turbulence model. A turbulence model closes the set of equations by relating the Reynolds stresses to the averages or to the gradients of the averages. For setting up a turbulence model a knowledge of the physics of the fluctuating motion affecting the correlations (the Reynolds stresses) is essential.

Establishing a turbulence model has been a long standing goal of researchers in fluid mechanics. While the challenges of this task are well known, we wish to recall here a point relevant in the experimental context. It is that the foundation of a turbulence model lies essentially in the experiment. For checking the validity of a turbulence model in a certain flow, it is necessary to subject the measured instantaneous data in this flow to *Reynolds-averaging*, and to examine if the properties of the Reynolds stresses evaluated herefrom are as hypothesised. This procedure is equivalent to *solving the equations first and then averaging the solution* in contrast to *averaging the equation first and then solving them*. The latter is implicitly followed in the approach of treating the flow problem through Reynolds-averaged equations. The experimental evidence available up to now indicates that the properties of the Reynolds stresses in the turbulent flow with embedded longitudinal vortices are not as hypothesised in any of the hitherto known turbulence models, which explains the inadequacy of the turbulence models for this flow.

**Data analysis of quantities in large-eddy simulation**

The difficulties encountered in treating complex turbulent flows through a turbulence model in the Reynolds-averaged equations are traceable to an outstanding property of the turbulent fluctuating motion, viz. that in its large scales it is strongly dependent on factors influenced by geometry and global flow features. This stands in the way of accounting for their effects on the *mean* motion through *universal* turbulence models. The approach through large-eddy simulation attemps to circumvent this difficulty by further division of the turbulent fluctuating motion into a large-scale motion and a rest. Conceptually, no averaging is done over the large-scale motion, only the influence of the *rest* having to be accounted for through a *subgrid-scale model*. Since this approach aims at covering the effects of the fluctuating motion over only the smaller scales and evidence up to the present indicates *universality* of some features of the small scale motion, a suitably formulated subgrid-scale model may be expected to be independent of the global flow features. Just to what extent this expectation is fulfilled for a certain flow has to be examined during validation.

The equations used for *large-eddy simulation* of a flow field may be found in relevant literature, see eg. Lesieur [16]. These are as follows:

$$\frac{\partial(\overline{u}_i+\tilde{u}_i)}{\partial x_i}=0$$

$$\rho\left[\frac{\partial \tilde{u}_i}{\partial t}+(\overline{u}_j+\tilde{u}_j)\frac{\partial(\overline{u}_i+\tilde{u}_i)}{\partial x_j}\right]=-\frac{\partial(\overline{p}+\tilde{p})}{\partial x_i}+\mu\frac{\partial^2(\overline{u}_i+\tilde{u}_i)}{\partial x_j\partial x_j}-\rho\frac{\partial(\langle\hat{u}_i\hat{u}_j\rangle)}{\partial x_j} \qquad (3)$$

where

$$u_i(t,x_j)=\overline{u}_i(x_j)+\tilde{u}_i(t,x_j)+\hat{u}_i(t,x_j) . \qquad (4)$$

In (3) the symbol $\langle\ \rangle$ denotes averaging over the small scales only. For a corresponding formulation for the temperature field in the flow the reader is referred to articles by Moin et al. [17] and Lilly [18]. A search through published literature indicates that the use of the large-eddy concept for transport of a passive scalar has not received attention comparable to its counterpart for momentum. In the present paper the authors restrict themselves therefore to the flow field only. In (3) $\overline{u}_i(t,x_j)$ is the Reynolds-averaged velocity, $\tilde{u}_i(t,x_j)$ and $\hat{u}_i(t,x_j)$ may be identified in that order with the large and small-scale contributions to the turbulent fluctuating motion. The quantity $\tilde{u}_i(t,x_j)$ is often referred to as a *coherent structure*. It may be interpreted as the time-dependent motion filtered out and left behind, when partial averaging, i.e. over the small scales only, has been done over the whole fluctuating motion.

It is clear from (3) that $\tilde{u}_i(t,x_j)$ is influenced by correlations of the small-scale motion components $\hat{u}_i(t,x_j)$ between themselves, and that these equations have to be closed by a sub-gridscale model. The subgrid-scale model in common use is the *Smagorinski model*, see eg. Lesieur [16] and Braun et al. [19].

The guiding principle for validation of this model for any class of flows may be stated in simple terms as follows:

> The measured data $u_i(t,x_j)$ in the flow have to be subjected to an analysis aimed at isolating the large-scale motion $\tilde{u}_i(t,x_j)$ from the rest. Parallel to the analysis, it should be examined if this solution for $\tilde{u}_i(t,x_j)$ is arrived at from the equations with the sub-grid-scale model.

While the underlying principle for data analysis may be stated in relatively simple terms as above, charting out a route to separate out the turbulent fluctuating motion into its large-scale components and the rest is a task of considerable complexity. It is a subject of vigorous ongoing research, and we refer to three more or less parallel approaches with many ramifications. They are:

1. VITA (Variable-Interval Time Averaging)
2. POD (Proper Orthogonal Decomposition)
3. wavelet analysis.

Literature on these methods has been published by several authors over two decades that is spread over several journals, and for an overview we refer the reader to survey articles, eg. by Morrison et al. [20], Berkooz et al. [21] and Farge [22].

In the present state of understanding, the relation between these approaches is not entirely clear. As a step towards clarifying the relation, the authors have conducted comparative tests between the VITA and wavelet analysis of selected signals. These will be described in the section titled "Structure identification studies". For the present we note an important condition to be fulfilled regardless of the approach that is adopted. This is that in order that the spirit of large-eddy simulation be properly captured in the analysis of experimental data, it is necessary to assure that the division of the turbulent fluctuating motion into its large-scale components and the rest is consistent with the definition of the averaging implicit in the derivation of equation (3) using (4). A particularly important step in the derivation is that the large-scale motion is not correlated with the motion at the smaller scales! Otherwise products of the type $\langle \tilde{u}_i \hat{u}_i \rangle$ would appear in (3). It is also worthy of note that a classical Fourier decomposition of the fluctuating motion with a continuous spectrum, followed by dividing the *large scales* from the *small scales* on the basis of wave numbers/frequency does not necessarily satisfy the condition of the two being not correlated with each other. Building in the requirement called for into the algorithmic procedures of the VITA or POD or wavelet analysis requires a much deeper insight into the characteristics of turbulent motion than is available today. The examples in the section "Structure identification studies" will make this point more clear. Here we merely wish to note that although significant advances have indeed been made both at solving the equations of large-scale simulation and in algorithms for identifying and isolating the large-scale motion in measured data, research at present has not reached a stage where the subgrid-scale model in a large-eddy simulation of a complex turbulent flow, such as the channel flow with embedded longitudinal vortices, may be conclusively validated. The section "Structure identification studies" has therefore to be read against this background.

## EXPERIMENTAL SET UP FOR PROBE MEASUREMENTS

Probe measurements of the velocity and thermal fields were conducted in two similar but separate facilities. Separation into the cold and heated facility was done to meet the engineering requirements of heating installation. It also facilitated better control of thermal conditions during the temperature measurements which was not necessary for velocity measurements. Flow field measurements were conducted in the cold facility. The heated facility was used for thermal field measurements.

### The cold facility

The experimental facility for flow field measurements is a low-speed channel flow rig of large aspect ratio with air as the working medium. Its salient dimensions are given in fig. 1. The winglets were mounted on one of the channel walls (dotted area in fig. 1). A computer operated traverse mechanism for the probes is installed on the channel wall opposite to the winglets, see Lau et al. [23].

The nominal channel height is $40mm$. The large aspect ratio of 1:18, together with measures taken for supressing separation in the section between the blower and settling chamber, ensured uniformity and two-dimensionality of the oncoming undisturbed channel flow ahead of the winglets.

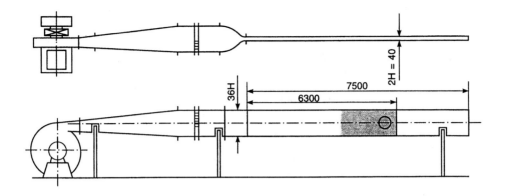

Figure 1: Channel flow facility (dimensions in mm)

## The heated facility

Salient dimensions of the facility for conducting measurements in the heated flow were the same as for the cold facility shown in fig. 1. An electrically conducting thin foil was cemented on the wall with winglets made out of thermally insulating material. Adjustment of the current passed through the foil regulated the heat input into the flow. The heating rate $\dot{q}_w$ in our experiments was set to $325 \pm 1.5 W/m^2$ taking into consideration heat flux losses of 4% through the back wall. This gave rise to a wall temperature of around $10K$ above the ambient, see also Neumann [24].

## Winglet configurations

The longitudinal vortices are generated by rectangular winglet pairs mounted on one of the channel walls. This basic geometry is common to all the B projects. Essential geometric parameters of the basic winglet pair are shown in fig. 2. The winglets protude up to the mid-channel height, i.e. $h = H$. This protusion, although rather large from the point of view of an engineering device for heat transfer enhancement, was chosen for the present experiments in order to generate longitudinal vortices large enough to be resolvable with our probe.

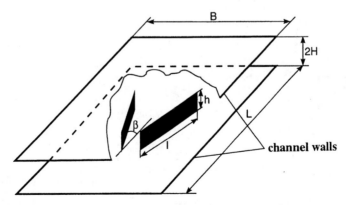

Figure 2: Periodic element of winglet configuration, 2H=40mm=2h, L=200mm=10h, l=80mm=4h, B=160mm=8h, $\beta=30^0$, thickness of the winglets s=1mm

259

In the present work the flow investigations have been carried out in altogether three different winglet configurations. They are as follows:

Configuration 1 is an isolated winglet pair, corresponds to fig. 2.

Configuration 2 is a single row of the winglet pairs arranged periodically in the spanwise direction only, see fig. 3. G, H, J, K and L are the planes in which the probe measurements were done.

Configuration 3 consists of the winglet pair arranged periodically in both spanwise and streamwise directions as shown in fig. 4. This corresponds to the basic winglet configuration of the research group. The planes shown as A and C, which are at $x = 41mm$ and $x = 107mm$ behind the plane of the winglet trailing edges of the 6th row, are the planes of probe measurement.

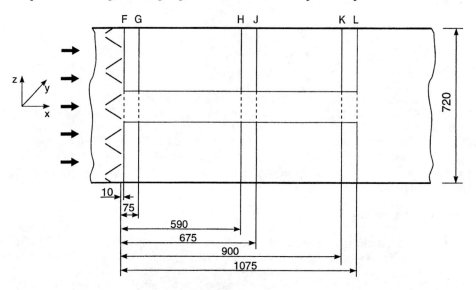

Figure 3: Configuration 2: row of winglets, G, H, J, K and L are planes of measurement

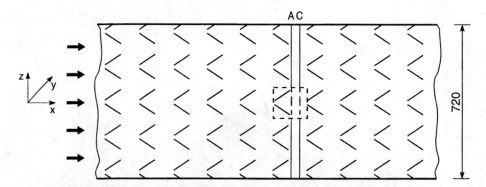

Figure 4: Configuration 3: periodic winglet configuration, A and C are planes of measurement

260

Both velocity and thermal measurements with probes as well as flow visualisation studies were conducted with configuration 3. With configuration 2 no thermal measurements were undertaken. However, the probe velocity measurements were supplemented by flow visualisation. The studies with configuration 1 were restricted to flow visualisation only.

## PROBE MEASUREMENT TECHNIQUES

### Instrumentation

The measurements were conducted with both X- and quadruple hot-wire probes, used successively. The probes were fabricated in our laboratory, because commercial probes with their mountings were found to be too large for installation in our channel facility. The wire arrangement of the quadruple hot-wire probe was the same as in Eckelmann et al. [25] and Schulz [26], see fig. 5. It consisted of two pairs of X-wires in planes perpendicular to each other. With all the wires operated as hot-wires the probe could measure the three velocity components and the Reynolds shear stresses simultaneously. For combined measurements of temperature and velocity we refer to the subsection under the corresponding heading later in this section.

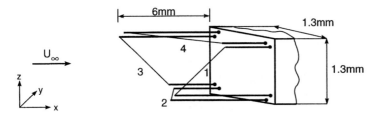

Figure 5: Quadruple hot-wire probe (lengths are not to scale)

Since in velocity measurement the three components are measured by four wires, there is a redundancy in measurement. With the present wire configuration the redundancy is in the u-component. The size of the probe is $1.3 \times 1.3 \, mm^2$, the wire diameter $5 \mu m$, the length of the wires is approximately $1.5 mm$. The wires of both the X-wire and the quadruple wire probe were driven independently of each other by the AN 1003 constant-temperature anemometer without linearizer marketed by AA Lab. During measurement the four hot-wire signals were recorded after passing them through the DAS20 A/D-converter of 12 bit resolution and the accompanying sample & hold box marketed by Keithley Instruments. The A/D-converter was installed in a 386 MS-DOS computer, while the sample & hold box was external.

### Calibration of the probe for flow measurements

In our calibration facility $\alpha$ and $\beta$ are the angles that are set, see fig. 6. The step in calibration procedure of obtaining voltages versus velocity was repeated for 169 pairs of $(\alpha,\beta)$-angles within the cone of semi-apex angle $28°$. The error in setting $\alpha$ and $\beta$ was estimated to be $0.5°$. The entire family of calibration curves, hot-wire voltage $E_i$ versus velocity $U$ for each wire, $i=1,2,3,4$, forms a hypersurface. $E_i$ can formally be written as $E_i = E_i(\alpha,\beta,U)$, where

$\alpha, \beta$ are the angles in fig. 6. Alternatively, $E_i$ may also be written as $E_i = E_i(\varphi, \gamma, U)$, where $\varphi$ and $\gamma$ are angles the velocity vector makes with the $x-$ and $z-$axis in the cartesian coordinate system in which the velocity components are generally sought in measurement. Examination of the calibration data showed that, while the voltage versus velocity for any pair $(\alpha, \beta)$ followed a power law, Jorgensen's equations with its cosine law does not capture satisfactorily the angular characteristics of the probe over its entire measuring range. In view of this difficulty it was decided to develop a look-up table scheme for purposes of measurement with this probe. This scheme comprises of two major steps, viz. generation of the look-up table and development of a search algorithm to assign the velocity components to the measured hot-wire signals.

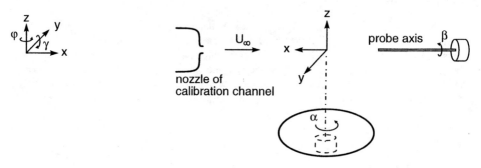

Figure 6: Calibration facility

## Look-up table algorithm

A look-up table, with a resolution much higher than in the straightforward calibration, was generated by using a functional relationship between the hot-wire signals and the three-dimensional velocity vector as seen by the probe. Matching of signals from all the four wires to within a specified error in each voltage is used to determine the three-dimensional velocity vector, because a redundancy in the u-component may be expected to lead to higher accuracy in measurement. An alternative would be to use signals from only three of the four wires to determine the three velocity components and repeat this for different sets of three wires. The authors found that the alternative does not always permit for our probe a unique inversion of the calibration data to measurement over the entire range.

The straightforward look-up table would start with about 2535 ($=169 \times 15$) velocity vectors and the corresponding four hot-wire voltages, $E_1, E_2, E_3, E_4$. All those arrays of seven quantities are sorted with respect to the magnitude of the voltages with a given sequence of the different wires. A 4th order polynomial for voltages vs. velocity vector was found to fit the calibration data for any $(\alpha, \beta)$, so that calibration at 15 velocities was found to be sufficient. From the 4th order polynomial voltages could be obtained for any intermediate velocity through interpolation. For dependency of the signals on the pair of angles, the hot-wire signals were interpolated linearly. The final look-up table contains about 287000 velocity vectors with the corresponding four hot-wire voltages $E_1, E_2, E_3, E_4$. It should be noted here that, to obtain the velocity components in the $x, y, z-$coordinate system, conversion from $(\alpha, \beta)$ to $(\varphi, \gamma)$ is necessary. Also, even after interpolation the look-up table is not equally spaced, neither in $(u, v, w)$ nor in $(E_1, E_2, E_3, E_4)$, but this is not a disadvantage as such.

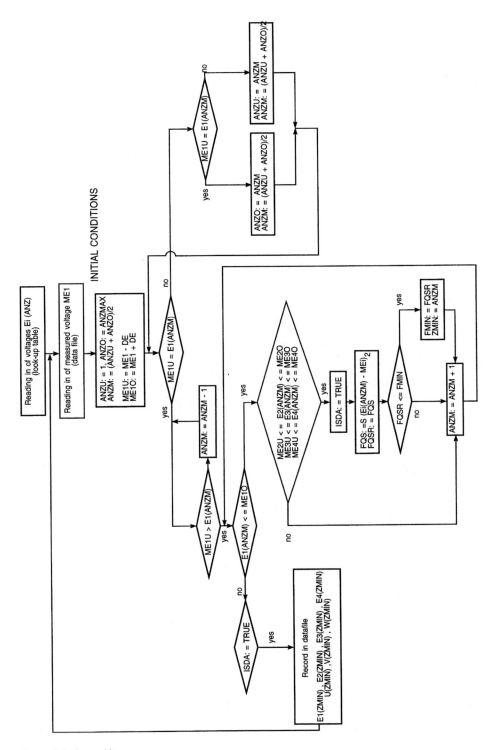

Figure 7: look-up table

The second step is assigning the components of the velocity vector to the quadruple of voltages as sketched in fig. 7. Matching the quadruple of voltages in measurement to the quadruple in the look-up table to within the specified error in each voltage generally led to more than just one set of possible velocity components. The non-uniqueness of inversion in theses cases was handled by chosing the set of velocity components according to the criteria of least square error. This set was then recorded on a file. When no matching of the quadruple in measurement with one in the look-up table was possible the measurement at this instant of time was discarded.

## Calibration of the probe for combined velocity and temperature measurements

The quadruple hot-wire probe sketched in fig. 5 was also used for combined temperature- and velocity measurements. One pair of wires forming an X-probe served for temperature measurement, whereas the second one was for measurement of the two velocity components in the planes of the wires. For the temperature measurements a $2.5\mu m$ wire was welded onto the prongs forming this X-wire. These wires were operated independently in the constant current mode (CCA-bridges of the AN1003- Anemometer by AA Lab) with a low heating current of $0.3mA$. Rewelding the wires onto the prongs with the wire diameters interchanged permitted measurement of the velocity components in the plane perpendicular to the previous one.

For the combined temperature- and velocity measurements the quadruple hot-wire probe with two cold wires was calibrated in two steps. The cold wires were first calibrated for temperature in the range from $20-50°C$. Since their input-output relationship was found to be linear, calibration at three temperatures was regarded suffucient. The thermal calibration facility consisted of a box immersed in a thermal water bath. The box contained the probe and a reference Pt 100 thermometer. The water temperature could be held constant to within $0.06K$ at several discrete levels.

In the second step the probe was calibrated at a fixed reference air temperature $T_c$ with respect to the velocity vector in a plane only within a wedge of semi-angle $30°$ in the calibration facility described above. For the angle $\alpha=0°$ the calibration for velocity was repeated at three air temperatures with the cold wires used as thermometers.

The calibration method uses a linear relation between the instantaneous temperature and the cold wire signals, with different reduction parameters for the wires as suggested by Meyer [27] and Klick [28]. For the two components of the velocity vector the method by Browne et al. [29] was followed.

The dependency of the cold wire signal $E_3$ on air temperature $T_m$ can be described by equation (5). The constant $A_3$ and the factor $\gamma_3$ were determined by calibration of the temperature wire at different air temperatures.

$$E_3 = A_3 + \gamma_3 T_m . \qquad (5)$$

Using equation (6) the measured hot-wire voltages $E_{mi}(i=1,2)$ at the air temperature $T_m$ were transformed to corresponding voltages $E_{ci}$ based on the reference temperature $T_c$. The coefficients $\gamma_1$ and $\gamma_2$ describe the air temperature dependency of the hot-wires. Their values are average values over the entire region of angles of calibration.

$$\frac{E_{mi}^2}{E_{ci}^2} = 1 + \gamma_i (T_c - T_m) \ . \tag{6}$$

We then applied the algorithm by Browne et al. [29] to the signals $E_{ci}$. In contrast to Browne et al. [29], who used a 3th order polynomial, we used a 4th order polynomial to describe the $U$ versus $E$ fits. For each of the hot-wires in the pair and for each yaw angle $\alpha$ we write

$$U = C_{1i} + C_{2i} E_{ci} + C_{3i} E_{ci}^2 + C_{4i} E_{ci}^3 + C_{5i} E_{ci}^4 \ . \tag{7}$$

For each $\alpha$ the five coefficients were determined using a least square fit procedure and stored, see Browne et al. [29] for details.

For reliability of measurement, the velocity vector has to lie entirely within the calibration cone of the probe at all instants of time. In order to meet this requirement the measurements were conducted in two runs. In the first run, which we call the preliminary one, an X-wire probe was employed to provide an estimate for the angle the local mean velocity vector makes with the mainstream $(x-)$ direction in the plane parallel to the channel walls $(x,z-\text{plane})$. In the second run, which we call the measurement run, the quadruple hot-wire was aligned in the $x,z-$plane with the local mean velocity vector thus estimated. This alignment was used for both velocity and temperature measurements. This procedure, although a little tedious, was found to be necessary to acquire reliable instantaneous flow data from the signals of the probe. The final data obtained are then in the form of discrete time series of the instantaneous velocity vector and temperature.

From the probe signals the components of the Reynolds stress tensor and of the heat flux vector, quantities that are of interest both for verification of turbulence models and their further development for a complex flow, have been obtained. The time series available from measurement with this probe also permit extraction of structural features of the fluctuating motion that have acquired importance in more recent work on turbulence modelling.

## LONG TIME AVERAGES

### FLOW QUANTITIES

In this section we present sample results of long-time averaged quantities measured in the turbulent channel with both the winglet configurations 2 and 3.

### CONFIGURATION 2 (a single spanwise row of winglet pairs)

#### Velocity field

In the following, for convenience at reading we refer to planes F and G in fig. 3 as the near field, to H and J as the far field, and to K and L as the distant field. In the plane F the fluctuations were so large that the instantaneous velocity vector was often found lying outside the calibration cone of the probe making evaluation of a meaningful average impossible. So we restrict ourselves to planes downstream of G.

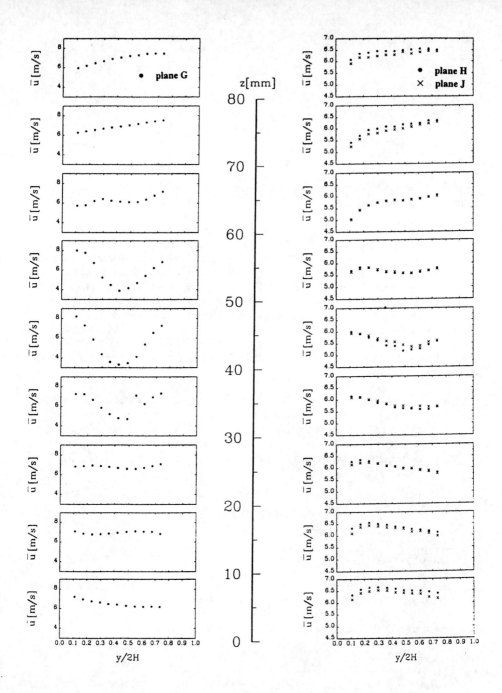

Figure 8: u component in planes G, H and J (configuration 2)

The measured velocity components in the planes G, H and J are shown in figs. 8 and 9. In plane G $\bar{u}$ shows a strong wakelike profile at the spanwise stations $z = 20, 30, 40$ and $50mm$. As one proceeds downstream, this characteristic of the profile becomes less pronounced in planes H and J, see fig. 8. Further downstream, in planes K and L, which are not shown in this paper, it is seen to have almost vanished.

In fig. 9 the $\bar{v}$ and $\bar{w}$ components of the mean velocity vector are shown as vectors to show the magnitude and direction of the secondary flow. Identifying the longitudinal vortex by non-vanishing $\bar{v}$ and $\bar{w}$ components, it is seen, that the vortex fills the entire cross section of the channel. The cross section of the vortices is elliptical. The core of a vortex is close to the middle of the channel, but it is on the side of the winglets at $y/2H = 0.42$, $y = 18mm$. The axes of both counterrotating vortices are mainly parallel over the entire measurement section in streamwise direction, that is $1065mm\,(54H)$.

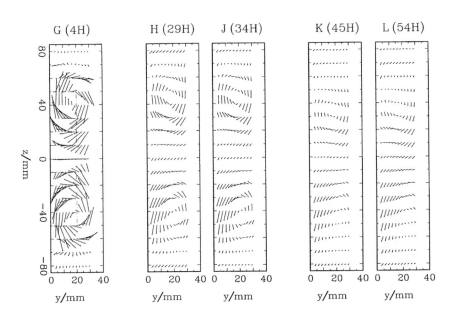

Figure 9: secondary velocity components behind a row of winglet (configuration 2)

## Turbulent kinetic energy

The turbulent kinetic energy is shown in fig. 10. In plane G, in the near field just behind the winglets, the maximum of the turbulent kinetic energy $k/u_{ref}^2 = 0.075$ is in the upwash region ($z = 60mm$), the minimum in the core of the vortex and - looking at the surrounding flow - in the downwash region, where the fluid with low turbulent kinetic energy from outside the vortex gets sucked into the vortex. Downstream, a redistribution of $k$ takes place. The maximum of $k$ moves towards the core of the vortex. From plane H to plane L the turbulent

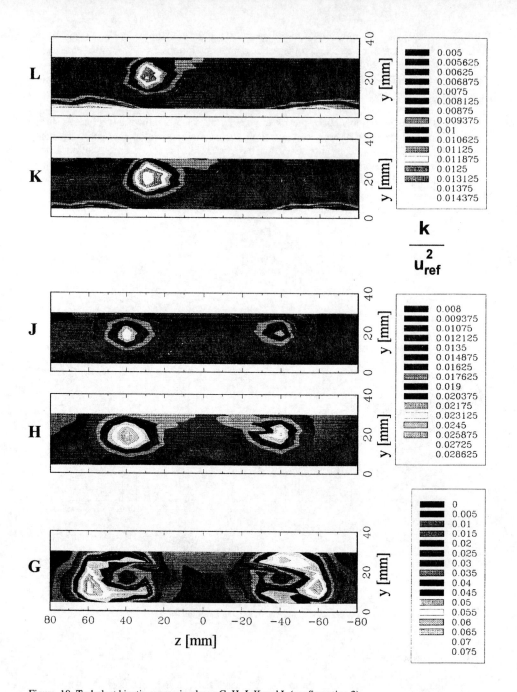

Figure 10: Turbulent kinetic energy in planes G, H, J, K and L (configuration 2)

kinetic energy $k$ decreases from $k/u_{ref}^2 = 0.0286$ to $k/u_{ref}^2 = 0.0144$. The plots of $\overline{u'^2}, \overline{v'^2}$ and $\overline{w'^2}$ show that a redistribution between the flucuations of the three components occur. Downstream, the major distribution to $k$ comes from $\overline{w'^2}$, which suggest that there may be oscillations in the spanwise direction occuring in the far and distant field $(30-54H)$.

## CONFIGURATION 3 (winglet pairs, periodic both spanwise and streamwise)

### Velocity field

The measured velocity components in plane C are shown in fig. 11. Identifying the longitudinal vortex by nonvanishing $\overline{v}$ and $\overline{w}$ components, it is seen that the vortex fills the entire cross section of the channel. Just like the flow behind a row of winglets, the $\overline{u}$ component in these planes has a wakelike profile at the spanwise stations $z = 20, 30$ and $40 mm$, see also Lau [30].

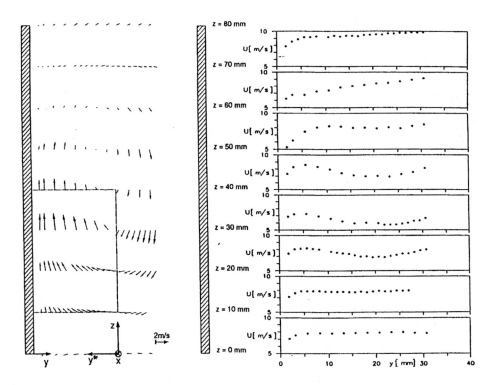

Figure 11: u, v and w components of the velocity in plane C (configuration 3), winglet position in projection shown by dashed line

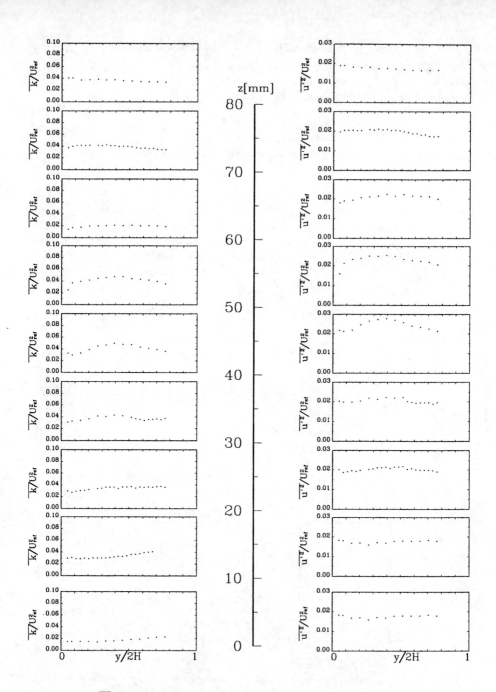

Figure 12: k and $\overline{u'^2}$ in plane C (configuration 3)

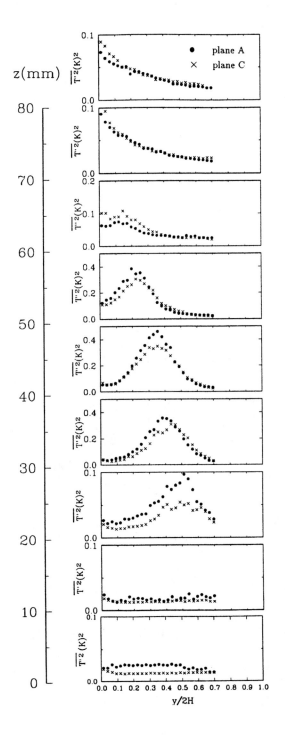

Figure 13: $\overline{T'^2}$ in planes A and C (configuration 3), 2H=40mm

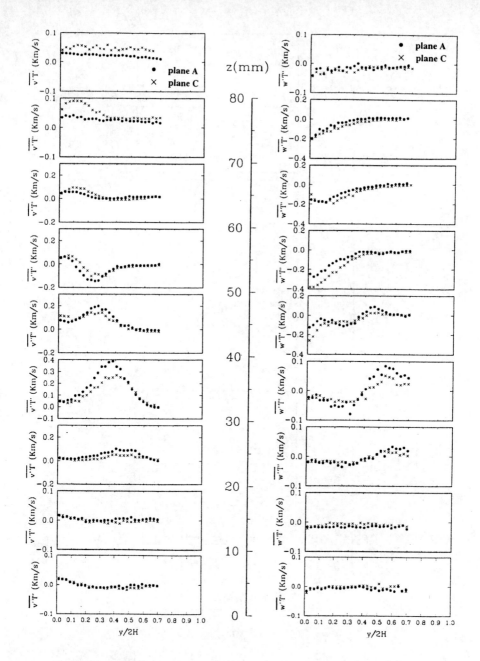

Figure 14: $\overline{v'T'}$ and $\overline{w'T'}$ in planes A and C (configuration 3), 2H = 40mm

**Turbulent kinetic energy**

The distribution of the turbulent kinetic energy is shown in fig. 12. In the plane C of the configuration under discussion, which corresponds to plane G of configuration 2, the maximum of the turbulent kinetic energy $k/u_{ref}^2 = 0.054$ is in the core of the vortex. This is similar to the distribution, that is found in the more far field behind the row of winglets in configuration 2. The minimum of $k$ can be found in the downwash region at $z = 0mm$, whereas in the region between the two vortices at $z = 70$ and $80mm$ $k$ is nearly constant at a higher level $k/u_{ref}^2 = 0.04$. Apparently, because of the winglets farther upstream in the periodic winglet configuration, the flow shortly downstream of any row of winglets of configuration 3 shows characteristics typical of the far field, but not of the near field of configuration 2. The geometry of the latter, we note, is not periodic in the streamwise direction.

## COMBINED TEMPERATURE AND VELOCITY MEASUREMENTS

The measurements of the thermal field using hot- and cold-wires were conducted only for configuration 3, see fig. 4.

Due to limitation of space we again restrict ourselves herein to presenting a sample selection of results on the long-time averages of some chosen quanities in the two planes A and C, x=41mm and x=107mm downstream of the trailing edge of the winglet. These are: $\overline{T'^2}$ and the $\overline{v'T'}$ and $\overline{w'T'}$ components of the heat-flux vector. For further results, e.g. the corresponding Reynolds shear stresses $\overline{u'v'}$ and $\overline{u'w'}$, see Lau [35].

Fig. 13 shows high values of $\overline{T'^2}$ within the core region of the vortex and, for larger z, in the upwash region, that is $y < 0.4 \times 2H = 16mm$, $z > 50mm$. The maximum of $\overline{T'^2} = 0.48 K^2$ is in the vortex core, that is for $y = 0.4 \times 2H = 16mm, z = 40mm$, whereas $\overline{T'^2}$ tends to zero for $z < 20mm$ and in the outer regions of the vortex.

The $\overline{v'T'}$- and $\overline{w'T'}$- components of the heat-flux vector are shown in fig. 14. $\overline{v'T'}$ has a maximum with $\overline{v'T'} = 0.4 Km/s$ in the vortex region for $z = 30mm, z = 40mm$. The sign of $\overline{v'T'}$ changes and $\overline{v'T'}$ acquires a relative minimum $\overline{v'T'} = -0.18 Km/s$, when there is a change of sign in the v-component, cf. fig. 11. Outside the vortex $\overline{v'T'}$ is approximately zero.

$\overline{w'T'}$ is close to zero in the regions for $z < 10mm$ and $z = 80mm$, which, as can be seen from fig. 11, lie outside the vortex within the planes of symmetry of the winglet configuration. Within the core of the vortex, $z = 30mm, z = 40mm$, $\overline{w'T'}$ has a strong gradient with a change of sign, where there is a change of sign in the w-component of the velocity, cf. fig. 11. Large negative values of $\overline{w'T'}$ are within the core of the vortex and in the upwash region $y < 0.4 \times 2H = 16mm, 40 < z < 70mm$.

# VISUALISATION STUDIES

Flow visualisation is an important tool in the study of complex flow patterns. The visualised images permit a qualitative description of the entire flow field. In many flows visualisation is an essential step at interpretation of measured data in term of flow characteristics such as structures, hence their importance for quantitative data evaluation too. A general overview of visualisaton studies can be found in the book of Merzkirch [32]. Since single-point measurements cannot give concrete hints on the instantaneous spatial extent of structures, flow visualisation is indispensible for rounding up hot-wire measurements such as in this work.

We investigated the flow pattern for the three different winglet configurations sketched in figs. 2, 3 and 4 for Reynolds numbers in the range from 300 up to 7000. At the lower end of the Reynolds number range the channel flow not disturbed by the winglets is laminar. Increasing the Reynolds number the flow first becomes transitional and then turbulent. Starting with configuration 1 with a single pair of winglets, we first present visualisation studies of the channel flow with 2 main embedded longitudinal vortices generated by both the winglets. This sheds light on the interaction of the channel flow with the vortices. The next configuration with a single spanwise row of winglet pairs (configuration 2) enables investigation of the interaction of more than two embedded longitudinal vortices. The flow visualisation study with winglet pairs periodic in both spanwise and streamwise directions (configurations 3) assists in gaining understanding of probe measurements in this configuration which is of interest to designers of heat exchangers. For reasons of space we restrict ourselves to presenting a sample of flow visualisation results for configurations 1 and 2 only.

## The facility

A special channel flow facility with water as the flowing medium was designed, see fig. 15. Its salient dimensions are: channel height $10mm$, span $200mm$. This large aspect ratio is chosen to realise two-dimensionality in the oncoming flow in the channel without winglets. Water is circulated by a $2.2kW$ pump in a closed system. Flow rate through the channel is regulated by a frequency divider, throttle valve and a bypass. The pump in this system could maintain a steady flow rate in the range from $0.02$ up to $4 l/s$, which corresponds to an average through flow velocity in the measurement section of $1 cm/s - 2 m/s$. Flow direction is vertically upwards. To prevent transmission of motor vibrations to the channel wall a soft and flexible tube connection was installed. Water is pumped into a compact box with a cross section of $60mm \times 200mm$ and height of $150mm$. To ensure a good quality of the oncoming flow precautionary

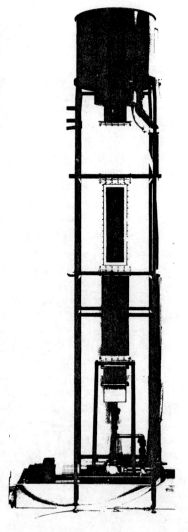

Figure 15: Water flow facility

measures were taken to suppress flow separation in the transition section from a circular to the rectangular cross section. This box is filled up to the flange with ceramic spheres of $10mm$ diameter packed as densely as possible on a perforated plate, which was fixed at a distance of $25mm$ from its entrance. A coarse screen in the flange keeps the spheres in place.

After passing through two honeycombs separated by a distance of $45mm$ and a fine screen in the settling chamber, water flows through a two-dimensional nozzle of contraction 6:1. There is an entrance section $800mm$ long and of the same cross section as the measurement section before the measurement section starts. The entrance section as well as the measurement section are $10mm \times 200mm$ in cross section and are made of plexiglas so that the flow can be observed in both. For purposes of cleaning the walls off the coating from the particles and of changing the winglet configuration, one of the channel walls in the measurement section is removable.

Winglets made out of Lexan® and sized $5mm \times 20mm \times 0.5mm$ are glued on to this plate. The winglet configurations are geometrically similar to those described earlier under this heading, however scaled with respect to the present channel height of $10mm$. After passing through another section the flow enters a diffuser where it increases in velocity. This diffuser leads into a circular trough of diameter $380mm$ and height $720mm$. The trough has four V-notches, which serve for volume flow measurement. Water flows over the V-notches into a second circular reservoir, that surrounds the first one. The water head above the tip of the V-notch is measured in a communicating plexiglas pipe with a scale, that can be read off from outside. The flow rate was measured by reading the water head above the V-notches which was calibrated for this purpose. The calibration curve of flow rate vs. water head above the V-notches was found to follow functional relationships similar to those proposed by Streeter [33]. The pipe connection at the bottom of the reservoir leads to the pump completing the circuit.

Several sources were available for illumination of the flow. Flood light generated by two $300W$ incandescent lamps and/or one of $1000W$ with rectangular cross section mounted at different positions serves for illumination within the measurement section. A He/Ne-laser (nominally $35mW$) was used to produce a laser light sheet via a small cylindrical lens. A photo camera or video camera was installed on a trestle standing on one side of the channel flow facility.

The photo camera was a Canon F-1 with a standard Canon objective with focal length $50mm$ and a macro objective Vivitat Macro with focal length $100mm$. Most of the photos have been taken at an exposure of $1/120s$ on a 400 ASA film (black/white). The video camera was a Sony video Hi8 Handycam.

For flow visualisation the circulating water is seeded with Iriodin® 120 Glanzsatin particles (titanium-dioxide coated glimmer) marketed by MERCK. These particles are saucer-shaped, roughly $10-25\mu m$ in diameter and with a thickness of $3-4\mu m$. Since the particles tend to stick onto the channel walls when there is no flow, sediment and therefore assemble in the pump which is at the lowest level of the facility, the facility has to be emptied into a $200l$ rain barrel after conducting the experiments and be filled again for further experiments.

For preparation of the figures presented in the following contrast has been enhanced with the aid of a computer program.

## Visualised images of the flow

The photos and the videos were taken in the near field and in the far and distant fields behind the winglet pair, and behind the row of winglets. The Reynolds number range for the study was 300 up to 7000. Particular attention was paid to the feature of steadiness or otherwise of the flow. This feature enables the evaluation of characteristic time and length scales of the flow. In this paper we restrict our presentation to studies at Re = 7000, which is the closest in our flow visulisation studies to the Reynolds number of our hot-wire measurements.

### Configuration 1 (a single pair of winglets)

Fig. 16 shows a photograph of the visualised flow in the near, far and distant fields behind a single pair of winglets at Re = 7000. It can be seen that the counterrotating vortices are nearly parallel over the entire distance ($50H$). Waviness of the vortex axes in the spanwise direction starts at $15H$. Comparison with videos shows the waviness is associated with oscillations in the spanwise direction. In the photo, which shows an instantaneous view, the waviness of the vortex axes shows up through the curvature of the bright or dark streaks caused by light reflection.

### Configuration 2

The visualised flow behind the row of winglet pairs at Re = 7000 is shown in fig. 17. Again it can be seen that the counterrotating vortices are nearly parallel over the entire distance ($50H$). Corresponding to the measured high fluctuations in the spanwise direction $\overline{w^2}$ in the far field at $x = 27H$, see Lau [31], the visualisation shows oscillations in this direction starting at about $x = 15H$.

Fig. 16: Flow visualisation, configuration 1, Re = 7000

Fig. 17: Flow visualisation, configuration 2, Re = 7000

# STRUCTURE IDENTIFICATION STUDIES

It has been recognised for quite some time now that the fluctuating motion in a turbulent flow is not totally random but comprises regions over which the motion is of a highly organised structure. These have come to be known as *coherent structures*. Evidence from studies of experiments and direct numerical solution in the *simpler* class of turbulent flows avialable up to now indicate that the size and shape of the coherent structures, while depending to some extent upon the flow, also seem to possess some universal features. For a recent review of the state of knowledge on this subject the reader may wish to consult Bonnet et al. [34] in which attention has been drawn to some hitherto unresolved questions. Problems are encountered at defining a *coherent structure* meaningfully, recognising one in an experiment or in a direct numerical simulation. Incorporating this knowledge into modelling and control of turbulence is yet to be done in a satisfactory manner. While it seems obvious that the characteristic features of a structure should be contained in the solutions of the "suitably averaged" equations of motion, at least as an approximation in some sense, it is not clear what precisely the nature of the averaging involved or the approximation is, and how to extract these from the equations of motion. This is the "important missing link" to which Bonnet et al. [34] refer to as the (absence) of a "clear connection between Navier-Stokes dynamics and CS (for coherent structure) dynamics". Due to this "missing link", questions concerning structures in a certain flow, such as whether they are present, on methods to recognise and isolate them, on their dynamical significance and the like, can only be addressed to the flow actually realised and measured in the experiment, or, as in more recent times, also to results obtained from direct numerical simulation. Conceptually, extracting characteristics of structures either from flow visualisation or measurements, or from direct numerical simulation, is equivalent to seeking the information in the solutions of the equations of motion under the boundary conditions prescribed by the experiment, as against in the equations themselves prior to solution. Incorporating the insight gained from such analysis of solutions into methods for solving the equations of motion is a goal of research work. However, a discussion of points concerned with this task falls outside the scope of the present paper.

An essential step in structure identification studies in any flow requires the answering of two basic questions.

1. The first is regarding the characteristic features of the coherent structure present in the flow in question. Features of coherent structures that have been discerned hitherto, whether in wall bounded or free shear flows, have all been restricted to the class of relatively *simple* turbulent flows like mixing layers, jets, wakes, parallel channel flows and boundary layers under zero or modest pressure gradient. In a complex turbulent flow such as the one at present, since we have no hint regarding the nature of the structures arising, we have no alternative to the tedious and uncertain procedure of checking if a certain anticipated structure is present in the flow. In charting out the route through such an unknown terrain a guiding principle is setting up *benchmark structures* with which the structures in the unknown flow can be compared. The *benchmarks* should be judiciously chosen from *simpler* turbulent flows in which the properties of the turbulent motion have been thoroughly examined with respect to their structural characteristics. The choice is indeed important and will be discussed later in this section.

2. The second question is regarding the method by which a coherent structure may be discerned. In this context it has to be noted that a successfull application of any of the methods proposed hitherto requires careful "matching" of the method to the characteristic sought. All these methods, VITA, POD or wavelet analysis, require for their success a

much deeper insight into the characteristics of turbulent motion than is available today. For example, for the VITA, this calls for definitions of the "event" to be detected and of the threshold. POD is carried out meaningfully on spatial correlations which are mostly not available from single point measurements. Uncovering a hidden structure by wavelet analysis hinges upon a choice of the mother function that is *properly matched* to the structural feature actually present. To reduce the possibility of missing a structure present due to the choice of a *poorly matched* procedure, investigations into the interrelationships between the methods for identifying structures are urgently called for. An essential part in such investigation of which is a certain "calibration" of the methods to examine what features of the coherent structures they respond to and uncover. Both the above questions are addressed in the present work of the authors. In the part of our study on the relationsships between the methods we have subjected two of the methods, viz. the VITA and the wavelet analysis, to carefully constructed academic signals. In the other part we have analysed our measured data by VITA-methods and wavelets and examined if these show indications of structures being present. For details we refer the reader to Lau [31].

## Comparison of structure identification in academic signals by VITA methods and wavelet analysis

Since recognition of the predominance of large-scale coherent structures in turbulent shear flows, the task of development of methods for identifying and analysing those structures in experiments has been accorded importance in turbulence research. The methods hitherto proposed and applied with some success fall broadly under the headings: Variable-Interval Time Averaging (VITA)-techniques, proper orthogonal decomposition (POD) and wavelet analysis. The success of any of these methods strongly depends upon sufficient knowledge of salient features of the structure to be sought. In an arbitrary turbulent shear flow such knowledge of the structure is generally not available in advance. However, despite this handicap, we may hope to deploy the structure identification algorithms more directly by subjecting these to rigorous tests of their outcome in academically generated canonical signals, which, in order to be not too far away from applicabilty to experiments, are meaningfully chosen to contain essential features of structures observed in real flows.

In our study we have subjected academic signals of several different kinds of complexity to tests for structure identification. The methods are VITA and wavelet analysis with Marr-, Morlet- and Daubechies-wavelets. The academic signals belong to the following classes:

a) signals with continuous change in frequency, $\sin(\cos t)$ and $\sin(t^2)$. The latter is shown in fig. 18.

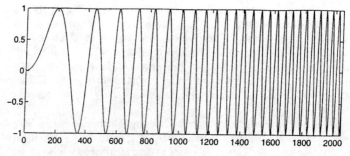

Figure 18: academic signal $\sin(t^2)$

b) signals obtained by addition and multiplication of computer generated noise with signals shown in figs. 19 and 20.

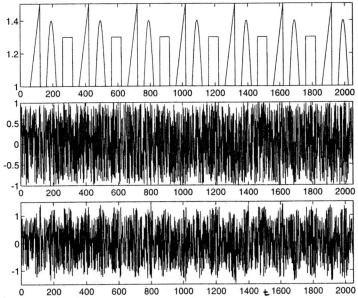

Figure 19: a) academic structures
b) computer generated noise
c) academic signal generated by multiplication of a) and b)

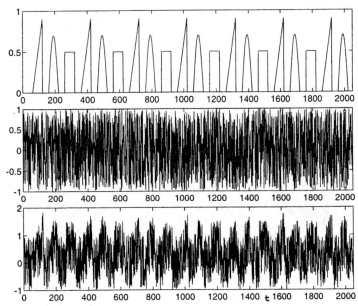

Figure 20: a) academic structures
b) computer generated noise
c) academic signal generated by addition of a) and b)

The class a) may be regarded as extensions of the more basic academic signals studied by Farge [22] and Liandrat and Moret Bailly [36]. Signals of the class b) are idealisations of structures typically found in, say, hot-wire signals in a turbulent shear flow. The goal of the structure identifications algorithm is to capture the salient features of all the signals, their complete recovery where possible being the ultimate aim.

A sample of results from our studies with the VITA method and the Marr wavelet are presented in figs. 21, 22, 23, 24 and 25. In the figures of wavelet analysis the shading is a measure of the coefficients of the wavelet terms with bright patches corresponding to larger magnitude of the same. For details we refer to Lau [31]. We sum up our findings from the comparative study as follows:

The VITA-algorithm, with modifications where necessary, is in a position to identify structural features of the following kind in academic signals:

i. VITA is not suitable for signals of the kind a).
ii. In a signal generated by additive superposition in b) above, if the feature characterizing the structure is a strong gradient, it may be detected with short averaging times.
iii. In a signal generated my multiplication in b), if the features characterizing the structure is a change in variance, it may be detected with properly tuned averaging times.

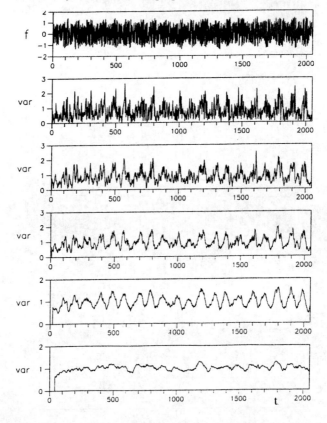

Figure 21: VITA-variances of academic signal f in fig. 19c
from top: signal f, VITA-variance for averaging times $T_m$=4, 8, 16, 40, 80

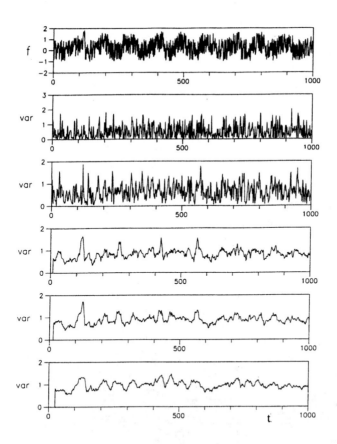

Figure 22: VITA-variances of academic signal in fig. 20c
from top: signal f, VITA-variance for averaging times $T_m$=2, 4, 20, 30, 50

The scope and limitations of the wavelet algorithms are as follows:

i. It is suited to capture the instantaneous frequency and its location for signals of the kind a).
ii. It can identify with a high degree of precision the relevant scales and their position in academic signals of the additive type in b). The structures are recoverable on inverse transformation if the frame condition is respected.
iii. A comparable precision is only partially achievable for signals fo the multiplicative type in b). The choice of the mother function is crucial in this context.

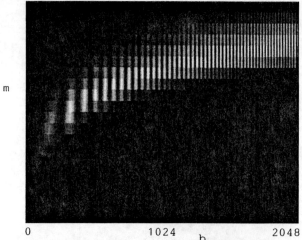
Figure 23: Wavelet coefficients (Marr wavelet) for analysed academic signal in fig. 18

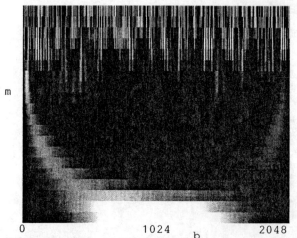
Figure 24: Wavelet coefficients (Marr wavelet) for analysed signal in fig. 19c (multiplicative)

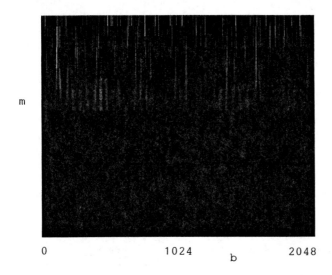

Figure 25: Wavelet coefficients (Marr wavelet) for analysed signal in fig. 20c

## Structure identification studies in the experimental data by VITA methods and wavelet analysis

In the flow under consideration at present, two kinds of source are envisable for the generation of the longitudinal vortices. One is the mechanism that comes into play when a wall-bounded shear layer encounters an obstacle that is finite in the spanwise direction. This leads to the formation of a *horse-shoe vortex* such as at a wing-body junction or at the root of a strut produding out of the boundary layer, see eg. figs. 92 and 93 in van Dyke [3]. The other is the mechanism that is effective in producing longitudinal vortices through rolling up of a shear layer separating off sharpened edges of a lifting body, such as of a delta wing in a uniform stream, see eg. fig. 90 in van Dyke [3]. In general, the vortices from the two kinds of source are of different size and strength and may interact with one another. The flow visualisation studies presented earlier indicate the possibility of the vortices either retaining their identity for merging to lose the same depending on the interaction mechanism. Physical considerations also suggest that vortices from the first kind of source are closer to the wall. In the flow presently under examination, both flow visualisation studies and probe measurements show that the vortices from the second kind of source are more dominant. This may be ascribable to the geometry of the winglets and their orientation. But, given these as at present, from the point of view of data analysis for structure identification, it is more meaningful to chose as the *benchmark flow* for comparison, not a wall bounded flow but the *mixing layer* that is uninfluenced by the walls. However the "mixing layer" should be a three dimensional mixing layer such as in Gründel [35] and not the classical plane mixing layer. Coherent structures in this flow have been analysed through the methods of POD.

In our present work we have employed the VITA and wavelet algorithms to examine if stuctures of the kind identifiable in principle through these methods are discernible in our experimentally realised turbulent channel flow with embedded longitudinal vortices. These analyses were done on the discrete time series of $u'w'$ from the hot-wire measurements at selected locations. In view of limitations of probe resolution we have restricted our studies mostly to the region further away from the channel walls. Again we refer the reader to Lau [32]. Our studies show that although the wavelet algorithm with the Marr-wavelet as mother function hints at the existence of structures, this cannot yet be regarded as conclusive.

## DISCUSSION

The sample of measured data presented under the heading "Long-time averages" already contains hints of the extraordinary complexity of the flow under study. Focussing our attention, for example, on the turbulent kinetic energy, the probe measurements show a significant increase in the contribution of the spanwise velocity fluctuations to turbulent kinetic energy as the flow evolves downstream. The nature of anisotropy of the fluctuating motion manifested through the difference in the spanwise and streamwise velocity fluctuations is different in this flow from, say, in extensively studied "simpler" turbulent shear flows. It is also worthy of note that in the flow of present interest this strong anisotropy is prevalent even in the mid-region of the channel, away from the walls, again standing in contrast to "simpler" turbulent flows. As a second example of complexity we may note the pronounced dissimilarity in shape between the measured distributions of $u'w'$ and $T'w'$. Features such as these would call for a surveillance of terms in the equations for the components of Reynolds stresses and the heat flux vector. However, it is not meaningful to carry out this task without due regard to due regard to the limitations imposed by experiment. It is therefore desirable to take stock of the sources of error in the evaluation of these terms from experimental data.

In the channel flow with embedded longitudinal vortices errors in measurement and data reduction arise over and above those in "simpler" turbulent shear flows. They can be bracketed together as limitations imposed by probe resolution, data acquisition and handling. Firstly, restrictions due to probe resolution are more severe in this flow than in the "simpler" flows. Secondly, not only are there steep gradients of the flow quantities in the region bridging the longitudinal vortex with the surrounding channel flow, but also the direction of the steepest gradient undergoes drastic changes within the flow field. It is spanwise over a part of the vortex but perpendicular to the walls over another. Due to data acquisition and handling considerations it is often inevitable that the data points obtainable in experiment are less dense spanwise than in the direction perpendicular to the walls. Therefore, if the long-time averaged data are not sufficiently smooth, as is the case in the flow in question, errors creeping in due to replacing derivatives by differences in estimating the production terms are larger than in, say, the simpler flows. As a consequence, the experimentally evaluated terms in the budgets of turbulent kinetic energy or Reynolds stress, or of heat flux are not easily comparable in accuracy with the corresponding terms in the "simpler" flows. For the same reason they are not well suited to test hypotheses in turbulence models in Reynolds averaged equations.

In view of such inherent limitations the authors have chosen to view the outcome of probe measurements against the background of flow visualisation. For this purpose Configuration 2 is well adapted since it permits acquisition of probe measured data with relative ease from the near field behind the row of winglets all the way into the distant field. A comparison of photographs of the visualised flow of figs. 16, 17 with the corresponding probe measurement data of figs. 11 shows the vortex location in the two to be essentially compatible with each other. The Reynolds number at which the probe measurements have been carried out is not identical with that for flow visualisation. But, since these are far away from the critical Reynolds numbers at which the flow patterns drastically change, the difference in Reynolds numbers may be regarded as acceptable for a qualitative comparison. The flow photograph shows the axes of the vortices in Configuration 2 to be essentially parallel, a further feature in agreement with the probe measurements.

While the hints of structures in the probe measured data are faint, they are more evident in flow visualisation. This is particularly so at lower Reynolds numbers, as would be expected. To facilitate more definite conclusions in respect of structures at higher Reynolds numbers, very thorough further investigations into its techniques and applicability are essential. Without these verifying the model hypotheses in large-eddy simulations of this complex flow is not feasible.

**Acknowledgements**

The authors thank the Deutsche Forschungsgemeinschaft for grant of financial support for this work. One of the authors, S. Lau, also wishes to thank the Jawaharlal Nehru Center for Advances Scientific Research in Bangalore, India, for grant of a scholarship that enabled her to carry out at their institution parts of the investigations reported in this paper.

# REFERENCES

[1] FIEBIG, M. (1995): "Embedded vortices in internal flow: heat transfer and pressure loss enhancement". International Journal of Heat and Fluid Flow, vol.16, 376-388.
[2] LIU, J.T.C., LEE, K. (1994): "Heat transfer in a strongly nonlinear spatially developing longitudinal vortex system". Physics of Fluids, vol. 7, 559-599.
[3] VAN DYKE, M. (1988): "An album of fluid motion". The Parabolic Press.
[4] CUTLER, A.D., BRADSHAW, P. (1993): "Strong vortex/boundary layer interactions, Part1, vortices high". Experiments in Fluids, vol. 14, 321-332.
[5] CUTLER, A.D., BRADSHAW, P. (1993): " Strong vortex/boundary layer interactions, Part 2, vortices low". Experiments in Fluids, vol. 14, 393-401.
[6] EIBECK, P., EATON, J.K. (1987): "Heat transfer effects of a longitudinal vortex embedded in a turbulent boundary layer". Journal of Heat Transfer, vol. 109, 16-24.
[7] MEHTA, R.D., BRADSHAW, P. (1988): "Longitudinal vortices imbedded in turbulent boundary layers, Part 2, vortex pair with ´common flow´ upwards". Journal of Fluid Mechanics, vol. 188, 529-546.
[8] SHABAKA, I.M.M.A., MEHTA, R.D., BRADSHAW, P. (1988): "Longitudinal vortices imbedded in turbulent boundary layers. Part 1, single vortex". Journal of Fluid Mechanics, vol. 55, 37-57.
[9] SHIZAWA, T., EATON, J.K. (1992): "Turbulence measurements for a longitudinal vortex interacting with a three-dimensional turbulent boundary layer". AIAA Journal, vol. 30, 49-56.
[10] ZHU, J.X. (1992): "Wärmeübergang und Strömungsverlust in turbulenten Spaltströmungen mit Wirbelerzeugern". Dissertation at Fakultät für Maschinenbau, Ruhr-Universität Bochum, Bochum, Fortschrittberichte VDI, Reihe 7: Strömungstechnik, Nr. 192, VDI-Verlag, Düsseldorf.
[11] BRAUN, H. (1996): "Grobstruktursimulation turbulenter Geschwindigkeits- und Temperaturfelder in Spaltströmungen mit Wirbelerzeugern".Dissertation at Fakultät für Maschinenbau, Ruhr-Universität Bochum, Bochum, Cuviller Verlag, Göttingen.
[12] KLINE, S.J., CANTWELL, B.J, LILLE, G.M. (1981): "Complex turbulent flows, computation and experiment". Proceeding of the 1980 Standford Conference, Thermosciences Divisio, Mechanical Engineering Department, Stanford University, Stanford.
[13] BRADSHAW, P. (1997): "Understanding and prediction of turbulent flow - 1996 -". International Journal of Heat and Fluid Flow, vol. 18, no.1, 45-54.
[14] TENNEKES, H., LUMLEY, J.L (1990): A first course in turbulence. The MIT Press (13th printing).
[15] GERSTEN, K., HERWIG, H. (1992): Strömungsmechanik. Vieweg Verlag, Braunschweig/Wiesbaden.
[16] LESIEUR, M. (1990): Turbulence in fluids. 2nd revised edition, Kluwer Academic Publishers, Dordrecht/Boston/London.
[17] MOIN, P., SQUIRES, K., CABOT, W., LEE, S. (1991): "A dynamic subgrid-scale closure method". Physics of Fluids A, vol. 3, 2746-2757.
[18] LILLY, D.K. (1992): "A proposed modification of the Germano subgrid-scale closure method". Physics of Fluids A, vol. 4, 633-635.
[19] BRAUN, H., FIEBIG, M., MITRA, N.K. (1995): "Large eddy simulation of duct flow with a periodically ribbed wall". Computational Fluid Dynamics Journal, vol. 4, no. 1, 79-88.
[20] MORRISON, J.F., TSAI, H.M., BRADSHAW, P. (1989): "Conditional sampling schemes for turbulent flow, based on variable-interval time averaring (VITA) algorithm". Experiments in Fluids, vol. 7, 172-189.
[21] BERKOOZ, G., HOLMES, P., LUMLEY, J.L. (1993): "The proper orthogonal decomposition in the analysis of turbulent flows". Annual Review of Fluid Mechanics, vol. 25, 539-575.
[22] FARGE, M. (1992): "Wavelet transforms and their applications to turbulence". Annual Review of Fluid Mechanics, vol. 24, 395-457.
[23] LAU, S., SCHULZ, V., VASANTA RAM, V. (1993): "A computer operated traversing gear". Experiments in Fluids, vol. 14, 475-476.
[24] NEUMANN, H. (1997): "Experimentelle Untersuchung von hydrodynamischen und thermischen Anlaufströmungen mit periodisch angeordneten Wirbelerzeugern". Dissertation at Fakultät für Maschinenbau, Ruhr-Unversität Bochum, Bochum.
[25] ECKELMANN, H., KASTRINAKIS, E., NYCHAS, S.G. (1984): "Vorticity and velocity measurements in a fully developed turbulent channel flow". In ´Turbulent and Chaotic Phenomena in Fluids´ (Ed. T. Tatsumi), Proceedings of the IUTAM Symposium, Kyoto 1983, 421-426.
[26] SCHULZ, V. (1989): "Der Wellencharakter der Turbulenzstruktur einer ebenen, räumlich gestörten Kanalströmung". Dissertation at Fakultät für Maschinenbau, Ruhr-Universität Bochum, Bochum, Fortschrittberichte VDI, Reihe 7: Strömungstechnik, Nr. 153, VDI-Verlag, Düsseldorf.
[27] MEYER, L. (1994): "Measurements of turbulent velocity and temperature in axial flow though a heated rod bundle". Nuclear Engineering and Design, vol. 146, 71-82.

[28] KLICK, H. (1992): "Einfluß variabler Stoffwerte bei der turbulenten Plattenströmung". Dissertation at Fakultät für Maschinenbau, Ruhr-Universität Bochum, Bochum, Fortschrittberichte VDI, Reihe 7: Strömungstechnik, Nr. 213, VDI-Verlag, Düsseldorf.

[29] BROWNE, L.W.B., ANTONIA, R.A., CHUA, L.P. (1989): "Calibration of X-probes for turbulent flow measurements". Experiments in Fluids, vol. 7, 201-208.

[30] LAU, S. (1995): "Experimental study of the turbulent flow in a channel with periodically arranged longitudinal vortex generators". Experimental Thermal and Fluid Sciences, vol. 11, no. 3, 255-261.

[31] LAU, S. (1996): "Messung und Analyse der Transportgrößen in der turbulenten Kanalströmung mit eingebetteten Längswirbeln". Dissertation at Fakultät für Maschinenbau, Ruhr-Universität Bochum, Bochum.

[32] MERZKIRCH, W. (1987): Flow visualisation. Academic Press Inc., Orlando.

[33] STREETER, V. (1966): Fluid mechanics. McGraw-Hill Book Company.

[34] BONNET, J.-P., LEWALLE, J., GLAUSER, M.N. (1996): "Coherent structures, past, present and future". Advances in Turbulence IV, Kluwer Academic Publishers.

[35] GRÜNDEL, H. (1995): "Strukturen in symmetrischen und asymmetrischen Scherschichten". Dissertation at Technische Universität Berlin, Verlag, Dr. Köster, Berlin.

[36] LIANDRAT, J., MORET-BAILLY, F. (1990): "The wavelet transform: some applications to fluid dynamics and turbulence". European Journal of Mechanics B/Fluids, vol. 9, no.1.

# VORTEX STRUCTURE, HEAT TRANSFER AND FLOW LOSSES IN TURBULENT CHANNEL FLOW WITH PERIODIC LONGITUDINAL VORTEX GENERATORS

H. Neumann, H. Braun, M. Fiebig
Institut für Thermo- und Fluiddynamik, Ruhr-Universität Bochum
D-44780 Bochum, Germany

## SUMMARY

Turbulent channel flow with embedded longitudinal vortex generators is studied experimentally and numerically. The base configuration consists of an array of small rectangular winglet pairs mounted on one wall with an angle of attack of $\beta=45°$. For $\beta=90°$ the winglets become continuous transverse ribs.

The experiments focus on developing flow and heat transfer in the entrance region over 14 winglet rows. For comparison two configurations are studied, one generating dominant longitudinal vortices ($\beta=45°$), and one generating transverse vortices ($\beta=90°$).

The LES is used to investigate the time-dependent flow structure and temperatures of the 45° winglet configuration in periodically fully developed channel flow.

For the experiments the Reynolds number range is $5 \cdot 10^3 \leq Re_H \leq 14 \cdot 10^3$. The attachement of the vortex generator configurations to one wall causes an asymmetry of the flow. The velocity maximum is displaced towards the smooth wall. This displacement of the velocity maximum is stronger for the base configuration with $\beta=45°$ than for the transverse rib configuration with $\beta=90°$. Starting with a laminar flow at the channel entrance velocity fluctuations are initiated and progressively amplified by the vortex generators. The flow at the ribbed wall becomes turbulent first, further downstream the turbulent fluctuations occur also at the smooth side. For the winglet configuration ($\beta=45°$) this occurs further upstream than for the transverse rib configuration ($\beta=90°$).

Measurements for 14 winglet rows show that the 45° winglets generate higher heat transfer ($Nu_{\beta=45°}/Nu_{\beta=90°}=1,46$) and pressure loss ($f_{app, \beta=45°}/f_{app, \beta=90°}=1,23$) than the 90° winglets.

The large eddy simulations (LES) are in very good aggreement with direct numerical simulations (DNS) and experimental results. For fully periodic turbulent channel flow at $Re_H=6000$ the 45° winglet configuration is investigated by the LES and compared with turbulent plane channel flow. The streaky structure of the flow is reinforced. The maximum of the velocity and temperature profiles are higher and shifted to the smooth wall. The turbulent kinetic energy and the turbulent transport is enhanced. Significant augmentation of heat transfer on the smooth and ribbed wall occur. The mean heat transfer on the winglets is more than a factor three higher than on the walls. The global heat transfer enhancement is a factor of 1.7 for a pressure loss penalty factor of 2.5. Reduction of the vortex generator length to 50% results in no significant global changes in heat transfer and pressure loss but changes the individual surface values up to 20%.

# INTRODUCTION

Reduced costs for material, manufacturing and energy are main goals of heat exchangers design. Heat transfer intensification is one tool to achieve these goals. One way to intensify heat transfer is the enhancement of convective heat transfer by flow manipulation, for example by vortex generators. Vortex generators (VGs) may cause heat transfer enhancement by (1) swirl, (2) flow destabilization, and (3) developing boundary layers. VGs may be distinguished by the type of vortices they mainly generate, namely transverse vortex generators (TVGs) and longitudinal vortex generators (LVGs), see Fiebig [3]. TVGs generate vortex systems with their vortex axes mainly transverse to the primary flow direction, while LVGs generate vortex systems with vortex axes mainly along the primary flow direction, see Figs. 1 and 2.

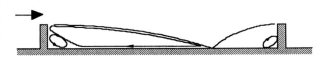

Fig. 1: *Transverse vortices between two rectangular ribs*

Fig. 2: *Pair of counterrotating longitudinal vortices generated by rectangular winglets*

Fundamental investigations of laminar and turbulent flows in flat channels with periodically embedded transverse and longitudinal vortices have been carried out by the research group „*Vortices & Heat Transfer*" at the Institut für Thermo- und Fluiddynamik at the Ruhr-University Bochum, see Fiebig [4]. For these studies a base configuration consisting of a periodically arranged array of rectangular winglets mounted on one wall has been chosen, see Fig. 3. The winglets have a height of e/H=0.5, a longitudinal pitch of Lp/e=10, a length of l/e=4, a thickness of δ/e=0.1, and an angle of attack of β=45°. The lateral pitch Bp of one period consisting of two diverging winglets in the flow direction is equal to Bp/e=8. The distance s between the leading edges of the winglets is s/e=(1-sin β)·l/(2·e). Grosse-Gorgemann [7] and Weber [30]

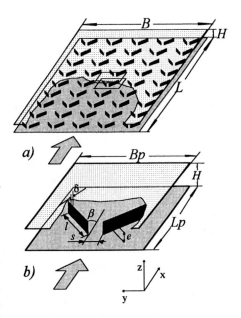

Fig. 3: *Base vortex generator configuration with the geometric parameters of one periodic element: e/H = 0.5; δ/e = 0.1; Lp/e =10; Bp/e = 8; l/e = 4; s/e = (1-sin β)·l/(2·e); β = 45°*

investigated flow structure and heat transfer in steady and self oscillating laminar flows for this base configuration for varying angle of attack and Reynolds number. For angle of attack $\beta=90°$ the configuration transforms into a channel with transverse ribs. A summary of their results is contained in this volume, Neumann, Hahne, Müller and Fiebig [20]. For hydrodynamically and thermally developed periodic flow with the base configuration of Fig. 3 and constant wall temperature ($T_W$=const.) their main results can be summarized as follows:

- The use of the base configuration (Fig. 3) causes a reduction of the critical Reynolds number and a stretching of the transition range compared to that of plane channel flow. With increasing angle of attack and increasing height of the VGs from e/H=0.25 to e/H=0.5 the critical Reynolds number decreases, while the fluctuation amplitudes and number of frequencies increase. For example the TVG configuration, i.e. the base configuration (Fig. 3) with $\beta=90°$, with a height of the VGs of e/H=0.5 reduces the critical Reynolds number by more than an order of magnitude from $Re_{crit,0}$ of 3470 for a plane channel flow to $Re_{crit}\approx 200$.

- The LVGs with $\beta\approx 45°$ cause higher heat transfer and lower pressure loss than the corresponding TVG configuration ($\beta=90°$). The continuously rising pressure loss with increasing angle of attack can be attributed to the increasing dominance of the form drag while the frictional resistance reaches a maximum value at $\beta\approx 45°$ in analogy to the heat transfer.

- The reduction of the VG height from e/H=0.5 to e/H=0.25 causes a reduction of heat transfer and pressure loss. However the reduction in pressure loss is considerably larger than the reduction in heat transfer, so the ratio of heat transfer to pressure loss is higher for the configuration with smaller e/H. Table 1 gives the heat transfer and pressure loss enhancement relative to plane laminar channel flow for the LVG and the TVG configuration and the two VG heights at $Re_{2H}=10^3$.

**Table 1:** *Heat transfer and pressure loss enhancement of the base configuration of Fig. 3 for different VG heights and angle of attacks at $Re_{2H}=10^3$ [30]*

| $Re_{2H}=10^3$ | $\beta=45°$ | | $\beta=90°$ | |
|---|---|---|---|---|
| e/H | 0.5 | 0.25 | 0.5 | 0.25 |
| $Nu/Nu_0$ | 2.76 | 2.14 | 2.68 | 1.75 |
| $f_{app}/f_{app,0}$ | 8.13 | 3.46 | 20.46 | 5.16 |

For higher Reynolds numbers ($10^3 \leq Re_{2H} \leq 5\cdot 10^4$) Riemann [23] investigated heat transfer and pressure loss for the base configuration of Fig. 3. But he studied a different boundary condition, namely periodically fully developed flow combined with a thermally developing condition. For his infrared technique he had constant heat flux ($\dot{q}_W$ = const.) on the ribbed wall for nearly one longitudinal pitch between two successive winglet rows. The opposite smooth wall was adiabatic. The results showed that heat transfer enhancement - the measured heat transfer for $\beta=45°$ related to the measured heat transfer for $\beta=0°$ - and corresponding pressure loss enhancement increase with increasing Reynolds number. Furthermore Riemann [23] varied the base configuration to compare in-line and staggered configurations as shown in Fig. 4. He showed that VG-pairs in a symmetric, in-line arrangement (Fig. 4a) generate a slightly higher heat transfer and lower pressure loss than VG-pairs in symmetric staggered arrangement. The parallel in-line configuration (Fig. 4c) causes a higher pressure loss of up to 40% compared to the parallel staggered configuration (Fig. 4d), while the heat transfer rates of both parallel configurations (Figs. 4c,d) are nearly the same. Güntermann [8] showed for the densest possible packaging without overlay of rectangular VG-pairs of Fig. 3 that in-line configurations cause higher heat transfer and lower pressure loss than in VG-pairs in staggered configurations.

All previous results have shown that symmetric, in-line configurations (Fig. 3 and 4a) generating pairwise counterrotating longitudinal vortices with the same rotation direction in successive periods result in higher ratios of heat transfer to pressure loss than comparable staggered or parallel configurations (Fig. 4b-d) or a comparable TVG configuration. Furthermore the use of VGs with a height of the order of the channel height ($e/H \geq 0.1$) is only suitable for laminar flows where an intensive mixture of near wall fluid and core fluid is induced by the vortices. For turbulent channel flows turbulent transport mechanisms dominate the heat transfer in the core and additional core mixing is not efficient for increased heat transfer. But the turbulent fluctuations decrease towards the wall, so the main thermal resistance of a turbulent flow lies in the viscous sublayer near the wall. Consequently for a high ratio of heat transfer to pressure loss in turbulent flows the height ratio of the VGs has to be much smaller than for laminar flows.

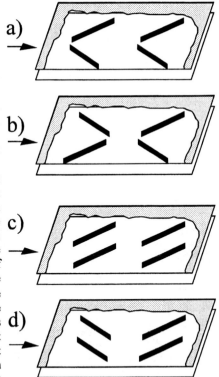

*Fig. 4: Different arrangement of the base configuration of Fig. 3 for comparison of in-line and staggered configuations*
*a) base configuration of Fig. 3: in pairs symmetric, in-line configuration; b) in pairs symmetric staggered configuration; c) parallel in-line configuration; d) parallel staggered configuration*

Here the configuration of Fig. 3 with a height ratio of the VGs of $e/H=0.05$ is studied experimentally and numerically, Fig. 5. The experimental part of this study consists of the investigation of the developing flow and heat transfer distribution in the entrance region of a channel, i.e. investigations of hydrodynamically and thermally developing flow with the winglets attached to one wall. The Reynolds number variation extends from $Re_H = 5 \cdot 10^3$ to $14 \cdot 10^3$. Velocity profiles and profiles of the velocity fluctuations for the LVG configuration of Fig. 5 and a comparable TVG configuration, shown in Fig. 6, are studied by hot wire anenometry. The developing local heat transfer distribution for the LVG and TVG configurations of Fig. 5 and 6 are examined by high resolution infrared thermography. The complex two-dimensional heat transfer distribution induced by the LVG configuration (see, [6, 7, 19, 29]) calls for a surface covering measurement technique with a high local resolution. In the numerical part of this paper a periodically fully developed turbulent channel flow with the configuration of Fig. 5 attached to one wall is studied at $Re_H = 6 \cdot 10^3$. The height of the VGs of $e/H=0.05$ corresponds to a roughness Reynolds number of $e^+ \approx 10$ to 20. Since our purpose is to investigate the influence of longitudinal vortices in the viscous sublayer

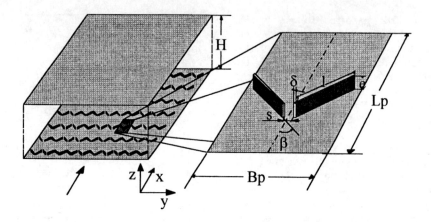

**Fig. 5:** Base vortex generator configuration of the present work with the geometric parameters of one periodic element: $e/H=0.05$; $Lp/e=10$; $Bp/e=8$; $l/e=4$; $\delta/e=0.05$; $s/e=(1-\sin\beta)\cdot l/e$

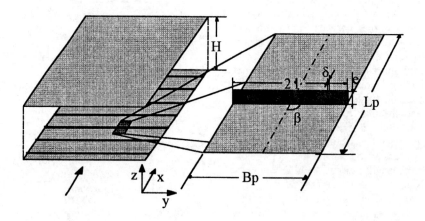

**Fig. 6:** Transverse rib configuration with: $e/H=0.05$; $Lp/e=10$; $Bp/e=8$; $l/e=4$; $\delta/e=0.05$; $s/e=(1-\sin\beta)\cdot l/e=0$

on heat transfer we need to simulate accurately the near wall flow structure. A numerical scheme with extremely refined grids near the wall can resolve the near wall vortical flow structure. Computations solving the Reynolds averaged Navier-Stokes equations with a turbulence model and law of the wall would not be sufficient, a direct numerical simulation (DNS) would be the proper tool. The extreme computational effort of DNS can be reduced by large eddy simulation (LES). In LES eddies larger than the grid size are directly simulated and a subgrid stress (SGS) model is used to take into account the turbulent stresses caused by the smaller eddies. This compromise has been used in the numerical part of this study.

# PART I

## EXPERIMENTAL INVESTIGATIONS OF DEVELOPING CHANNEL FLOW WITH PERIODICALLY ARRANGED VORTEX GENERATORS

### Experimental Setup

The experiments are performed in the closed circuit wind tunnel shown in Fig. 7 which is constructed and described in detail by Tiggelbeck [28]. In this wind tunnel air flows through the test section from bottom to top. The volume flux is determined by the pressure drop at the orifice measured by a pressure cell (10 Torr MKS Baratron).

For the investigations of the configurations of Figs. 5 and 6 a special test section shown in Fig. 8 is constructed. The heat transfer coefficient is determined by an infrared (IR) technique, see Fig. 8a. The test section is divided by a 670 mm long plate into two identical channels with a ratio of the channel width to height B/H=4.1. The test plate made of glass is pasted on both sides with thin heating steel foils (Inconell 600, 50μm) to achieve a constant heat flux boundary condition. The plate is fixed by four copper rails which are pressed together with several threaded rods. The contact pressure of the copper rails with the steel foils accomplished by the threaded rods guarantees good electrical contact. For the power supply by direct current the steel foils are series-connected. The difference of the resistance of both steel foils is less than 1.5%. To guarantee identical temperature fields for the investigations of the VG configurations winglets are attached to both sides of the plate at the same positions. The VGs are glued to the heating foils. As a consequence the heat flux on the VG surfaces is unknown and the heat transfer coefficient can only be determined on the heating foils between the VGs correctly. The heat transfer coefficients at the winglet positions are not accurate and the span and area averaged values are also slightly incorrect by the incorporation of these values, i.e. the heat transfer coefficients are slightly enhanced by these values. But the influence of these pixel on the heat transfer coefficients is small because the characteristic

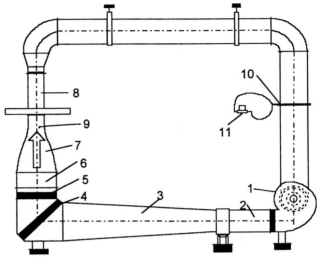

1. fan
2. expansion
3. steep diffuser
4. sheet-metal deflector
5. precombustion chamber with straightener
6. mixing chamber with honey combs
7. nozzle
8. test section (320*160*800 mm$^3$)
9. thermocouple
10. orifice
11. pressure cell

*Fig. 7: Wind tunnel*

length of one pixel is about 2.5 the winglet thickness. A typical infrared frame represented one periodic element - Lp·Bp= 40mm·32mm=1280mm² - by 89·38=3382 pixel. So one pixel extension is equal to 0.45mm·0.84mm in x·y-direction. With a winglet length of l=16mm and a thickness of δ=0.17mm the temperature of the two winglets of one periodic element influence at most 144 pixel of one period (4.26% of the pixel of one periodic element) assuming the worst case with an angle of attack of β=0° and an influence on two pixel in the span direction of each winglet. For a transverse rib (β=90°) the numbers of influenced pixel is reduced to a maximum of 76 pixel, i.e. 2.25% of the pixel one periodic element.

To enhance the emissivity of the heated wall the steel foils are coated with a black paint (Nextel Velvet Coating 2010). The radiation properties of this laquer is described by Lohrengel [18]. A CaF$_2$-window is installed in the opposite smooth wall to provide infrared-optical visibility to the heated surface. The whole test section is insulated by styrofoam plates with a thickness of 15 mm to avoid heat losses through the unheated walls, i.e. to achieve adiabatic conditions. For the recording of the IR-frames the insulation of the IR-window is removed.

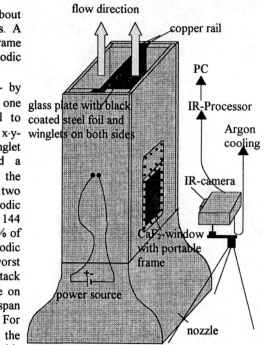

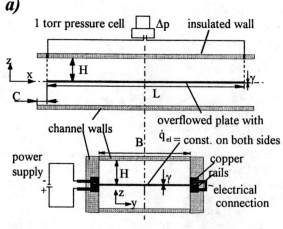

*Fig. 8:* a) test section with IR-camera for the investigation of the winglet configurations with $e/H = 0.05$
b) simplified cross section drawing of the test section; $H = 78$ mm; $B = 320$ mm; $L = 670$ mm; $C = 55$ mm; $\gamma = 4$ mm

## Data Analysis of the Measurements

Hot-Wire Measurements

The hot-wire equipment consists of a DISA type 55P31, tungsten-wire, single-wire probe, and a PSI model 6100 constant temperature anemometer. The data were sampled via a fast Keithley-Metrabyte DAS-20 analog/digital converter with a Keithley-Metrabyte simultaneous-sample-and-hold SSH-4 module. The data were directly stored on a PC hard disc at a sample frequency of 2048 Hz. The principles of the used hot-wire anemometry are described in detail by Neumann et al. [20] and Weber [30].

Turbulent channel flows are characterized by velocity fluctuations superimposing the time averaged velocities:

$$u(x,y,z) = \bar{u}(x,y,z) + u'(x,y,z,t) \quad (1)$$
instantaneous = time averaged + velocity fluctuation.
velocity             velocity

The time averaged momentum equations contain the Reynolds shear stress and the energy equation the turbulent heat flux [11]. These additional terms enhance heat and momentum transfer. To describe the intensity of the velocity fluctuations the standardized root-mean-square value for the velocity component in the flow direction is used:

$$u_{rms} = \frac{\sqrt{\overline{u'^2}}}{\bar{u}_m} . \quad (2)$$

Here $\bar{u}_m$ describes the spatial (index m) and temporal ( $\bar{\phantom{u}}$ ) mean velocity.

The calculation of the Gaussian standard deviation shows that an instantaneous veloctiy of two meters per second, corresponding to $Re_H=10^4$, can be determined with an accuracy of $\Delta u / u = \pm 4.2\%$. The time averaged velocity consisting of 6145 single values has an accuracy of $\Delta \bar{u} / \bar{u} = \pm 0.06\%$ and the standardized root-mean-square value (eq. (2)) an accuracy of $\Delta u_{rms} / u_{rms} = 0.77\%$. For lower velocities the standard deviations increase and for higher velocities they decrease.

Pressure Drop and Volume Flux Measurements

The flow losses are experimentally determined by measuring the static pressure drop. Following Kakac et al. [12] the apparent friction factor $f_{app}$ is defined from the pressure drop as:

$$f_{app} = \frac{2 \Delta p}{\rho \bar{u}_m^2} \frac{A_f}{A} . \quad (3)$$

The pressure loss is measured with a 1 Torr MKS Baratron pressure cell at the smooth channel wall at the beginning and the end of the plate (see Fig. 8b). For higher statistical precision three pressure taps are arranged over the channel width.

The volume flow $\dot{V}$ is determined from the pressure drop measured at an orifice, see Fig. 7. The Reynolds number based on the channel height is defined as:

$$Re_H = \frac{\bar{u}_m H}{\nu} = \frac{\dot{V}}{\nu B} . \quad (4)$$

The accuracy of the apparent friction factors is determined with the Gaussian standard deviation and depends strongly on the Reynolds number. The standard deviation of the apparent friction factor is given in the results as error stripes around the measured values.

## Heat Transfer Measurements

The heat transfer coefficient is characterized by the Nusselt number:

$$Nu_H = \frac{\alpha(x,y) H}{\lambda} \tag{5}$$

where the local heat transfer coefficient is defined by

$$\alpha(x,y) = \frac{\dot{q}_{con}(x,y)}{T_W(x,y) - T_B(x)} . \tag{6}$$

Here $\dot{q}_{con}$ is imposed and constant, $T_W(x,y)$ measured, and $T_B(x)$ determined from the electrical heating. Constant heat flux is realized by electric heating of the steel foils by a Hewlett Packard Modell 6011 power supply. The infrared-camera (AVIO TVS 2200 IR-camera system) measures the local wall temperature $T_W(x,y)$. It has to be corrected for emission and reflection at the $CaF_2$-window. Here we follow the correction given by Neumann, Lorenz, and Leiner [21]. Their theoretical view based on a radiation balance to determine the true wall temperature from the displayed wall temperature at the IR-camera is confirmed by a calibration. The bulk temperature $T_B(x)$ is deduced from an energy balance in the flow direction x, where x=0 corresponds to the leading edge of the plate, see Fig. 8b. Neglecting the very small amounts of radiation to the exit and entrance the bulk temperature is

$$T_B(x) = \frac{\dot{q}_{el} \, x}{\bar{u}_m \, H \, \rho \, c_p} + T_E . \tag{7}$$

The entrance temperature $T_E$ is measured by a thermocouple at the nozzle exit, see Fig. 7. The electrical heat flux is obtained by measuring the electrical current and voltage drop at the steel foils.

$$\dot{q}_{el} = \frac{U \, I}{A_W} . \tag{8}$$

The electric heat flux minus the radiation flux is the convective flux.

$$\dot{q}_{con}(x,y) = \dot{q}_{el} - \dot{q}_{rad}(x,y) \tag{9}$$

Tangential heat conduction in the steel foils can be neglected because of their high resistance. The radiation heat flux is estimated by the radiation exchange of the heated wall (index W) with the opposite unheated wall (index R), both unheated side walls (index S), the entrance cross section (index E; y-z plane with x=0), and the outlet cross section (index A; y-z plane with x=L).

$$\dot{q}_{rad} = \dot{q}_{WR} + 2\dot{q}_{WS} + \dot{q}_{WE} + \dot{q}_{WA} \tag{10}$$

The radiation exchange between two gray surfaces 1, 2 of constant temperatures $T_1$, $T_2$ is [29]:

$$\dot{q}_{12} = \frac{\sigma \, \varepsilon_1 \, \varepsilon_2 \, \varphi_{12}}{1 - (1-\varepsilon_1)(1-\varepsilon_2) \, \varphi_{12} \, \varphi_{21}} (T_1^4 - T_2^4) . \tag{11}$$

The emissivities of all the surfaces except the entrance and the exit are assumed equal to 0.9 according to [1,29]. For the entrance and exit cross section reflections can be neglected, and $\varepsilon_E = \varepsilon_A = 1$. The strong contraction in the nozzle and the expansion in the diffusor are the reason for neglegible reflections into the test section from the entrance and exit sections. The radiation temperature at the entrance and exit are assumed to be $T_E$ and $T_A$, where $T_A = T_B(x=L)$. The view factors $\varphi_{12}$ are determined by standard methods, see [29]. The view factors are $\varphi_{WR} = \varphi_{RW} = 0.7076$; $\varphi_{WS} = 0.1007$; $\varphi_{WE} = \varphi_{WA} = 0.0455$; $\varphi_{SW} = 0.4030$. The temperatures of the unheated walls are not known and must be estimated to determine the radiative heat flux $\dot{q}_{rad}$. The side walls and the wall opposite to the plate are heated by radiation exchange with the electrically heated wall and cooled by convection. Because the test section is well insulated the 'opposite'

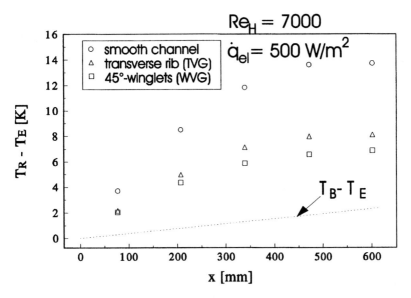

*Fig.9: Temperature values of the unheated wall minus the entrance temperature measured by thermocouples*

wall is in equilibrium between radiation and convection. Neglecting the indirect small radiation exchange by the side walls the energy balance for the 'opposite' wall is

$$\frac{\sigma \, \varepsilon_W \, \varepsilon_R \, \varphi_{WR}}{1-(1-\varepsilon_W)(1-\varepsilon_R) \, \varphi_{WR} \, \varphi_{RW}} (T_W^4 - T_R^4)$$
$$= \sigma \, \varepsilon_W \, \varphi_{WE} \left( (T_R^4 - T_E^4) + (T_R^4 - T_A^4) \right) + \alpha_R \, (T_R - T_B). \tag{12}$$

In equation(12) $\alpha_R$ is the mean convective heat transfer coefficient of the 'opposite' wall and $T_R$ is assumed constant. We measured $T_R$, see Fig. 9, and also estimated the influence of $\alpha_R$ on $T_R$, and of $T_R$ on the thermal boundary condition $\dot{q}_{con}$ of the ribbed wall. A change of the mean Nusselt number $Nu_R$ by 30% resulted in a change of $T_R$ by 0.5 K. A change of 0.5 Kelvin in the mean temperature of the 'opposite' wall resulted in a change in the thermal boundary condition $\dot{q}_{con}$ by less than one percent. The correction of $\dot{q}_{el}$ by $\dot{q}_{rad}$ was however not negligible and varied between 9% and 12% for the considered Reynolds number range. The Nusselt number $Nu_R$ was estimated from the laminar flat plate relation. The measured 'opposite' wall temperatures for $Re_H = 7 \cdot 10^3$ are shown in Fig. 9. The electric heating was 500 W/m². For the smooth wall the heat transfer coefficient is smaller than for the ribbed wall. Hence the surface temperature of the smooth wall is higher than for the ribbed wall. As a consequence the radiation heating of the 'opposite' smooth wall is higher for the smooth plate and lower for the heat transfer surface (45° winglets). With the measured mean temperature of the unheated wall $\dot{q}_{rad}$ in equation (10) has been evaluated with equation (11).

Riemann [23] equated the temperature of the unheated wall opposite to the heated wall with the entrance temperature. Lorenz and Leiner [17] who investigated a ribbed channel under similar conditions with the IR-thermography equated the temperature of the unheated wall opposite to the heated wall with the mean bulk temperature between the inlet and outlet. Both Riemann [23] and Lorenz et al. [17] made no measurements and gave no reasons to verify their assumption concerning their choice for the temperature of the unheated wall, but their test sections were not insulated. The difference between the bulk temperature and the measured wall temperature $T_R$ shows that the assumptions of Riemann [23] and Lorenz et al. [17] are not valid for the insulated test section investigated here.

For the span averaged Nusselt numbers the data of two periods are averaged to reduce the influence of local variations caused by manufacturing inaccuracies of the winglet configuration. The spanwise and area averaged heat transfer coefficients are defined by equations (13) and (14).

$$\overline{\alpha}(x) = \frac{\int_{y=0}^{y=Bp} \dot{q}_{con}(x,y)\,dy}{\int_{y=0}^{y=Bp} (T_W(x,y) - T_B(x))\,dy} \qquad (13)$$

$$\overline{\alpha} = \frac{\int_{x=0}^{x=Lp}\int_{y=0}^{y=Bp} \dot{q}_{con}(x,y)\,dy\,dx}{\int_{x=0}^{x=Lp}\int_{y=0}^{y=Bp} (T_W(x,y) - T_B(x))\,dy\,dx}. \qquad (14)$$

The accuracy of the Nusselt numbers are determined with the Gaussian standard deviation. Table 2 shows the mean relative standard deviation of the Nusselt numbers for the first 14 periods for different Reynolds numbers and different configurations.

*Table 2: Relative standard deviation of the mean Nusselt numbers of the first 14 Periods for different Reynolds numbers and configurations*

| configuration | $Re_H$ | $Nu_H$ | $\Delta Nu/Nu$ [%] |
|---|---|---|---|
| smooth channel | 5022 | 27.17 | 0.878 |
| | 10018 | 37.37 | 1.208 |
| | 14036 | 57.98 | 2.210 |
| **TVG-configuration** | 5009 | 45.92 | 1.182 |
| | 10019 | 79.37 | 2.409 |
| | 14026 | 103.9 | 3.673 |
| **LVG-configuration** | 5002 | 67.35 | 1.894 |
| | 10019 | 115.88 | 2.925 |
| | 14041 | 149.34 | 3.996 |

## Experimental Results

The results of the hot-wire measurements downstream of the first (x/H=0.45) and the twelfth (x/H=6.09) vortex generator row (the positions of the HDA-probe are schematically shown in Fig. 10) are depicted in Fig. 11 and 12.

Figure 11 shows the velocity profiles of the velocity component in the flow direction. The asymmetric profile of the plane channel flow is caused by the asymmetric channel geometry (Fig. 8b). At the channel wall which is a continuation of the nozzle wall (z/H=1), the boundary layer development continues, while at the other wall (z/H=0), the blunt flat plate, the boundary layer devolops from the stagnation flow. The vortex generator configurations change the velocity profiles to a more asymmetric form in the downstream direction. Along the flow path the maximum velocity is displaced more and more towards the smooth wall (z/H=1). This displacement of the velocity maximum to the smooth wall is larger for the winglet configuration ($\beta$=45°) than for the transverse rib configuration ($\beta$=90°).

Starting with a laminar steady-state flow at the channel entrance the velocity fluctuations are initiated and progressively amplified by the vortex generators. The flow at the ribbed wall becomes turbulent while the flow at the smooth wall remains still laminar as shown in Fig. 12. Finally the turbulence reaches the smooth side. This extension of the turbulence from the vortex generator side to the opposite smooth wall is stronger for the winglet configuration ($\beta=45°$) than for the transverse rib configuration ($\beta=90°$).

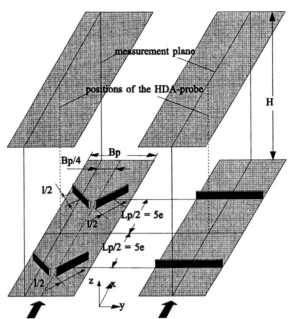

*Fig. 10: Measurement plane and the position of the HDA-probe respectively between two successive vortex generator rows*

Figure 13 shows the apparent friction factors versus Reynolds number for the plane channel, the transverse rib ($\beta=90°$; Fig. 6), and the winglet configuration ($\beta=45°$; Fig. 5). The static pressure drop measured at the smooth wall (Fig. 8b) shows that longitudinal vortex generators ($\beta=45°$) cause higher pressure losses than the transverse rib configuration ($\beta=90°$). The ratio of the apparent friction factors of both is nearly constant $f_{app,\,\beta=45°}/f_{app,\,\beta=90°}=1.23$ in the Reynolds number range $4\cdot10^3 \leq Re_H \leq 15\cdot10^3$. This higher pressure loss of the longitudinal vortex generators ($\beta=45°$) compared with the transverse rib configuration ($\beta = 90°$) can be explained by the stronger asymmetry of the velocity profile and the faster spreading of the velocity fluctuations from the ribbed to the smooth side, see Figs. 11 and 12.

Figure 14 shows the span averaged Nusselt numbers in the entrance region of the plane channel for different velocities or $Re_H$-numbers respectively. For comparison the correlations for a laminar and turbulent flow $Nu_{x,\,lam}$ and $Nu_{x,\,turb}$ of an infinitely thin plate are shown. Because of the blunt leading edge of the plate the flow separates near the leading edge of the plate at a critical Reynolds number based on the thickness of the plate. A recirculation zone originates from this separation with a resulting enhanced heat transfer downstream. Kottke et al. [15] determined the critical Reynolds number as:

$$Re_{\gamma,\,crit.} = 360\left(\frac{\gamma}{2\,x_\gamma}\right)^{-1.4} \qquad (15)$$

where $x_\gamma$ describes the length of the profile along the plate from the leading edge to the point of the full plate thickness $\gamma=4$ mm at x=0. Here $x_\gamma$ is equal to $x_\gamma=2.5$ mm. So the resulting critical Reynolds number is equal to $Re_{\gamma,\,crit.}=492$ or $Re_H=9594$ respectively. Accordingly for $Re_H \geq 10^4$ the heat transfer is influenced by the flow separation near the leading edge and different heat transfer correlations result for $Re_H \leq 10^4$ and $Re_H \geq 10^4$ valid for $Re_x \geq 2\cdot 10^5$, see Fig. 14.

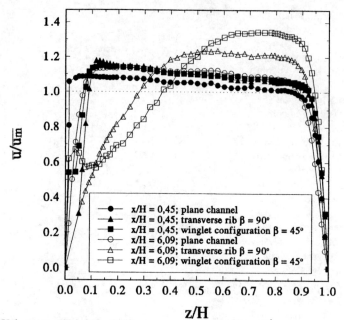

**Fig. 11:** Velocity profiles from HDA-measurements at $Re_H = 10^4$

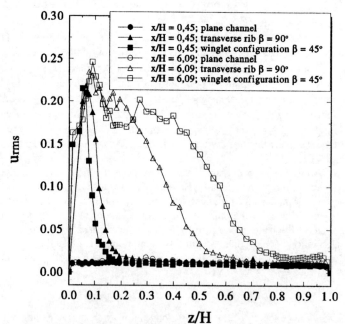

**Fig. 12:** Profiles of the standardized velocity flucuations from HDA-measurements at $Re_H = 10^4$

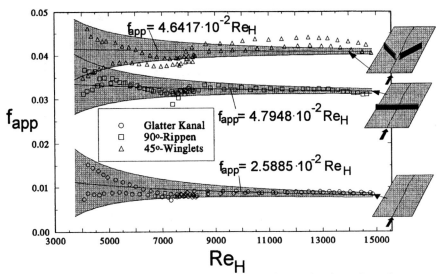

*Fig. 13:* Apparent friction factor versus Reynolds number of the plane channel entrance (L/H=8.6) and the vortex generator configurations with β = {45°, 90°} (Figs. 5, 6); *the grey stripes describe the standard deviation of the measurements*

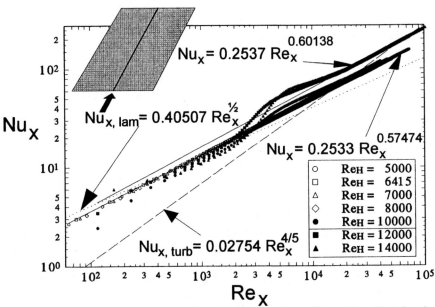

*Fig. 14:* Span averaged Nusselt numbers ($Nu_x$) versus Reynolds number ($Re_x$) for the entrance region of the plane channel flow

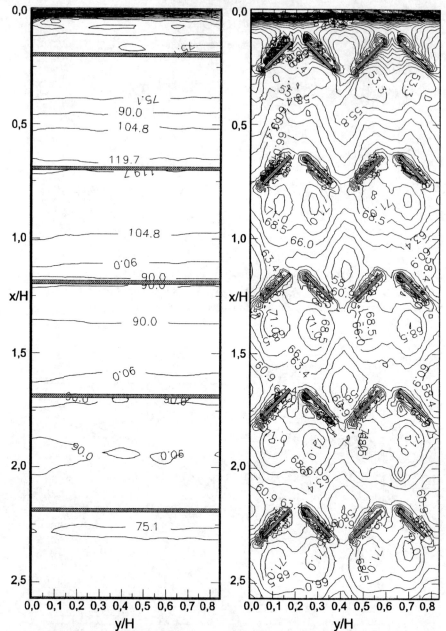

*Fig. 15:* Isolineplot of the local Nusselt number distribution of the transverse rib configuration ($\beta = 90°$) at $Re_H = 10^4$

*Fig. 16:* Isolineplot of the local Nusselt number distribution of the winglet configuration ($\beta = 45°$) at $Re_H = 10^4$

For the entrance region and $Re_H=10^4$ Figs. 15 and 16 show isolines of the local Nusselt number distribution for the transverse rib configuration ($\beta=90°$) and the winglet configuration ($\beta=45°$), respectively. For the transverse rib configuration ($\beta=90°$) the nearly parallel isolines document the one dimensional heat transfer distribution which can be completely described by the span averaged Nusselt numbers. For the winglet configuration ($\beta=45°$) a complex two dimensional heat transfer distribution can be seen. The local heat transfer distribution around each winglet indicates three kinds of vortex structures:

- The pressure difference between upstream and downstream side of the winglet leads to flow separation along the top edge of each winglet. The unstable shear layers then develop into the dominant longitudinal vortices shown in Fig. 2.

- A horseshoe-like vortex is generated in front of the leading edge of each winglet and enhances the heat transfer directly upstream of each winglet.

- The flow separates from the leading and trailing edges of the winglets and generates transverse vortices with their vortex axes vertical to the heat transfer surface.

For both vortex generator configurations ($\beta = \{45°, 90°\}$) the heat transfer distribution is qualitatively periodic downstream of the third vortex generator row in the sence that the local positions of heat transfer minima and maxima within one period are the same for each period downstream of the third vortex generator row. A quantitatively fully developed heat transfer distribution was not reached within the measured 14 periods, i.e. the mean Nusselt number of each period decreases downstream of the third vortex generator row.

Figures 17 and 18 show the span averaged Nusselt number of the first four vortex generator rows of the transverse rib configuration ($\beta=90°$) and the winglet configuration ($\beta=45°$) at different Reynolds numbers. In Fig. 17 high local heat transfer values appear at the transverse rib positions. As mentioned before the Nusselt numbers on the VGs themselves could not be determined accurately because of the unknown heat flux. The shown Nusselt numbers at the rib positions resulted with the usual evaluation, constant heat flux, bulk temperature and measured pixel temperature. Here the latter is a mean value of the wall and VG temperature. For the transverse rib configuration the high Nusselt numbers at the rib positions result because the rib temperature is considerably lower than the neighbouring wall temperature. For the transverse rib configuration only one or two span averaged values are influenced by the VG temperature. For the 45° winglet configuration (Fig. 18) the winglet positions are spread over 14 pixel and 14 span averaged Nusselt values are influenced. No corrections are made in Figs. 17 and 18 for this. The effect on the mean Nusselt number is very small because the winglets cover only 0.5% of the heat transfer area.

The span averaged Nusselt numbers show that the heat transfer distribution is qualitatively periodic developed downstream of the third vortex generator row. Comparison of Fig. 17 with Fig. 18 shows that the heat transfer of the winglet configuration ($\beta=45°$) is higher than the heat transfer of the transverse rib configuration ($\beta=90°$).

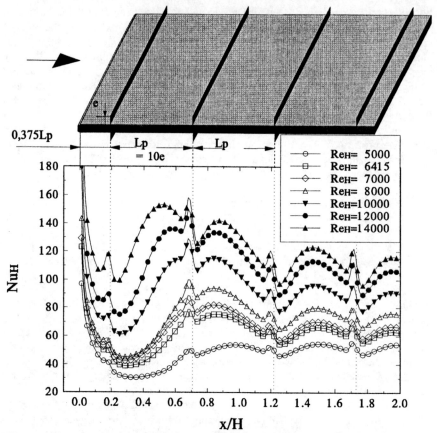

**Fig. 17:** *Spanwise averaged Nusselt numbers ($Nu_H$) versus the flow path $x/H$ for the first four vortex generator rows of the transverse rib configuration ($\beta = 90°$) at different Reynolds numbers ($Re_H$)*

Figure 19 shows the mean Nusselt numbers of successive periods for the vortex generator configurations with $\beta=\{45°, 90°\}$ and different velocities or $Re_H$-numbers. The mean x-coordinate of each period is used as characteristic length of the Nusselt and Reynolds number. For each configuration a correlation of the form:

$$Nu_x = a \, Re_x^{[b+c\ln(Re_x)]} \qquad (16)$$

exists. Table 3 shows the coefficients a, b and c of equation (16) for both vortex generator configurations.

**Table 3:** *Coefficients a, b and c of Equation 17 to determine the meanNusselt numbers of the single succesive periods of the vortex generator configurations with $\beta=\{45°, 90°\}$*

| Configuration | a | b | c |
|---|---|---|---|
| transverse rib configuration ($\beta=90°$) | $3.8931\cdot10^{-5}$ | 2.2262 | -0.06855 |
| winglet configuration ($\beta=45°$) | $1.4295\cdot10^{-5}$ | 2.4826 | -0.08044 |

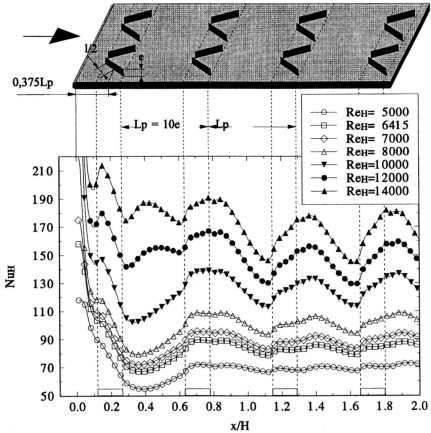

**Fig. 18:** *Spanwise averaged Nusselt numbers ($Nu_H$) versus the flow path $x/H$ for the first four vortex generator rows of the winglet configuration ($\beta = 45°$) at different Reynolds numbers ($Re_H$)*

The area averaged Nusselt numbers and apparent friction factors as functions of Reynolds number are shown in Fig. 20 for the investigated entrance region ($0 \leq L/H \leq 8.6$). The maximum of the heat transfer enhancement at $Re_H \approx 10^4$ coincides with the critical Reynolds number $Re_\gamma$. For all Reynolds numbers the longitudinal vortex generators generate a higher heat transfer and pressure drop compared to the corresponding transverse rib configuration. The mean heat transfer enhancement of the winglet configuration is equal to $Nu_{\beta=45°}/Nu_s = 3.6$ while for the transverse rib configuration this ratio is equal to $Nu_{\beta=90°}/Nu_s = 2.47$. So the mean heat transfer ratio of $Nu_{\beta=45°}/Nu_{\beta=90°}=1.46$ and the mean pressure drop ratio of $f_{app,\beta=45°}/f_{app,\beta=90°}=1.23$ shows that the ratio of heat transfer and pressure drop is more favourable for the LVGs of Fig. 5 than for the TVGs of Fig. 6.

The periodically fully developed region where the heat transfer distribution is identical for successive periods could not be obtained in the experiments. A wind tunnel with considerably higher L/H would be needed. Furthermore the expenditure of manufacturing a vortex generator array to obtain the periodically fully developed flow

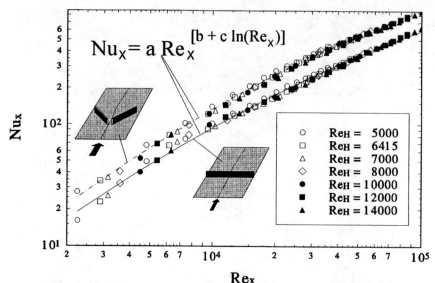

**Fig. 19:** *Mean Nusselt numbers $Nu_x$ of successive periods in the flow direction of the transverse rib configuration ($\beta=90°$) and the winglet configuration ($\beta=45°$) corresponding to different $Re_x$ for different velocities or $Re_H$-numbers respectively*

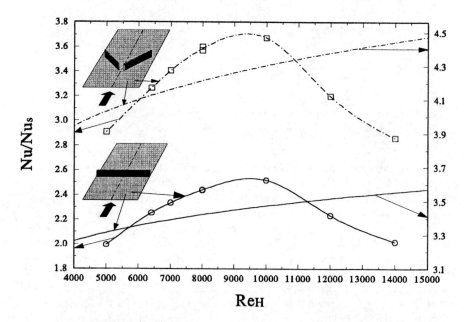

**Fig. 20:** *Standardized area averaged Nusselt numbers of the first 14 periodic elements and apparent friction factors for the entrance region ($L/H = 8.6$)*

region would be much higher. The periodically fully developed region was simulated with the large eddy simulation (LES).

# PART II

# NUMERICAL INVESTIGATION OF FULLY DEVELOPED CHANNEL FLOW WITH VORTEX GENERATORS

## Large Eddy Simulation

The large eddy simulation (LES) of a turbulent flow consists of solving filtered Navier-Stokes and energy equations in order to simulate directly the eddies which are larger than the filter size. In a finite difference scheme the grid size defines the filter. In LES the influence of turbulent fluctuations smaller than the grid size are taken into account by incorporating a subgrid stress model in the averaged equations. The averaged equations are:

$$\frac{\partial \overline{u}_i}{\partial x_i} = 0 \tag{17}$$

$$\frac{\partial \overline{u}_i}{\partial t} = -\frac{\partial \overline{p'}}{\partial x_i} - \frac{\partial \overline{u}_i \overline{u}_j}{\partial x_j} + \frac{\partial \overline{\tau}_{ij}}{\partial x_j} + \frac{\partial}{\partial x_j}\left(\frac{1}{Re}\frac{\partial \overline{u}_i}{\partial x_j}\right) + \delta_{i1}\beta \tag{18}$$

$$\frac{\partial \overline{T}}{\partial t} = -\frac{\partial \overline{u}_j \overline{T}}{\partial x_j} - \frac{\partial \overline{q}_{ij}}{\partial x_j} + \frac{\partial}{\partial x_j}\left(\frac{1}{Re\,Pr}\frac{\partial \overline{T}}{\partial x_j}\right). \tag{19}$$

$\overline{\tau}_{ij}$ results from the filtering of the incompressible Navier-Stokes equations. It includes the cross terms, the Leonard stresses, which have the same order as the truncation error, and the subgrid-scale stresses. This term and the analogeous term $\overline{q}_{ij}$ of the energy-equation (19) have to be modeled.

The modelling of $\tau_{ij}$ is perfomed with an eddy viscosity formulation, Smagorinsky [26]

$$\tau_{ij} - \frac{1}{3}\delta_{ij}\tau_{kk} = 2\,\nu \overline{S}_{ij} \tag{20}$$

$$\nu_{ij} = (C_s\,\Delta)^2\,\left|S_{ij}\right|. \tag{21}$$

$\Delta$ is the characteristic length scale, which is proportional to the volume of the mesh cell. $\overline{S}_{ij}$ is the strain rate of the filtered velocity components.

$$\overline{S}_{ij} = \frac{1}{2}\left(\frac{\partial \overline{u}_i}{\partial x_j} + \frac{\partial \overline{u}_j}{\partial x_i}\right). \tag{22}$$

$C_s$ is calculated with the dynamical approach of Germano [6] for one- or two-dimensional geometries. The characteristic of the dynamical approach is the determination of the variable $C_s$ in dependence of the state of the flow field. That means $C_s$ can vanish, if the state of the flow can be resolved by the chosen grid resolution. The influence of the subgrid-scale model and with it the values of $C_s$ are approximated by comparing the calculated flow field with a filtered flow field. The filtered flow field is determined by interpolating the results of the simulation onto a coarser grid. With the assumption that the local values of $C_s$ are the same on both grids the equation system is closed and $C_s$ is determined by the least square approach. For further details see Germano [6].

At the walls the Schumann boundary condition [25] is used. Although this approach was applied first over 20 years ago (eq. (23)), this formulation is one of the most reliable ones to determine the correct wall shear stress for large-eddy simulations (see Piomelli et al. [22]).

$$c_\tau = \frac{\overline{\tau}_w}{\overline{u}|_w} = \frac{\overline{\overline{\tau}}_w}{\overline{\overline{u}}|_w} . \qquad (23)$$

This synthetic boundary condition changes automatically into the natural boundary condition, if the grid point nearest to the wall is positioned in the viscous sublayer of the turbulent flow. In this case the wall stress is calculated by Newton's law of shear.

The modelling of $q_{ij}$ is analogous to the modelling of $\tau_{ij}$.

$$-\overline{q}_{ij} = \frac{2}{Pr_t} \overline{\upsilon}_{ij} \frac{\partial T}{\partial x_j} . \qquad (24)$$

The results of a calculation with this simple model have been compared with the direct simulation of Horiuti [9]. The test case is a fully developed channel flow with the thermal boundary condition of constant wall temperature. The agreement of heat transfer at the walls and turbulent heat transfer over the channel height was excellent (see Braun [2] for details).

For the energy equation a modified Schumann boundary condition is used (eq. (25)):

$$c_q = \frac{\overline{q}_w}{\overline{T}_1|_w} = \frac{\overline{\overline{q}}_w}{\overline{\overline{T}}_1|_w} . \qquad (25)$$

In the streamwise direction periodic boundary conditions for velocity and temperature field are used. This means the velocity field at the end of the computational box is the same as the velocity field at the entrance. The same procedure is performed with the pressure and temperature field, but they have to be corrected with the averaged gradients multiplied by the periodic length.

For the LES the programm system FRACAS was developed to solve the incompressible Navier-Stokes equations and the energy equation without or with the subgrid stress model by Smagorinsky [26] and Lilly [16]. The numerical procedure is similar to the fractional step method of Kim and Moin [13]. The discretization is second order in space and time, which is realised with central differences in space and a combination of the Adams-Bashforth and the Crank-Nicholson scheme for the time discretisation.

The application of the fractional step algorithm leads to a Poisson-equation for pressure, which is solved by the strongly implicit procedure (SIP) of Stone [27], see eq. (26).

$$\frac{1}{\Delta t} \frac{\partial^2 \overline{p'}}{\partial x_i^2} = \frac{\partial \overline{u_i^*}}{\partial x_i} \quad \text{with} \quad \overline{u}_i^* = \overline{u}_i^{n+1} + \Delta t \frac{\partial \overline{p'}}{\partial x_i} . \qquad (26)$$

## Numerical Results

Geometry, Resolution and Boundary Conditions

The base configuration of Fig. 5 allows to combine the advantages of second order discretization, high grid resolution, and a geometry of technical relevance. The computational domain 0.5·H*0.4·H*1·H is resolved by 102*102*62 grid points. Variable mesh sizes are used in the streamwise and normal direction. For the ribbed wall the nearest grid point is positioned at $z^+=1$. Therefore natural wall boundary conditions are used at the ribbed wall (no slip and constant temperature); at the smooth wall Schumann's boundary conditions are employed. The thermal boundary condition is always constant at all solid walls. In addition to the base geometry of Fig. 5 a configuration with half the winglet length and lateral pitch is studied at the same Reynolds number.

The developed LES code FRACAS is tested by comparing its results with DNS calculations for fully developed channel flow and with experiments for channel flow with transverse ribs on one wall.

## Test Cases

Fully developed channel flow; $Re_H=6000$

In fully developed turbulent channel flow heat transfer is enhanced in comparison to laminar flow by natural longitudinal vortex structures called streaks. They are the major instantaneous structures which transport fluid from the centerline region to near wall areas. Streaks are generated near the wall, spiral up and away from the walls and lose their structures in the main flow. In Fig. 21 the streaks are visualised by surfaces of iso-vorticity in a fully developed channel flow at $Re_H=6000$. Figure 22 shows a cross sectional view of the corresponding temperature field. The vortical structures lead to an exchange of heated near-wall flow and cold core flow. It is interesting that the vortices of turbulent channel flow occur not only as counter-rotating pairs. Robinson [24] has observed the same behaviour of these structures in turbulent boundary layers. The highest values of secondary flow can be found between $0.05 \cdot H$ to $0.15 \cdot H$ in the normal direction from the walls. Their distance from the wall corresponds to the position of highest turbulent kinetic energy (see Fig. 25). These long-lived structures have a large extension in the streamwise direction. For the calculation of convective heat transfer problems a good accuracy of the simulation of the flow field is a prerequisite for a sufficient accuracy of the temperature field calculation. The fully developed channel flow is the standard test case for most turbulence models. The time averaged flow is one-dimensional, but the structure of the instantaneous flow field is 3-dimensional with characteristic time-dependent coherent structures. For the present study the results of the direct numerical simulation of Horiuti [9] are used as reference. The Reynolds number is 6000 with $Re_\tau= 180$ identical with the calculations of Kim, Moin and Moser [14]. The results of both direct simulations correspond very well so it is guaranteed that the used data base is of high quality. Horiuti employed a spectral procedure with 2.1 million grid points, the computational domain was $6.4 \cdot H * 3.2 \cdot H * 1 \cdot H$.

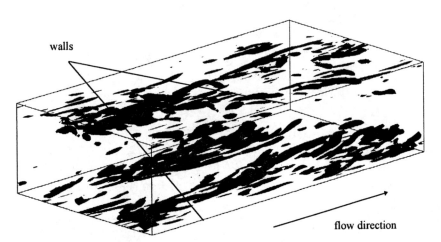

***Fig. 21:*** *Turbulent streaks of fully developed channel flow visualised by areas of equal vorticity; $Re_H=6000$*

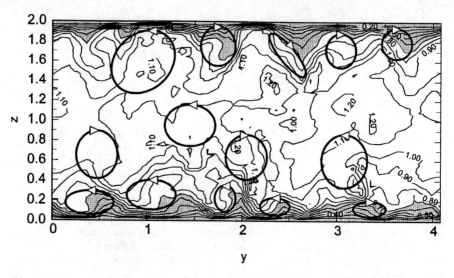

*Fig. 22: Cross-sectional view of the instantaneous temperature field; Fully developed turbulent channel flow ($Re_H=6000$); Arrows mark rotation direction of turbulent streaks*

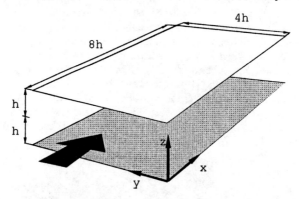

*Fig. 23: Computational domain for the large-eddy simulation of fully developed channel flow*

*Table 4: Grid specifications*

|  | nx | ny | nz | Dx | Dy | $Dz_{max}$ | $Dz_{min}$ | $Dz^+_w$ |
|---|---|---|---|---|---|---|---|---|
| 25 grid points | 102 | 52 | 27 | 0.08 h | 0.08 h | 0.08 h | 0.08 h | 7.2 |
| 30 grid points | 102 | 52 | 32 | 0.08 h | 0.08 h | 0.11 h | 0.028 h | 2.5 |
| 40 grid points | 102 | 52 | 42 | 0.08 h | 0.08 h | 0.07 h | 0.033 h | 3.0 |

Here, three LES calculations with different grids are performed. The different grid sizes are shown in Table 4 and the computational domain is shown in Fig. 23. To limit the computational effort only the grid spacing in the normal direction is varied. The coarse grid is equidistant, while for the two other test cases variable grid spacing is used. The spread numbers in the normal direction with respect to the ribbed wall are 1.11 with 32 grid and 1.07 with 42 grid points These positions near-wall grid points in the viscous sublayer and allows the use of natural boundary conditions (Newton`s wall equation).

The comparison of the time-averaged velocity and temperature profiles shows good agreement of the three large-eddy simulations with the direct numerical simulation of Horiuti, Fig. 24. The differences of the centerline velocities are less than 1% and the differences of the centerline normalized temperatures are less than 3%. More sensitive to inaccurate profiles are wall gradients represented by dimensionless friction factor $C_f$ and Nusselt number $Nu_H$. The values of $C_f$ are 2.5% too high for the coarsest grid, while the simulations with the fine grids underpredict the $C_f$-values. The over and underprediction may be connected with the different formulation of the wall boundary condition. With Schumann's boundary condition $C_f$ is overestimated (25 grid points), while with natural boundary condition $C_f$ is underestimated (30 and 40 grid points). The tendency of the Nusselt numbers $Nu_H$ is not influenced by a change of the boundary condition. The difference of the calculation on the coarsest grid to the direct numerical simulation is 13.2%. It decreases to 1.8% with 40 grid points in the normal direction. The comparison of the global wall gradients also shows the improvement of the results by using fine grids with natural boundary conditions.

*Table 5: Comparison of different flow values; fully developed flow $Re_H = 6000$*

| Case | $C_f$ | $u_c$ | $E_{max}$ |
|---|---|---|---|
| DNS (Horiuti) | 0.00785 | 1.157 | 0.01682 |
| 40 grid points | - 1.8 % | + 0.4 % | - 0.8 % |
| 30 grid points | - 3.8 % | + 0.5 % | - 1.6 % |
| 25 grid points | + 2.5 % | + 1.0 % | - 4.8 % |

*Table 6: Comparison of different temperature field values; fully developed flow $Re_H = 6000$*

| Procedure | $Nu_H$ | $T_c$ | $u'T'_{max}$ |
|---|---|---|---|
| DNS (Horiuti) | 4.865 | 1.120 | 0.0283 |
| 40 grid points | - 2.7 % | + 1.3 % | - 7.1 % |
| 30 grid points | - 6.9 % | + 0.9 % | - 4.2 % |
| 25 grid points | - 13.2 % | + 2.5 % | - 27.2 % |

Indicative of the performance of different grids is also the resolved turbulent kinetic energy and turbulent heat transport $\overline{w'*T'}$. The coarsest grid with 25 grid points in the normal direction produces a higher resolved turbulent kinetic energy in the core region than the direct numerical simulation, see Fig 25. Probably this is a consequence of modelling the viscous sublayer with synthetic boundary conditions. The calculations with the finer grids with natural boundary conditions produce a better agreement with the DNS. The resolved turbulent transport of momentum and heat is an indicator of the importance of the subgrid model for the near wall turbulent transport. With the two finer grids approximately 80-90% of turbulent transport are calculated directly, the other 10-20% are calulated by the eddy viscosity model.

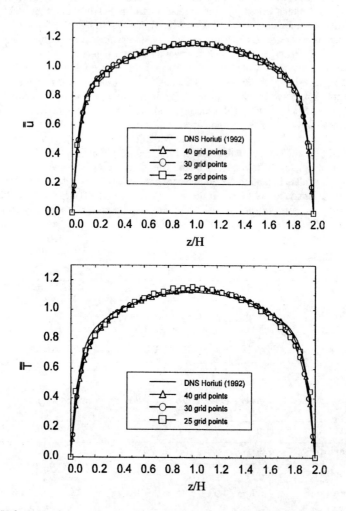

*Fig. 24: Velocity and temperature profiles for grid resolutions normal to the wall and LES results with different DNS results; fully developed turbulent channel flow $Re_H = 6000$*

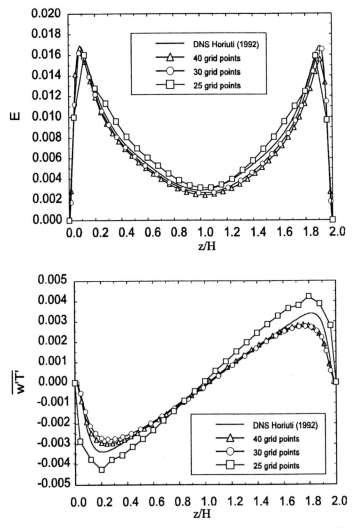

*Fig. 25: Turbulent kinetic energy profiles and normal turbulent heat flux profiles for grid resolutions normal to the wall and DNS results; fully developed turbulent channel flow $Re_H = 6000$*

Channel Flow with Transverse Ribs; $Re_H = 1500$

There are no sufficient data bases available for channel flow with separation and turbulent heat transfer. For testing the LES for seperated flow with heat transfer an experiment for ribbed channel flow was carried out by Weber [30], which used the analogy between heat and mass transfer. The rib height was a quarter of the channel height. The Reynolds number was 1500 based on the channel height. In the LES 353.808 grid points were employed to resolve one period. Grid refinement was used in the vicinity of the rib in the normal and streamwise direction (see Fig. 27). It should be noted, that fully developed plane channel flow is laminar at this Reynolds number, but that the transversely ribbed channel flow was turbulent.

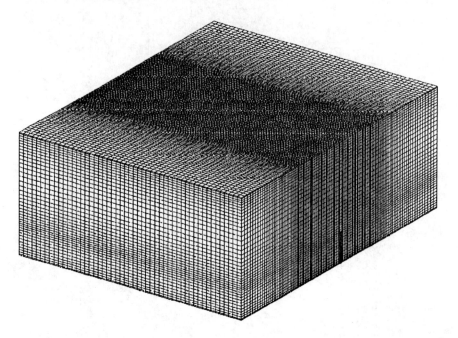

*Fig. 26: Computational domain for the LES of periodically arranged ribs (e/H=0.25), $Re_H = 1500$*

*Table 7: Comparison of LES and experiment for periodically ribbed channel flow; $Re_H=1500$*

| Procedure | $f_{app}/f_{app0}$ | $Nu_{H, ribbed}$ | $Nu_{H, smooth}$ | $Nu_{H, total}$ |
|---|---|---|---|---|
| **Experiment Weber [30]** | 17.30 | 11.425 | 7.24 | 9.335 |
| **LES (H/4 rib)** | + 1.2 % | − 2.3 % | + 2.2 % | − 0.5 % |

Weber [30] measured the global pressure loss and local time and span-averaged mass transfer for constant wall concentration with the ammonia absorption method (AAM). The local heat transfer followed from the analogy between heat and mass transfer. The thermal boundary condition for the experiment and the numerics was constant wall temperature with adiabatic transverse ribs (90°). The comparison between experiment and LES can be deduced from Fig. 27 and Table 7. The maximum local Nusselt number differences occur on the wall downstream of each rib, see Fig.27. This is the area of the reversed flow of the time-averaged vortex. The local Nusselt numbers differ up to 10% based on the averaged global Nusselt number. The agreement of the wall-averaged heat transfer is very good, the deviations amount to 2.3% and 2.2% on the ribbed and smooth wall, respectively. The pressure loss differs by only 1.2% (see Table 7). The agreement between experimental and LES values is better than the accuracy of the experiment. It is also an indication that the numerical procedure will give reliable results for separated flows with heat transfer.

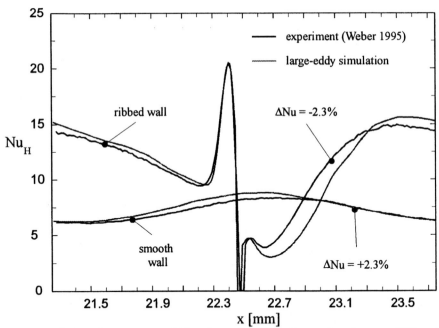

*Fig. 27:* Time and span averaged Nusselt numbers $Nu_H$; Comparison between LES and experiment; $Re_H=1500$; constant wall temperature and adiabatic ribs

Channel Flow with 45°-Winglet Configuration; $Re_H=6000$

To enhance heat transfer and turbulent transport of fully developed channel flow in which longitudinal vortices naturally occur, it is necessary to intensify the near-wall vortex structure and to decrease the viscous sublayer. LES for the two vortex generator configurations shown in Fig. 28 were performed. The first one called RWE8 is the base configuration of Fig. 5 with lateral pitch 8e. For the second configuration the vortex generator length is divided in half to investigate the influence of changing the aspect ratio of the vortex generators. It is called RWE4 because the lateral pitch is now 4e. The geometry parameters of both configurations are presented in Table 8. The dimension of the computational domain was $0.5 \cdot H * 0.4 \cdot H * 1 \cdot H$, see Fig. 28. 102*102*62 grid points were employed with variable grid spacing in the streamwise and normal direction to concentrate most of the grid points around the vortex generators. Periodic boundary conditions are used at the open borders of the computational domain. The thermal boundary condition of constant wall temperature is used for the walls and the winglet surfaces. The Reynolds number ($Re_H=6000$) is the same as in the fully developed channel flow described before, so it is used as reference case.

In Fig. 29 time- and space-averaging over one periodic element is applied to the velocity and turbulent kinetic energy. These ensemble-averaged profiles can be compared to the profiles of fully developed channel flow (VE), also shown in Fig. 29. The maximum of the velocity is shifted to the smooth wall. It is enhanced from 1.15 (fully developed turbulent channel flow) to 1.37 for both configurations. Near the smooth wall the velocity gradients are increased and the $C_f$-Value is approximately 66% (RWE8) and 52% (RWE4) higher than for turbulent channel flow. At the ribbed wall of the channel the mean wall gradient is also enhanced. Near the top of the winglets ($Z \sim 0.05$ H) an inflection point of the velocity profile can be recognized, which is caused by the winglets. The global apparent friction factor which include the form drag of both geometries is enhanced by a factor of 2.50 (RWE8) and 2.45 (RWE4) compared to fully developed channel flow, see Fig.29.

**Table 8:** Geometric parameters of the configurations related to rib height e

|  | RWE8 | RWE4 |
|---|---|---|
| Channel height H/e | 20 | 20 |
| Periodic length Lp/e | 10 | 10 |
| Periodic width Bp/e | 8 | 4 |
| Vortex generator length l/e | 4 | 2 |
| Vortex generator width d/e | 0.1 | 0.1 |

The vortex generators are also turbulence generators as can be seen from the turbulent kinetic energy profiles of Fig. 29. The turbulent kinetic energy is enhanced near the ribbed wall. The maximum is increased by a factor of 2.5 (RWE8) and 2.3 (RWE4) compared to fully developed channel flow. This is approximately the same order of enhancement as the apparent friction factor. Fluctuating shear layers behind the vortex generators are the reason for that increase. The positions of the minimum turbulent kinetic energy values correlate with the

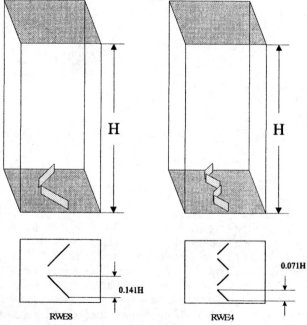

*Fig. 28: Vortex generator configurations*

maximum of the mean velocity, see Fig. 29. The enhancement of turbulent kinetic energy near the smooth wall is correlated with the shift of the maximum velocity and the increased wall gradients.

The ensemble-averaged temperature profiles in Fig. 30 are very similar to the corresponding ensemble-averaged velocity profiles described before. An inflection point and a shift and enhancement of the maximum value occur also for each temperature profile. For the geometry RWE8 the shift of the velocity maximum ($u_{max}=1.37$) from the centerline is 0.15·H, while the shift of the temperature maximum ($T_{max}=1.22$) is 0.07·H.

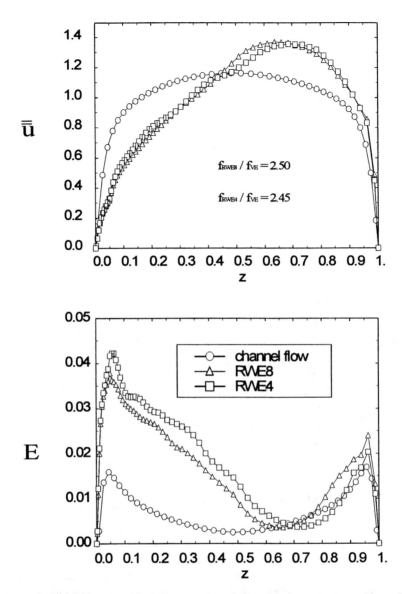

*Fig. 29: Ensemble-averaged velocity and turbulent kinetic energy profiles of the configurations RWE8 and RWE4 compared with a fully developed channel flow*

The lesser shift and maximum value affect the global Nusselt number on the smooth wall. While the friction factor is increased by 65%, the Nusselt number enhancement is only 50%. The turbulent flow of the geometry RWE4 shows the same behaviour: $u_{max}=1.37$ and $T_{max}=1.22$ result in $C_f/C_{f0}=1.52$ and $Nu/Nu_0=1.28$ at the smooth wall. The normal near-wall turbulent transport is enhanced for both configurations. For the ribbed wall the maximum values occur at the position $z=0.09 \cdot H$ and are 2.7 times higher as for fully developed turbulent channel flow. At the smooth walls the normal turbulent transport is less enhanced. The overall heat transfer enhancement in comparison to fully developed

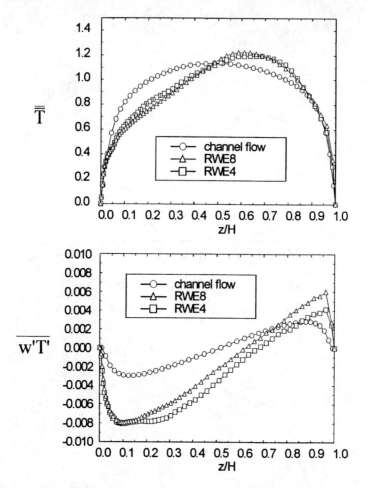

***Fig. 30:*** *Profiles of Ensemble-averaged temperature and turbulent heat flux normal to the wall, RWE8 and RWE4 f configurations and fully developed channel flow; $Re_H=6000$*

channel flow is 70% for the RWE8 configuration and 63% for the RWE4 configuration, while the apparent friction factor is increased by 150% (RWE4) and 145% (RWE8). For the smooth wall and the ribbed wall (but without the VGs) Fig. 31 shows the span and time-averaged Nusselt number distributions in the stream direction for one period for both configurations and the fully developed channel flow. The vortex generators extend in the streamwise direction from $x=0.19 \cdot H$ to $x=0.33 \cdot H$ for the RWE8 configuration and from $x=0.215 \cdot H$ to $x=0.285 \cdot H$ for the RWE4 configuration. Figure 31 shows for the ribbed wall without VGs that in front of each vortex generator a small Nusselt number maximum exists. It is caused by the horseshoe-vortex in front of each winglet. The maximum is higher for the RWE4 configuration because the number of winglets are doubled in comparison to the RWE8 configuration. Further downstream the Nusselt number of both geometries drops below the value of the fully developed channel flow. Behind the winglets the Nusselt number rises through the enhanced thermal transport by the generated longitudinal vortices. Before the flow reaches the next row of vortex generators it is deflected away from the wall and the heat transfer decreases. On the

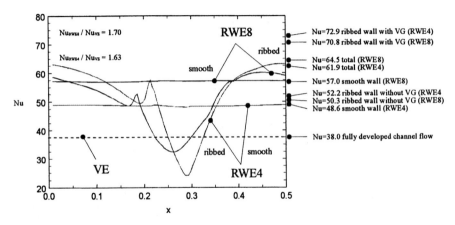

*Fig. 31: Span and time averaged Nusselt number distribution of the configurations RWE8 and RWE4; $Re_H=6000$*

smooth wall the Nusselt number distribution is not influenced by the local effects of the ribbed wall. The time-averaged Nusselt numbers are practically constant in the span and stream direction.

At the right side of Fig. 31 the average Nusselt number for the two configurations and their individual surfaces are listed. The designation 'ribbed wall without VG' denotes the wall with the winglets but without the winglets themselves. The global Nusselt numbers of both configurations differ by less than 3%. For the RWE8 configuration the Nusselt number of the ribbed wall without VG is about 10 % lower than the Nusselt number of the smooth wall. For the RWE4 configuration it is the other way around, the Nusselt number of the smooth wall is about 4% lower than the Nusselt number of the ribbed wall without VG. The VG's increase the overall Nusselt number of the ribbed wall by roughly 40 % for both configurations (~ 50 to ~ 70), even though their surfaces account for only 20% of the surface. The high VG heat transfer ($Nu_{VG}$ ~ 170 for both configurations) is probably the most striking result.

The dimensionless height of the vortex generators is $z^+=18$ relative to the shear stress velocity of the fully developed channel flow at the same Reynolds number $Re_H=6000$. Their side edges are in the turbulent area of the flow. Therefore large turbulent thermal transport approaches their surfaces and generates the high average Nusselt number.

For both configurations the two dimensional Nusselt number distributions are shown in Fig. 32 for the ribbed wall without VG. The large regions of low Nusselt numbers on the ribbed wall are interesting to note. Behind the vortex generators areas with time-averaged separated flow occur, in which the temperature gradients are lower than for fully developed turbulent channel flow. In the streamwise direction the separation regions merge into the upwash regions of the time-averaged longitudinal vortices. Here the Nusselt numbers are also lower than the reference values (see also Fig. 31).

At the smooth wall the Nusselt numbers are enhanced for both configurations due to the shift of the temperature maximum to that wall. The comparison of the RWE4 and RWE8 configuration shows that the Nusselt numbers on the smooth wall is 20% higher for the RWE8 configuration than for the RWE4 configuration.

The complex flow structure becomes apparent in the instantaneous vortex structure. In Fig. 33 vortical structures are visualised with the method of Jeong & Hussain [10]. The time-dependent flow pattern is completely different from the time-averaged one. Longitudinal vortices come mainly into being at the edges of the vortex generators and are lifted away

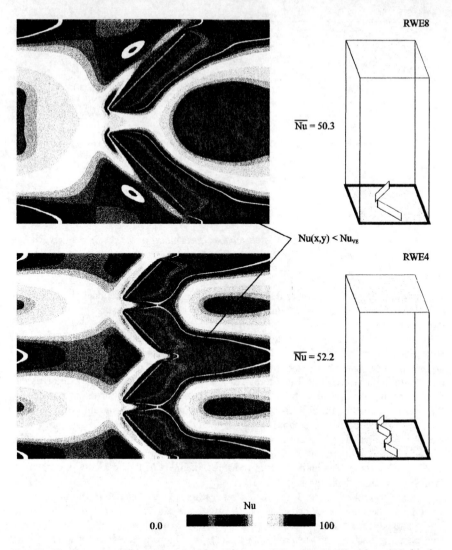

***Fig. 32:*** *Time averaged Nusselt number distributions of the ribbed wall without VG for the RWE8 and RWE4 configurations; $Re_H=6000$*

from the wall into the main stream. So a vortex is not reinforced by the next vortex generator row, there a new one is generated. It is interesting to note, that two co-rotating vortices can be generated by one rectangular winglet.

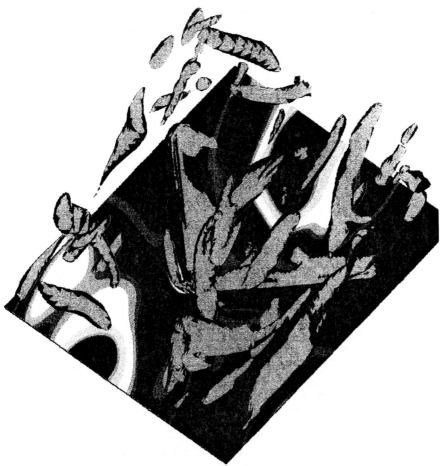

*Fig. 33: Instantaneous turbulent flow structure and heat transfer visualised by the vortex identification method of Jeong and Hussain [30]; $Re_H=6000$*

## CONCLUDING REMARKS

Channel flows have been studied with periodic winglet vortex generators. Flow field, temperature field, heat transfer, and flow losses have been investigated experimentally and numerically. The winglet vortex generators were slender rectangular plates attached to one wall in pairs at an angle of attack of ±45° (LVG-configuration) to the main flow direction. The base configuration had a winglet height e of 5% of the channel height H, and a length ratio of l/e=4, and longitudinal and lateral pitch of Lp =10e and Bp =8e, respectively. The configuration transformed into a transverse rib configuration when the angle of attack β was changed from 45° to 90° (TVG-configuration).

The experimental part investigated the entrance region of a channel formed by a blunt flat plate as one channel wall and the continuation of the nozzle as the other wall. The Reynolds numbers based on the channel height ranged from $5 \cdot 10^3$ to $14 \cdot 10^3$. The LVG and TVG configurations were compared. The local Nusselt numbers were determined for constant heat flux at the wall with the vortex generators. The other wall was insulated. An infrared technique was used to deduce the heat transfer coefficients. The main conclusions of the experimental study are:

- The wall with vortex generators introduces an asymmetry in the velocity profile which becomes more pronounced with flow length. The displacement of the velocity maximum is larger for the LVG ($\beta=45°$) than for the TVG ($\beta=90°$) configuration.

- The vortex generators initiate velocity fluctuations (turbulence) which are progressively amplified by successive vortex generator rows. At first the flow near the ribbed wall becomes turbulent while the main flow still remains laminar. Further downstream the turbulence extends more and more into the core and finally reaches also the smooth wall. The spreading of the turbulence from the vortex generator side to the opposite smooth side is faster for the LVG ($\beta=45°$) than for the TVG ($\beta=90°$) configuration.

- The LVG ($\beta=45°$) configuration generates a higher pressure drop and heat transfer than the corresponding TVG ($\beta=90°$) configuration. The ratio of the apparent friction factors is $f_{app,\ \beta=45°}/f_{app,\ \beta=90°}=1.23$, the ratio of the Nusselt numbers is $Nu_{\beta=45°}/Nu_{\beta=90°}=1.46$ for the investigated Reynolds number range ($5\cdot10^3$ to $14\cdot10^3$).

The numerical part used the large eddy simulations (LES) to investigate the identical LVG ($\beta=45°$) geometry for periodically developed turbulent flow and constant wall temperature on all walls, (smooth wall, ribbed wall and vortex generators). In addition a LVG configuration with half the length to height ratio, l/e =2 and identical frontal area, was investigated. It was named RWE4 because its lateral pitch Bp/e=4 while the base configuration with Bp/e=8 was named RWE8. The Reynolds number was $Re_H = 6000$. The mayor conclusions of the numerical simulations are:

- The LES were in very good agreement with (1) direct numerical simulations (DNS) for fully developed turbulent channel flow, and (2) local heat transfer measurements for fully periodic channel flow with transverse ribs.

- The streaky structure of the turbulent boundary layer is reinforced by longitudinal vortices.

- The vortex generators lead to higher values and to a shift of the velocity and temperature maximum to the smooth wall.

- The maximum turbulent kinetic energy and maximum turbulent heat flux normal to the wall is enhanced by more than a factor of two by the vortex generators in the wall region in comparison with fully developed channel flow.

- The enhanced turbulent kinetic energy, thermal energy, and momentum transport lead to enhanced heat transfer and friction. For the vortex generators the mean Nusselt numbers are more than three times the wall values.

- The potential of longitudinal vortex generator configurations to enhance heat transfer with low flow losses has been confirmed for turbulent flow.

- Reduction of the separation areas by reducing the angles of attack of the vortex generators should further reduce the flow losses.

- The reduction of the vortex generator length from 4e to 2e and lateral pitch from 8e to 4e has no significant overall effect because the frontal area remains the same. But the heat transfer on the smooth wall is reduced by 20 %.

## NOMENCLATURE

| | | |
|---|---|---|
| A | [m] | surface area |
| $A_f$ | [m] | cross section |
| a | [m²/s] | thermal diffusivity |
| a | [ - ] | constant |
| Bp | [m] | lateral periodic pitch |
| b | [ - ] | constant |
| C | [m] | distance between the end of the nozzle and the beginning of the overflowed plate |
| $C_s$ | [ - ] | subgrid model constant |
| c | [ - ] | constant |
| $c_\tau$ | [ - ] | Schumann's coefficient |
| E | | turbulent kinetic energy |
| e | [m] | vortex generator height |
| f | [Hz] | frequency |
| H | [m] | channel height |
| h | [m] | channel half height |
| I | [A] | current |
| Lp | [m] | longitudinal periodic pitch |
| l | [m] | length of the vortex generators |
| p | [Pa] | absolute pressure |
| p' | [Pa] | local periodic pressure |
| $\dot{q}$ | [W/m²] | heat flux |
| $S_{ij}$ | [ - ] | deformation tensor |
| s | [m] | distance between the leading edges of the vortex generators |
| T | [K] | temperature |
| U | [V] | voltage drop |
| u | [m/s] | velocity |
| $X_R$ | [ - ] | nondimensional distance between the edge of the rib and the reattachement line |
| x, y, z | [m] | coordinate in flow direction |

*greek symbols*

| | | |
|---|---|---|
| α | [W/(m²·K)] | heat transfer coefficient |
| β | [°] | angle of attack |
| β | [Pa] | fixed pressure gradient |
| δ | [m] | thickness of the vortex generators |
| ε | [ - ] | emission coefficient |
| γ | [m] | thickness of the overflowed plate |

| | | | |
|---|---|---|---|
| φ | [ - ] | | insolation coefficients |
| λ | [W/(m·K)] | | heat conduction coefficient |
| ν | [m²/s] | | kinematic viscosity |
| $\nu_{ij}$ | [m²/s] | | eddy viscosity |
| ρ | [kg/m³] | | density |
| σ | [W/(m²·K⁴)] | | Stefan-Boltzmann-constant |
| $\tau_W$ | [N/m²] | | wall shear stress |
| $\tau_{ij}$ | [N/m²] | | subgrid-scale stresses |

*subscripts*

| | |
|---|---|
| A | outlet |
| B | bulk temperature |
| E | entrance |
| R | non heated wall opposite the heated wall |
| S | non heated side walls |
| s | plane channel flow |
| W | heated wall |
| con | convective |
| el | electric |
| rad | radiation |
| 0 | plane channel flow |
| m | spatial averaged |
| ‾ | time averaged |
| ' | temporal variabel |
| - | large scale |
| = | time averaged large scale |

*dimensionsless parameters*

$$f_{app} = \frac{2\Delta p}{\rho \bar{u}_m^2} \frac{A_f}{A} \quad \text{apparent friction factor}$$

$$Nu_H = \frac{\alpha H}{\lambda} \quad \text{Nusselt number; the Index indicates the used characterisitic length}$$

$$Pr = \frac{\nu}{a} \quad \text{Prandtl number}$$

$$Re_H = \frac{\bar{u}_m H}{\nu} \quad \text{Reynolds number; the Index indicates the used characterisitic length}$$

# REFERENCES

[1] Baehr und Stephan: Wärme- und Stoffübertragung; Springer Verlag 1994, ISBN 3-540-55086-0.

[2] Braun, H.: "Grobstruktursimulation turbulenter Geschwindigkeits- und Temperaturfelder in Spaltströmungen mit Wirbelerzeugern", Dissertation Ruhr-Universität Bochum, 1996.

[3] Fiebig, M.: "Vortex Generators for Compact Heat Exchangers", J. of Enhanced Heat Transfer, Vol. 2, Nos. 1-2, pp. 43-61, 1995.

[4] Fiebig, M.: "Vortices as Tools to Influence Heat Transfer; Overview of the Results of the DFG-Research Group 'Vortices and Heat Transfer'".

[5] Germano, M.: "Turbulence: the filtering approach", Journal of Fluid Mechanics 238, 325-336, 1992.

[6] Germano, M., Piomelli, U., Moin, P. and Cabot, W. H.: "A dynamic subgrid-scale eddy viscosity model" Physics of Fluids A 3, 1760-1765, 1991.

[7] Grosse-Gorgemann, A.: "Numerische Untersuchung der laminaren oszillierenden Strömung und des Wärmeüberganges in Kanälen mit rippenförmigen Einbauten"; VDI-Fortschrittsberichte, Reihe 19: Wärmetechnik/Klimatecknik, Nr. 87, ISBN 3-18-308719-7, 1996.

[8] Güntermann, T.: "Dreidimensionale stationäre und selbsterregt-schwingende Strömungs- und Temperaturfelder in Hochleistungs-Wärmeübertragern mit Wirbelerzeugern"; VDI-Fortschrittsberichte, Reihe 19: Wärmetechnik-/Kältetechnik, Nr. 60, ISBN 3-18-146019-2, 1992.

[9] Horiuti, K., Miyake, Y., Miyauchi, T., Nagano, Y.: "Establishment of the direct numerical simulation data bases of turbulent transport phenomena", Department of Mechanical Engineering University of Tokyo, http:///www.thtlab.u-tokyo.ac.jp, 1992.

[10] Jeong, J., Hussain, F.:"On the identification of a vortex", Journal of Fluid Mechanics 285, 69-94, 1995.

[11] Jischa, M.: "Konvektiver Impuls-, Wärme- und Stoffaustausch; Vieweg Verlag, Braunschweig 1982, ISBN 3-528-08144-9.

[12] Kakac, S., Shah, R. K. aand Aung, W.: "Handbook of Single-Phase Convective Heat Transfer", Wiley & Sons 1987.

[13] Kim, J., Moin, P.: "Application of a fractional-step method to incompressible Navier--Stokes equations", Journal of Computational Physics 59, 308-323, 1985.

[14] Kim, J., Moin, P., Moser, R.: "Turbulence statistics in fully developed channel flow at low Reynolds number", Journal of Fluid Mechanics 177, 133-166, 1987.

[15] Kottke, V., Blenke, H. and Schmidt, K. G.: "Einfluß von Anströmprofil und Turbulenzintensität auf die Umströmung längsangeströmter Platten endlicher Dicke", Wärme- und Stoffübertragung, Vol. 10, pp. 159-174, 1977.

[16] Lilly, D.K.: "The representation of small-scale turbulence in numerical simulation experiments", Proceedings of the IBM Scientific Computing Symposium on Environmental Sciences, IBM Form No. 320-1951, 99-164, 1967.

[17] Lorenz, S. and Leiner, W.: "Flow Structure and Local Heat Transfer in an Asymmetric Grooved Channel in the Turbulent Regime".

[18] Lohrengel, J.: "Gesamtemissionsgrad von Schwärzen", Wärme- und Stoffübertragung 21, pp. 311-315, 1992.

[19] Neumann, H.: "Experimentelle Untersuchungen an periodisch angeordneten längs- und querwirbelerzeugenden Rauhigkeitselementen bei hydrodynamischem und thermischem Anlauf", Dissertation Ruhr-Universität Bochum, Institut für Thermo- und Fluiddynamik, 1997.

[20] Neumann, H., Hahne, W., Müller, U. and Fiebig, M.: "Heat Transfer and Flow Losses in Transition from Longitudinal to Transverse Vortices in Steady and Oscillating Channel Flow".

[21] Neumann, H., Lorenz, S. and Leiner, W.: "Infrarot-Thermographie durch Fenster bei niedrigen Temperaturen", Wärme- und Stoffübertragung 29, pp. 219-225, 1994.

[22] Piomelli, U., Ferziger, J., Moin, P., Kim, J.: "New approximates boundary conditions for large-eddy simulations of wall-bounded flows", Physics of Fluids A 6, 1061-1068, 1989.

[23] Riemann, K. A.: "Wärmeübergang und Druckabfall in Kanälen mit periodischen Wirbelströmungen bei therrmischem Anlauf", Dissertation Ruhr-Universität Bochum, Institut für Thermo- und Fluiddynamik, Lehrstuhl für Wärme- und Stoffübertragung, 1992.

[24] Robinson, S.K., Kline, S.J. and Spalart, P.R.: "Quasi coherent structures in the turbulent boundary layer: Part II. Verification and new information from a numerically simulated flat-plate layer", in Kline, S.J. et al., "Near-wall turbulence", Hemisphere, 218-247, 1988.

[25] Schumann, U.: "Subgrid scale model for finite difference simulations of turbulent flows in plane channels and annuli", Journal of Computational Physics 18, 376-404, 1975.

[26] Smagorinsky, J.S.: "General circulation experiments with the primitive equations. I. The basic experiment", Monthly Weather Review, Vol. 91, 99-164, 1963.

[27] Stone, H.L.: "Iterative solution of implicit approximation of multidimensional partial differential equations", SIAM Journal of Numerical Analysis, Vol. 5, 530-558, 1968.

[28] Tiggelbeck, S.: " Experimentelle Untersuchungen an Kanalströmungen mit Einzel- und Doppelwirbelerzeuger-Reihen für den Einsatz in kopakten Wärmetauschern", VDI-Fortschrittsberichte, Reihe 19: Wärmetechnik/Klimatecknik, Nr. 49, ISBN 3-18-144919-9, 1991.

[29] VDI-Wärmeatlas; 7. Auflage 1994.

[30] Weber, D.: "Experimente zu selbsterregt instationären Spaltströmungen mit Wirbelerzeugern und Wärmeübertragung" Cuvillier Verlag Göttingen, 1996, ISBN 3-89588-633-5.

# LONGITUDINAL VORTICES IN BOUNDARY LAYER HEAT TRANSFER AUGMENTATION*

J.T.C. Liu

Division of Engineering and the Center for Fluid Dynamics,
Brown University, Providence, Rhode Island 02912, USA

## SUMMARY

We use nonlinearly developing longitudinal vortices that originated from initial upstream weak Görtler vortices on semi-infinite concave walls as the "prototype" vortices to discuss the heat transfer problem. Some experimental aspects of longitudinal vortices and their effects on heat transfer are reviewed. Recent theoretical and computational results are reviewed and used to interpret experimental observations. The analogy between heat and mass transfer with that of longitudinal momentum transfer is discussed, which could be used for nonintrusive quantitative interpretation of streamwise velocity distribution from flow visualization studies. Similarity parameters are pointed out and this is dictated by the strong dependence of flow and heat transfer on initial upstream conditions. The mechanism for heat transfer enhancement is discussed as are issues for further research.

## INTRODUCTION

Studies of heat transfer enhancement and its applications can loosely be classified according to the method of longitudinal vorticity generation: One is the generation of longitudinal vorticity of sufficient strength for heat transfer enhancement through insertion of vortex generators in the flow [1]-[3]. In this case the basic flow may not necessarily be inherently unstable. The other is the excitation of the inherent instability of the basic flow configuration, such as the flow over a concave wall, which give rise to longitudinal vorticity elements due to centrifugal instability [4]-[9]. In the former, longitudinal vorticity is generated in the wake behind the generators, which persist through long distances downstream compared to the scale of the generator, but nevertheless, decay downstream. In the case of instabilities giving rise to longitudinal vorticity, in appropriate parameter ranges the disturbance flow and its heat transfer enhancement amplify downstream. These two aspects of heat transfer enhancement by longitudinal vorticity, which play an important role in the development of heat transfer surfaces for compact heat exchangers [1] and in the heat transfer problem in turbomachinery, are reviewed recently by Crane [10], Floryan [5], Jacoby and Shah [11], Peerhossaini [12] and more recently by Fiebig [1]-[3]. Saric [6] surveyed work on Görtler vortices and pointed out the need for studying the effects of upstream and free stream disturbances on such vortices. Such studies are important to the heat transfer problem [9]. Floryan [5] discussed some aspects of the heat transfer under longitudinal vortices.

## EXPERIMENTAL EFFORTS AND APPLICATIONS

Significant increases in surface heat-transfer rates under Görtler vortices were measured by McCormack, et al. [13]. At a downstream station from the leading edge an increase in local the Nusselt number was between 100% to 150% over that of the flat plate values. At this early date, point of view taken was that surface heat transfer rates were correlatable with local values of the Görtler number and the downstream dependence of heat transfer was not considereded. The strong dependence of downstream development of heat transfer on initial, upstream

---

* Dedicated to Professor Martin Fiebig on the auspicious occasion of his 65th Anniversary

conditions is only a recent development [9].

Kottke [14] addresses both mass- and heat-transfer effects over a concave wall with a variety of upstream turbulence-generating grids. Although flow-field measurements were not performed and questions arise concerning whether the longitudinal vortices developed from small perturbations or were directly related to the grid-generated turbulence, Kottke raised a number of important issues concerning the dependence of heat transfer enhancement on the spanwise wave length and initial velocity disturbance levels, with maximal enhancement coming from velocity disturbances initiated at the most amplified level according to the linear theory.

Several experiments now confirm the strong effectiveness of Gortler vortex-influence on heat transfer [15],[16]. In general, the spanwise-averaged heat transfer only became strongly enhanced when the vortex motion has become strongly nonlinear. It is thus not surprising, in retrospect, that a linear theory [13] produced zero-averaged heat transfer enhancement. More recent experiments on the effect of longitudinal vortices on heat transfer were performed by Crane and Sabzvari [16] in a water channel and Crane and Umur [17] in a wind tunnel. Their measurements were taken under active nonlinearly developing longitudinal rolls that developed from upstream Görtler vortices of small amplitudes. In this sense, their measurements and the computational results [9] share similar flow fields, although quantitative details about the flow field were not well documented in these experiments. Both theoretical and experimental results attribute the spanwise averaged heat transfer enhancement to the large spanwise regions of high heat-transfer rates surrounding the vortex downwash region. More refined experiments, with well documented flow fields directed towards the heat transfer problem (this important aspect was pointed out in [9]), are in progress at the Laboratoire de Thermocinétique, ISITEM in Nantes [18]-[21].

Recent theoretical [7],[9] as well as experimental results [16],[17] thus conclusively confirm that within parameter ranges for the initiation of most amplified Görtler vortices, local surface heat-transfer rates in the downstream nonlinear region can ttain values several times that of the undisturbed local flat-plate values on concave walls. This trend is also indicated by a weakly nonlinear analysis of the incipient onset of nonlinear Görtler vortices and their effect on heat transfer [22].

## DRAG RISE VS. HEAT TRANSFER ENHANCEMENT

Inserting vortex generators in the flow for heat transfer enhancement would give rise to considerable drag and pressure loss in heat transfer devices relative to the heat transfer enhancement achieved (e.g., [2]), whereas the relative heat transfer enhancements and drag rises are comparable for Görtler-type longitudinal vorticity system [9]. However, the heat transfer characteristics of the Görtler-type longitudinal vorticity system in open flows depend strongly on upstream initial conditions of the disturbance velocity field [9] and are thus difficult to control predictably. The understanding of response of the Görtler-type longitudinal velocity field to its upstream and environmental condition is essential to well-controlled heat and mass transfer enhancement. On the other hand, vortex generators [1]-[3] ensures the well-controlled presence of heat and mass enhancement vortices.

## UPSTREAM AND FREE STREAM DISTURBANCES

In turbomachinery flow environment, upstream and free stream disturbances are omni present [23]. Specific issues of heat transfer on the pressure side of turbine blades attributable to Görtler vortices have been discussed by Crane [10]. Estimates using actual conditions of the gas turbine operating environment indicate that the Görtler number would be in the range where such longitudinal vortices could be initiated in the amplified region [15]. But questions are raised as to the survivability of Görtler-type longitudinal vortices under conditions of high free stream turbulence levels encountered in the turbomachinery environment [10],[24]. On the other hand, according to Crane [10], measured heat transfer on the pressure side of cascade blade arrangements do not necessarily follow the results two-dimensional boundary layer computations without an explicit structure and role of longitudinal vortices.

Kestoras and Simon [25] performed experiments in a curved channel. They examined the transport effects of momentum and heat in the concave wall boundary layer and over a recovery, straight wall section. Of interest here is that, on the basis of their measurements, they conclude that in presence of relatively high free stream turbulence, the Görtler-type longitudinal vortex could not be sustained. It would be of considerable interest to study the effect of fine-grained turbulence within the boundary layer on the survivability of longitudinal vortices.

More recently, Wei and Miau [32],[33], through controlled upstream turbulence properties, experimentally related the turbulence length scales and intensity to that of stretched vorticity elements found near a two-dimensional stagnation region. They found by spatial correlations that the dimensions of the longitudinal vorticity elements near the stagnation region are well correlated with the integral scales of the upstream turbulence for a range of Reynolds numbers and free stream turbulence levels. The stagnation region stretched vortices are essentially aligned in a direction perpendicular to the stagnation line. The dimensions of the stretched vortices are smaller than the stagnation region dimension, which in turn, is larger than the local stagnation region boundary layer thickness. Wei and Miau [32],[33] showed that vorticity stretching was initiated far upstream and that the stretching mechanism is essentially an "inviscid" one. These recent quantitative findings are of an enormous impact in the connection between (1) the vortical region under which the boundary layer develops and (2) a characterizable upstream turbulence environment, identifiable with extractable properties of realistic upstream environments of turbomachines. Experimental measurements of controlled upstream vortical disturbances were reported earlier in Van Fossen and Semoneau [34] and recently by Van Fossen, et al. [35]. These studies point to the need for a systematic study of the mechanistic effect of external organized disturbances on the possible stimulation of longitudinal vortices within the boundary layer.

## THEORETICAL AND COMPUTATIONAL RESULTS

Heat transfer under an active, amplifying and spatially developing longitudinal vorticity system was studied by Liu and Lee [9], based on the spatial nonlinear momentum problem in Lee and Liu [8]. These authors considered the isolated concave wall problem, with the longitudinal vorticity system initiated on the upstream concave part of the wall according to experimental spanwise wave length and intensity (e.g., Swearingen and Blackwelder [36]). Only the streamwise velocity was measured experimentally, which is essentially curve-fitable by the mode shape of the linear theory (e.g., Floryan and Saric [4]). The linear theory gives, in addition, the spanwise and normal velocity shapes for their initial conditions. In this situation, the effects of upstream, stretched vortices as a result of free stream turbulence, as well as turbulence in the boundary layer, was not addressed. Inspite of this idealization, an important conclusion was that the initial disturbance velocity level was an important parameter in controlling the downstream surface heat transfer rate in such an active longitudinal vorticity system. The local heat transfer rate could attain a value as high as 400% of the local flat wall surface heat transfer rate without the longitudinal vorticity system.

Because of the sensitivity of surface heat transfer rates to the upstream, initiating velocity disturbance levels [9], it is crucial that a quantitative understanding be obtained in terms of realistic upstream disturbances and the effect of a blunt leading edge, such as in turbomachinery cascade. Shigemi, et al. [26] infact, found that Görtler-type longitudinal vortices were not confined to the concave section on the pressure side of their cascade airfoil, but that they in fact "crept upstream" towards the blunt leading edge. On the other hand, longitudinal vorticity stretching and amplification in the forward blunt stagnation region have been known for some time (e.g., [27]-[31]). It is thus conceivable that the longitudinal vortices on the downstream concave surface were stimulated by the slective leading edge amplification or stretching of upstream three-dimensional vorticity elements. The quantitative effects of upstream/free stream turbulence effects as well as the effect of turbulence in the boundary layer itself (Wei and Miau [32],[33]; Kestoras and Simon [25]) remain to be assessed from a mechanistic point of view in order to furnish quantitative as well as qualitative information for the management of heat transfer environment in presence of longitudinal vortices.

In recent measurements, Ajhak, et al. [21] observed that downstream of the wake-disturbance field of a row of thin cylinders placed perpendicularly to the leading edge stagnation line of their experimental model of a concave wall, disturbances with spanwise wavelength half that of the upstream disturbance were detected. In a kinematic calculation, the present author was able to show that this harmonic generation can be attributed to the nonlinear effects of "vortex" stretching and advection in the stagnation region. Insights from studies of upstream vorticity and entropy perturbations obtained through linearization about an otherwise basic potential flow about an airfoil [37] would be helpful in nonlinear dynamical studies of vortex stretching and selection effects in the interesting case when the leading edge appears "blunt" relative to the scale of the upstream disturbances [21],[32],[33].

## HEAT TRANSFER ENHANCEMENT MECHANISM

The disturbance velocity of the longitudinal vorticity element in the direction of the wall tends to bunch the local iso-temperature or iso-concentration lines towards the wall, giving rise to stronger normal-to-wall temperature or concentration gradients in regions of downwash [9]. The opposite is true in regions of upwelling. Net enhancement is achieved when local spanwise region of high heat transfer dominates, irrespective of how the longitudinal vorticity system was generated. This situation is brought about by the nonlinear development of the longitudinal vortices [8],[9], in which the "eye" of the longitudinal vorticity system would move closer together in the upwash region, thus concentrating the "weakened" streamwise velocity and temperature gradients into a smaller spanwise region. This migration of the "eye" leaves, at the same instance, a more extended spanwise downwash region with stronger streamwise and tempersture gradients.

Except for details, the spanwise bunching of weaker nornal-streamwise velocity gradients, hence of temperature gradients, and the spreading of stronger normal-velocity and normal-temperature gradients in the spanwise region can also be inferred kinematically from just the inviscid, but nonlinear, effects of the wall on upstream periodic streamwise vortices [38].The distinct spanwise focasing of the iso-streamwise velocity lines in the upwash region and the spread in the downwash region is clearly shown, which is caused by the migration of the "eye" of the system attributed to the mutual induction effect of the vorticity system [38]. However, the increasing closeness of isotherms towards the wall in the ever expanding downwash region as the flow develops nonlinearly downstream is clearly due to the intensification of the longitudinal vorticity flow field by the centrifugal instability mechanism.

## INJECTION AND SUCTION AT THE WALL

The issue relating to free stream turbulence effects on film cooling effectiveness was addressed in a series of papers by Bons, et al. [39], Schauer and Bons [40] and Bons, et al. [41]. They addressed primarily flow configurations with sharpe leading edges. General correlation of a wide range of experimental data for turbulent heat transfer is recently given by Maciejewski and Anderson [42]; concave wall geometrical effects were included in the correlation but the effect of blunt leading edge was not addressed. Basic research needs in gas turbine heat transfer control was recently reviewed and addressed by MacArthur and Rivir [43].

The severe thermal environment in which blades in modern gas turbine engines operate necessitates both internal and external cooling (e.g., Eckert [44], Subramanian, et al. [45]). In external cooling, the coolant is injected through holes or through porous matrix on the blade surface. Thus basic information about the interaction between spanwise distribution of coolant injection and bands of spanwise wave lengths of unstable longitudinal-vortex structures would be helpful in avoiding inadvertent stimulation of such vortices, which in turn, enhances heat transfer towards the wall [9]. Alternatively, an "anti-phase" spanwise injection or suction pattern relative to the longitudinal vortices may well lead to a decrease of heat advection towards the wall. Such limited injection or suction patterns directed at cancellation effects may be seen as a possibility towards reduction of mass flow needed for coolant injection. The use of suction in controlling the spawise wave length and structure of the longitudinal vortices was addressed experimentally by Myose and Blackwelder [46],[47] for which theoretical and computational studies would be helpful.

## FLOW VISUALIZATION AND HOT WIRE MEASUREMENTS

The amazing likeness between the down stream development of smoke lines in the normal and spanwise cross sectional plane [48],[49] and the hot wire measurements of the iso-streamwise velocity [36] do warrant some comments. Aihara, et al. [48] and Ito [49], among others, used spanwise oriented smoke lines to seed an air boundary layer on a concave wall. These were released upstream at different locations in the direction normal to the wall. The advected concentration patterns in cross sectional planes appeared as spanwise-sinusoidal undulations in the incipient linear stages of Görtler vortices. As the flow developed into the nonlinear stages, the concentration patterns resemmbled the shape of "mushrooms" in the cross sectional plane. The stem of the mushrooms correspond to the upwash region while the downwash region contributed to a bunching of iso-concentration lines towards the wall in between adjacent mushroom structures.

Detailed hot wire measurements within longitudinal vortices along the concave wall showed similar development of mushroom shapes in the iso-streamwise velocity lines prior to further breakdown. Computational results for both temporal [50],[51] and spatially developing boundary layers [8],[52] show that indeed the iso-streamwise velocity patterns in the cross sectional plane do develop according to the hot wire measurements and that the iso-therms, under similar boundary and upstream-initial conditions also develop into mushroomlike structures [9]. In these problems, the conditions for incompressible flow are met, that is, low Mach number flow and relative small temperature loading.

The scaling for the momentum problem [4]-[6],[53],[54] resulted in the streamwise momentum equation for the total streamwise velocity to be devoid of the disturbance-pressure gradient for thin boundary layers relative to the radius of the concave wall. If, in addition, the external flow has zero streamwise pressure gradient, which is the case for the temporal and spatial development problems as well as the experiments just discussed, then the streamwise momentum equation for the total velocity involves only the balance between nonlinear advection and viscous diffusion in the cross sectional plane.

Liu and Sabry [7] pointed out that this nonlinear advection-diffusion process is precisely in the same form as that of the heat transfer problem for an incompressible fluid except that the relative extent of temperature diffusion in the cross section planes would be scaled by $\sqrt{Pr}$, where Pr is the Prandtl. This similarity also extends to the overall mass fraction from the binary, reactionless species continuity equation in which the corresponding extent of diffusion would be scaled by the Schmidt number as $\sqrt{Sc}$.

The identical similarity between the overall streamwise velocity, temperature and species concentration is only possible for identical dimensionless upstream and boundary conditions in addition to Pr = Sc = 1. The upstream smoke injection is actually performed at one normal location for each experimental run [48],[49]; the superimposed composite picture nevetheless does mimick that of the uniform upstream streamwise velocity. For diffusion in air, Sc ≈ 1, the composite smoke lines in the cross sectional plane as they evolve downstream [48],[49] do indeed mimick, in extent and structure, the iso-streamwise velocity "mushroom structures"

[8],[50]-[52]. Computations for the scalar field for Pr = Sc ≈ 1 also confirm these ressemblences [9].

This similarity [7] reminds us of the Crocco-Busemann particular integrals in laminar boundary layer theory, however, the present smimilarity for the three-dimensional flow field is a limited one in that it relates the advected scalr field to only the streamwise velocity. The conclusions one can draw are nevertheless quite interesting for a gaseous media (Pr ≈ Sc ≈ 1 ): (1) the smoke visualization studies can now be related to the streamwise velocity and quantitative, nonintrusive diagnostics of at least the streamwise velocity component could be devised from optical measurements of the advected scalar field, and, (2) this analogy enables us to obtain from the Reynolds averaged skin friction (which is the spanwise average in this problem), the heat and mass transfer rates at the wall, leading immediately to a Reynolds analogy between skin friction coefficient the heat and mass transfer coefficients. The latter leads to the conclusion that under the present system of longitudinal vorticity elements, the relative heat transfer enhancement would be of the same order as the drag rise.

## SCALING PARAMETERS

The parameters for heat transfer naturally include those inherited from the momentum problem. Experimental data for the advective flow field is usually presented in the Görtler number vs dimensionless wave number diagram [6],[36]. One form of the Görtler number is written as

$$Gö_\theta = Re_\theta \sqrt{(\theta/R)},$$

where $Re_\theta$ is the free stream Reynolds number using the local momentum thickness $\theta$ as length scale; R is the radius of the concave surface. The wave number k is then made dimensionless by the local momentum thickness. For developing boundary layers, the downstream evolution of an experiment is more appropriately represented by rewriting the Görtler number in terms of the local dimensionless wave number as

$$Gö_\theta = \Lambda_k (k\theta)^{3/2},$$

where

$$\Lambda_k = Re_R (kR)^{-3/2}$$

is a wave number parameter, $Re_R$ is the surface-radius Reynolds number. In laboratory experiments [36], it appears that once the physical wave number k is fixed by upstream effects, it remains robust and persists downstream. In this case, the wave number parameter remains fixed for a surface with constant radius in a given experiment (Bemmalek and Saric [51] discuss the variable surface curvature problems). Thus experimental points for developing boundary layers follow positively-sloped lines in a log-log plot, with different lines for different values of the wave number parameter. However, it is the wave length parameter

$$\Lambda_\lambda = Re_R (\lambda/R)^{3/2}$$

that is commonly used to represent experiments; typically it takes on values of about $10^2$ to $10^3$ (see Swearingen and Blackwelder [36]). The maximum amplification line according to the local linear theory [4]-[6] is very nearly parallel to the line of about 210 in numerical value (see Crane [10], Floryan and Saric [4], Floryan [5], and Saric [6]).

The nonlinear development is actually an upstream-initial value problem, as all the developmental boundary layer experiments indicate [36] and, as such, the initial Görtler number

$$Gö_{\theta_o} = Re_{\theta_o} \sqrt{(\theta_o/R)},$$

becomes an important parameter of the problem, where the subscript zero denotes the initial momentum thickness at a distance $X_o$ from the leading edge where the disturbance is initiated. The wave length parameter and the initial form and amplitude of the disturbance then specifies the momentum problem. The influence of upstream and environmental disturbances, which was discussed earlier, would introduce further parameters.

The scalar transport problem in addition involves the Prandtl and Schmidt numbers and the initial forms and amplitudes of the temperature and concentration disturbances. For thin developing velocity and scalar boundary layers relative to the wall radius, the initial mean flow corresponds to that of the laminar boundary layer. Experimental measurements for a fixed wave length parameter begins at lower values of the Görtler and wave numbers in which the weak initiating disturbance is in the amplified region and "terminates" more or less at several initial boundary layer thicknesses downstream where a spanwise-modulated turbulent boundary layer develops (e.g., [36]).

The observed transition of mean skin friction in this problem from the local laminar value to that of turbulent flow [36] can be attributed to a large extent to the single initial longitudinal vortex mode and its modification of the mean flow in the nonlinear region [8],[50]-[52]. In this case, the use of a single mode for significant local heat transfer enhancement becomes practicable. The hastening of the transition process is strongly dependent on controlling the increase of initial velocity disturbance amplitude and by which process the hastening of heat transfer enhancement could be accomplished. As was found in [9], the scalar transport problem is indeed a passive one, with the influence on heat transfer dominated by the intensification of the advecting velocity field due to centrifugal instability.

## CONCLUDING REMARKS

Originally our studies of nonlinear development of longitudinal vorticity elements on concave walls were strongly motivated by heat transfer problems associated with turbomachinery components. During the course of performing our studies, we were most beneficially influenced by the body of work on vortices and heat transfer associated with Professor Martin Fiebig and his colleagues at the Institut für Thermo- und Fluiddynamik, Lehrstuhl für Wärme- und Stoffübertragung, Ruhr-Universität Bochum for which one of their aims was directed towards heat transfer enhancement in compact heat exchangers. This has also stimulated our thinking on ways in which longitudinal vorticity elements generated by body forces might also become useful in compact heat exchangers.

We conclude this brief review of our point of view by dedicating this paper to Professor Martin Fiebig on the occasion of his youthful Sixty Fifth Anniversary and to wish him well in his continued productive research which will no doubt continue to stimulate us all.

## ACKNOWLEDGEMENTS

I greatly enjoyed the discussions on vortices and heat transfer with Professsor Martin Fiebig on many occasions, especially during his sabbatical leave at Brown University in 1996. I have greatly benefited from discussions with Professors V. Vasanta Ram and N. K. Mitra as well as with the many research students associated with Wirbel- und Wärmeübertragung during my visits in Bochum. I am also indebted to Professors R. R. Mankbadi and A. S. Sabry of the Mechanical Power Engineering Department, Cairo University for continued collaboration and discussion. The partial support of NSF Grant INT-9602043 is acknowledged.

## REFERENCES

[1] FIEBIG, M.: "Vortex generators for compact heat exchangers", J. Enhanced Heat Transfer, 2 (1995) pp.43-61.

[2] FIEBIG, M.: "Embedded vortices in internal flow: heat transfer and pressure loss enhancement", Int. J. Heat and Mass Transfer, 16 (1995) pp. 376-388.

[3] FIEBIG, M.: "Vortices and Heat Transfer", Z. angew. Math. Mech., 76 (1996) pp.1-16.

[4]   FLORYAN, J. M., SARIC; W. S.: „Stability of Görtler vortices in boundary layers", AIAA J., 20 (1982) pp. 316-324.

[5]   FLORYAN, J. M., "On the Görtler instability of boundary layers", Prog.Aerospace Sci., 28 (1991) pp. 235-271.

[6]   SARIC, W. S.: "Görtler vortices", Annu. Rev. Fluid Mech., 26 (1994) pp. 379-409.

[7]   LIU, J. T. C., SABRY, A. S.: "Concentration and heat transfer in nonlinear Görtler vortex flow and the analogy with longitudinal momentum transfer", Proc.Royal Soc. London, A432 (1991) pp. 1-12.

[8]   LEE, K., LIU, J. T. C.: "On the growth of mushroom like structures in nonlinear spatially developing Görtler vortex flow", Phys. Fluids, A4 (1992) pp.95-103.

[9]   LIU, J. T. C., LEE, K.: "Heat transfer in a strongly nonlinear spatially developing longitudinal vorticity system", Phys. Fluids, 7 (1995) pp. 559-599.

[10]  CRANE, R. I.: "Boundary layers and transition on concave surfaces", in "Fundamental aspects of boundary layers and transition in turbomachines", Von Kármán Inst. Fluid Dynamics Lecture Series 1991-06 (1991).

[11]  JACOBY, A., SHAH, H.: "Heat transfer surface enhancement through the use of longitudinal vortices: A review of recent progress", EUROMECH 327, August 25-27, Kiev, (1994) pp. 6-7.

[12]  PEERHOSSAINI, H.: "Effect of curvature on transport properties of boundary layers: A review of recent progress and needs for future research", EUROMECH 327, August 25-27, Kiev, (1994) pp. 26-29.

[13]  McCORMACK, P. D., WELKER, H., KELLEHER, M.: "Taylor-Görtler vortices and their effect on heat transfer", ASME J. Heat Transfer, 92 (1970) pp.101-112.

[14]  KOTTKE, V.: "Taylor-Görtler vortices and their effect on heat and mass transfer", in"Proc. int. heat transfer conf.", 3, Hemisphere, Washington, DC.1986, pp.1139-1144.

[15]  MARTIN, B. W., BROWN, A.: "Factors influencing heat transfer to the pressure surfaces of gas turbine blades", Int. J. Heat and Fluid Flow, 1 (1979) pp. 107-114.

[16]  CRANE, R. I., SABZVARI, J.: "Heat transfer visualization and measurements in unstable concave-wall laminar boundary layers", ASME J.Turbomachinery, 111 (1989) pp. 51-56.

[17]  CRANE, R. I., UMUR, H.: "Concave-wall laminar heat transfer and Görtler vortex structure: Effect of pre-curvature boundary layer and favorable pressure gradients", ASME Paper No. 90-GT-94 (1990).

[18]  BAHRI, F., PEERHOSSAINI, H.: "Eigenfunctions of Görtler Vortices in the highly non-linear region: An experimental approach", in "Heat transfer enhancement by Lagrangian chaos and turbulence", H. PEERHOSSAINI, A. PROVENZALE, ed. Dec. 12-15, ISITEM, Nantes 1994, pp. 171-180.

[19]  BAHRI, F., KESTORAS, M., PEERHOSSAINI, H., AJAKH, A.: "An experimental investigation of the growth of Görtler vortices: the fluid mechanics study in view of the heat transfer study", EUROMECH 327, August 25-27, Kiev 1994, pp. 72-77.

[20]  PEERHOSSAINI, H., BAHRI, F.: "Experiments on mode decomposition in nonlinear Görtler instability," 9th Couette-Taylor Workshop, August 7-10, Boulder 1995.

[21]  AJAKH, A., KESTORAS, M., PEERHOSSAINI, H.: "Experiments on the Görtler instability: Its relation to transition to turbulence", ASME Fluids Engineering Division Annual Meeting, July 7-11 San Diego 1996.

[22]  SMITH, S. T., HAJ-HARIRI, H.: "Görtler vortices and heat transfer: A weakly nonlinear analysis",

Phys. Fluids, A5 (1993) pp. 2815-2825.

[23] MAYLE, R. E.: "Boundary layers in turbomachines" in "Fundamental aspects of boundary layers and transition in turbomachines", Von Kármán Inst. Fluid Dynamics Lecture Series 1991-06 (1991).

[24] BROWN, A., MARTIN, B. W.: "Flow transition phenomena and heat transfer over the pressure surfaces of gas turbine blades," ASME J Engineering for Power, 104 (1982) pp. 360-367.

[25] KESTORAS, M. D., SIMON, T. W.: "Turbulence measurements in a heated concave boundary layer under high free-stream turbulence", ASME J.Turbomachinery, 118 (1996) pp. 172-180.

[26] SHIGEMI, M., JOHNSON, M. W., GIBBINGS, J. C.:, "Boundary layer transition on a concave surface", Inst. Mech. Engrg., Paper C262/87 (1987) pp. 223-229.

[27] KESTIN, J., AND WOOD, R. T.: "Enhancement of stagnation line heat transfer by turbulence", Prog. Heat and Mass Transfer, 2 (1969) pp.249-253.

[28] HUNT, J. C. R.: "A theory of turbulent flow around two-dimensional bluff bodies", J. Fluid Mechs., 61 (1973) pp.625-706.

[29] BEARMAN, P. W., MOREL, T.: "Effects of free stream turbulence on the flow around bluff bodies", Prog. Aerospace Sci., 20 (1983) pp. 97-123.

[30] SADEH, W. Z., SUTERA, S. P., MAEDER, P. F.: "An investigation of vorticity amplification in stagnation flow", Z. angew. Math. Mech., 21 (1970) pp. 717-742.

[31] SADEH, W. Z., BRAUER, H. J.: "A visual investigation of turbulence in stagnation flow about a circular cylinder", J. Fluid Mechs., 99 (1980) pp. 53-64.

[32] WEI, C. Y., MIAU, J. J.: "Stretching of free stream turbulence in the stagnation region", AIAA J., 30 (1992) pp. 2196-2203.

[33] WEI, C. Y., MIAU, J. J.: "Characteristics of stretched vortical structure in two-dimensional stagnation flow", AIAA J., 31 (1993) pp. 2075-2082.

[34] VAN FOSSEN, G. J., SIMONEAU, R. J.: "A study of the relationship between free-stream turbulence and stagnation region heat transfer", ASME J. Heat Transfer, 109 (1987) pp. 10-15.

[35] VAN FOSSEN, G. J., SIMONEAU, R. J., CHING, C. Y.: "Influence of turbulence parameters, Reynolds number, and body shape on stagnation-region heat transfer", ASME J. Heat Transfer, 117 (1995) pp. 597-603.

[36] SWEARINGEN, J. D., BLACKWELDER, R. F.: "The Growth and breakdown of streamwise vortices in the presence of a wall", J. Fluid Mech., 182 (1987) pp. 255-290.

[37] GOLDSTEIN, M. E.: "Unsteady vortical and entropic distortions of potential flows round arbitrary obstacles", J. Fluid Mech. 89 (1978) pp. 433-444.

[38] GOLDSTEIN, M. E., LEIB, S. J.: "Three-dimensional boundary-layer instability and separation induced by small-amplitude streamwise vorticity in the upstream flow," J. Fluid Mech., 74 (1993) pp. 741-765.

[39] BONS, J. P., MACARTHUR, C. D., RIVIR, R. B.: "The effect of high free stream turbulence on film cooling effectiveness", Int. Gas Turbine and Aeroengine Congress and Exposition, The Hague, June 13-16. ASME Paper No. 94-GT-51 (1994).

[40] SCHAUER, J. J., BONS, J. P.: "Film cooling jet mixing with free stream turbulence", ICHMT Int. Symp. on Turbulence, Heat and Mass Transfer, Instituto Superior Técnico, August 9-12, 1994.

[41] BONS, J. P., RIVIR R. B., MACARTHUR C. D.: "The effect of unsteadiness on film cooling effectiveness", AIAA Paper No. 95-0306 (1995).

[42] MACIEJEWSKI, P. K., ANDERSON, A. M.: "Elements of a general correlation for turbulent heat transfer", ASME J. Heat Transfer, 118 (1996) pp. 287-293.

[43] MACARTHUR, C. D., RIVIR, R. B.: "Basic research needs in gas turbine heat transfer", 30th National Heat Transfer Conf., August 5-8, Portland (1995).

[44] ECKERT, E. R. G.: "Analysis of film cooling and full-coverage film cooling of gas turbine blades", ASME J. Engineering Gas Turbine and Power 106 (1984) pp. 206-213.

[45] SUBRAMANIAN, C. S., LIGRANI, P. M., GREEN, J. G., DONER,W., KAISUWAN, P.: "Development and structure of a film-cooling jet in a turbulent boundary layer with heat transfer", in "Rotating machinery transport phenomena" J. H. JIM and W.-J. YANG, eds. Hemisphere 1990, pp. 53-68.

[46] MYOSE, R. Y., BLACKWELDER, R. F.: "Controlling the spacing of streamwise vortices on concave walls", AIAA J., 29 (1991) pp. 1901-1905.

[47] MYOSE, R. Y., BLACKWELDER, R. F.: "Control of streamwise vortices using selective suction," AIAA J., 33 (1995) pp. 1076-1080.

[48] AIHARA, Y., TOMITA, Y., ITO, A.: "Generation, development and secondary instability of Görtler vortices", in "Laminar turbulent-transition" V. V. Kozlov, ed. Springer-Verlag, Berlin 1985, pp. 447-454.

[49] ITO, A.: "Breakdown structure of longitudinal vortices along a concave wall", J. Japan Soc. Aero. Space Sci., 33 (1985) pp. 166-173.

[50] SABRY, A. S., LIU, J. T. C.: "Longitudinal vorticity elements in boundary layers: nonlinear development from initial Görtler vortices as a prototype problem", J. Fluid Mech., 231 (1991) pp. 615-663.

[51] LIU, W., DOMARADZKI, A.: "Direct numerical simulation of transition to turbulence in Görtler flow", J. Fluid Mech., 240 (1993) pp. 267-309.

[52] BENMALEK, A., SARIC, W. S.: "Effect of curvature variations on the nonlinear evolution of Görtler vortices", Phys. Fluids, 6 (1994) pp. 3353-3367.

[53] HALL, P.: "The linear development of Görtler vortices in growing boundary layers", J. Fluid Mech., 130 (1983) pp. 41-58.

[54] HALL, P.: "The nonlinear development of Görtler vortices in growing boundary layers", J. Fluid Mech., 193 (1988) pp.243-266.

Providence, Rhode Island, USA
7 March 1997

# A NEW LOOK INTO THE TURBULENT MIXING IN VISCOUS FLUIDS: IMPLICATION TO HEAT AND MASS TRANSFER

B.A. Kolovandin, I.A. Vatutin

A.V. Luikov Heat and Mass Transfer Institute Academy of Sciences of Belarus

15 P. Brovka Street, Minsk 220072, Belarus

## SUMMARY

In the lecture, the fundamentals of turbulent mixing theory based on the supposition that main mechanism of turbulent mixing at high turbulence Reynolds numbers is attributed to persistent vortex structures are outlined. By virtue of the fact that an individual burgon, during its forming, is able to involve a scalar admixture contained in nonpremixed medium, the mixture fraction parameter is chosen as a *parameter of influence*. In this case, the thermodynamic parameters of the turbulence are the functionals depending on the mixture fraction parameter $z^{(\omega)}$, which is a random function of space and time. The techniques for the deducing of the differential equations governing the statistical moments conditioned by the parameter $z^{(\omega)}$ at a level $\eta$ is developed. One-dimensional differential equation governing the diffusion in $\eta$-space of the conditioned variance of $s_i = \left(\partial z^{(\omega)} / \partial x_i\right)\big|_{z^{(\omega)} = \eta}$ is derived, under particular restrictions imposed on the kinematics of burgons-mixers. The authors consider the model proposed as an approach to the general, "discrete-wave", theory of turbulence.

## INTRODUCTION

Over last decade, owing to the advances in direct numerical simulation (DNS) of turbulence and numerical visualization of the flow field in terms of vorticity, the vortical structures in the form of Burger's vortices (burgons) were discovered to be the principal elements of turbulence. These structures was shown to determine both the vorticity field of turbulence at rather large turbulence Reynolds numbers [1-5] and the field of a scalar contaminant transported by turbulence [6-9]. Direct numerical simulations with the computation grid of high resolution have revealed the global vorticity structures as filaments consisting of individual vortex tubes moving in the liquid of weaker vorticity as an undivisible material object. These elementary and global vortex structures playing a crucial role in the turbulent momentum transport allow the better understanding of intermitting turbulent velocity vector realizations, the decomposition of a realization in the mean component and the fluctuations, the physical meaning of the spectral analysis of velocity records, etc. Indeed, an individual vortex tube can be regarded as an liquid "particle" of the finite volume moving randomly during its lifetime with velocity vector perpendicular to its axis. When traveling before the collision with other tubes, an individual burgon generates velocity fluctuations in its wake. The interference of the wakes induced by an ensemble of vortex tubes results in the random velocity fluctuations ("noise") whose records reveal random samples including the regular events due to the velocities of individual vortex tubes.

One of important feature of the vortex structures discovered by DNS-technique is their ability to involve, during the "rolling-up" phase, the neighbouring fluid of weaker vorticity giving rise to the tube-like formations with vorticity distribution across the tubes. This inherent feature can, probably, to shed the light on the mechanism of turbulent mixing at large turbulence Reynolds numbers. Indeed, in the case when the environment includes the "spots" of a scalar contaminant, the entrainment of a spot into the core of individual rolling-up vortex tube has to proceed, producing the continuous profile of a scalar substance across the tube. These vortex tubes may be regarded as the elementary "vortex tube-mixers", mostly contributing to the production of the local spatial gradients of a scalar substance in turbulence at large turbulence Reynolds numbers.

Fundamental peculiarities of the beginning of vorticity forming in statistically homogeneous turbulent flow field, which follow from theoretical investigations [10, 11] and numerical studies [2, 7], are as follows. Pre-vortex structure occurs due to the strain localization in some spatial "blobs". These "blobs" are deformed during time becoming a near planar vortex sheets oriented in eigenvector plane $\vec{e}_2(\vec{x},\tau) - \vec{e}_3(\vec{x},\tau)$ normally to axis $\vec{e}_1(\vec{x},\tau)$.

A sheet-like vortex structure formed by local strain is rolled up due to shear instability into a vortex tube of finite length. Because of molecular diffusion of spiral layers inside the vortex tube and their opposite rotation, a solid core of strong vorticity is formed, with the vorticity vector $\vec{\omega}(\vec{x},\tau) = rot\,\vec{w}$ being directed most probably to the eigenvector $\vec{e}_2(\vec{x},\tau)$ corresponding to intermediate extension of the vortex sheet which is rolled up into a vortex tube. After the roll-up phase estimated by characteristic time $\tau_\omega \sim |\vec{\omega}|^{-1}$, the viscosity destroys the spiral structure inside the vortex tube producing continuous core with solid body rotation. Outside the core, the layer of strong viscosity influence (the "rate of dissipation layer") is formed, with instantaneous velocity profile being typical of cylindric Burger's vortex (computer graphics image of an individual vortex tube simulated numerically in isotropic turbulence is given in Fig. 1).

One of the remarkable peculiarities of dynamics of tube-like vortex structures is their merging into vortex filament moving as an indivisible material object, whose elements move with velocities directed normally to the tube axes (phase coherence). According to the data of Vincent and Meneguzzi [3], the lifetime of such a vortical filament is at least two times larger than that of individual vortex tube (time coherence). In this case, the locus of vorticity vectors of individual vortex tubes which form a vortex filament may be regarded as material vortical surface.

Isoscalar surfaces being formed near vortex tubes during their build-up are those which seem to make the main contribution to the *fine* turbulent mixing. Direct and indirect data available nowadays on isoscalar formations near vortex tubes [6-9, 12, 13] enables one to conceive the following picture of isoscalar "vortical" layers formed in the domain with preliminary (*discretely*) distributed passive scalar, e.g. an admixture being transported by the flow and not affecting the velocity field itself. During the vortex tube forming from strained sheets the "capture" of scalar "spots" by rolling-up vortex structures is probable. While strained sheet rolls up into the vortex tube, "a scalar ... is wrapped around the vortex

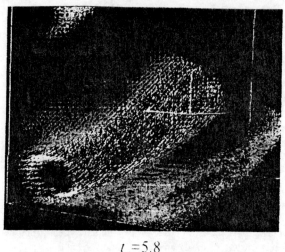

$t = 5.8$

Fig.1. The numerical visualization of the element of an individual vortex tube (Passot et al., 1995, under professor's Meneguzzi permission).

structure... with the scalar gradient pointing in the compressive strain direction" [7]. Kerr [6] have presumed that the "halo" he discovered near some of vortex tubes was a layer of a local peaks of a scalar gradient. Some tentative inferences following from DNS-data given by Kerr [6], Ashurst et al. [7], Ruetsch & Maxey [8, 9] can be summarized as follows:
- instantaneous profile of a scalar formed in radial cross section of an individual tube-mixer has a maximum gradient at tube periphery;
- vector of scalar gradient $\vec{g} = \nabla c$ is most probably directed normally to vorticity vector $\vec{\omega} = rot\,\vec{w}$, in the direction of contraction eigenvector $\vec{e}_1(\vec{x}, \tau)$.

An intuitive sketch presented in Fig. 2, reflects the following peculiarities discovered by Kerr [6] on the distribution of a scalar involved into a vortex tube-mixer: no scalar gradients exceeding a threshold value has been found in the plane with specified direction of maximum value of $|\vec{g}|$ after the plane rotation by 90°. This enables us to associate the vector of scalar gradient with the *surface* unlike the vorticity vector being associated with *material vortex line* (first the relation of the scalar gradient dynamics with the evolution of a surface was shown theoretically by Reid [10]).

A reasonable question arises on the relation of *material vortex surface* dynamics and that of *isoscalar surface* associated with vortex structures. One of possible scenarios of relative motion of the material vortex surface and associated isoscalar surface is based on Gibson's relationship [13] for relative velocity of a scalar surface, $w_\alpha^*(\vec{x},\tau) = w_\alpha^{(c)}(\vec{x},\tau) - w_\alpha(\vec{x},\tau) = -D\left(\partial^2 c \, \partial x_\alpha^2 / |\partial c/\partial x_\beta|\right) e_{\alpha\beta}$, where $e_{\alpha\beta}$ is cosine between some selected direction $\alpha$ and direction $\beta$ of maximum value of $|\vec{g}|$. If suppose the radial distribution of a scalar to be adequate to that presented in Fig. 2, then the velocity vectors of inner and outer (relative to the *"inflection point"* surface where $(\partial^2 c/\partial x_\alpha^2) = 0$) isoscalar

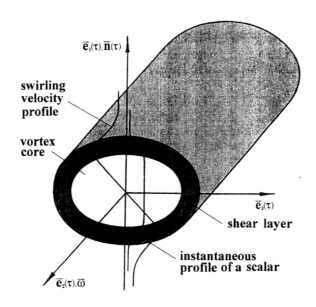

Fig.2  The sketch of vortex tube swirling velocity profile and the profile of a scalar involved in the tube.

surfaces are oppositely directed toward one to another, so that the thickness of the layer of maximal gradient of mixed scalar will decrease. In this case, the forming of some "double" scalar gradient layer with jump of scalar concentration in radial direction is possible. Such structural element of a scalar field can probably be realized, providing that the mechanism of scalar gradient layers stated above is adequate to the real process of turbulent mixing.

Since nowadays detailed numerical studies on elementary isoscalar "vortical" surface dynamics comparable by the grid resolution with that on vortex structure dynamics are not available, one may conclude that forming of the *scalar gradient layer* associated with burgon-mixer proceeds simultaneously with the forming of burgon itself. In this case, isoscalar "vortical" surface of a certain level is regarded as an ensemble of individual isoscalar "vortical" surfaces, each of them being generated by individual vortex tube. An element of such isoscalar "vortical" surface propagates normally to the tube axis in the direction of the contraction eigenvector $\vec{e}_1(\vec{x}, \tau)$ (vortex tube velocity vector normal to its axis does not coincide in general with elementary isoscalar surface velocity vector). In this connection, the characteristic size of *scalar gradient layer* is determined by the statistics of vector $g_i = \partial c / \partial x_i$, and the distance between a pair of elementary layers related to the same burgon-mixer is likely to be proportional to Corrsin-Oboukhov microscale $\lambda_c^2 = 6 \langle c^2 \rangle / \langle g_i^2 \rangle$, where $c$ is the pulsation of mixed scalar.

While approaching turbulence description based on the concept of vortical surfaces formed in turbulent flow at $R_l \gg 1$, it is necessary to identify statistical ensemble of events related to vortex tubes regarded as main structural elements of turbulent vorticity field. As to the parameter identifying the events related to vortex tubes, it should follow from the method of the "marking" of vortex tubes employed.

Although the vortex tubes and associated "vortex" scalar gradient structures can be viewed as rare events occupying rather small fraction of the total domain where turbulence is considered, it has been argued on the basis of DNS-approach, that these structures contribute significantly to the total vorticity ( as well as to the rate of strain and the rate of dissipation) and the variance of scalar spatial gradients playing the crucial role in the processes of turbulent mixing. Therefore, the models of the turbulent mixing based on the concept of turbulence which ignore the vortex structures are expected to be replaced by the corresponding models taking into account the contribution of the vortex structures and associated scalar-gradient formations governing the processes of momentum and heat / mass transfer in developed turbulence at $R_l >> 1$.

In this paper, an attempt to develop a model of turbulent mixing which is based conceptually on the forming of the vortex tubes and associated scalar-gradient structures in the developed turbulence is undertaken.

In the particular, we propose the formulation of dynamic correlational model of turbulence *influenced* by an inherent property of the vortex tube-like structure - its ability to mix a scalar contaminant transported in turbulent flow at high turbulence Reynolds numbers.

## MIXTURE FRACTION PARAMETER AS VORTICAL SURFACE "MARKER"

The degree of two substances mixing is known [14] to determine by mixture fraction parameter

$$z(\vec{x},t) = (c_p(\vec{x},\tau) - c_l(\vec{x},\tau) + c_l^0)/(c_p^0 + c_l^0), \tag{1}$$

where $c_p(\vec{x},\tau)$ and $c_l(\vec{x},\tau)$ are concentrations of two substances being mixed in the unit mixture volume; $c_p^0$ and $c_l^0$ are the values of parameters related to "fresh" substances, i.e. prior to their mixing. In turbulent medium values $z(\vec{x},\tau)$, $c_p(\vec{x},\tau)$ and $c_l(\vec{x},\tau)$ are random functions of space and time; a record of $z(\vec{x},\tau)$ function sample is a continuous solid signal ranged in the band $0 \le z(\vec{x},\tau) \le 1$.

As known, the source terms in the equations for reagent concentration balance each other in the case of equilibrium chemical reaction with two reagents, the parameter $z(\vec{x},\tau)$ being a conservative scalar. If a process of pure mechanical mixing of two scalars with no chemical reaction is considered, the difference $c_p(\vec{x},t) - c_l(\vec{x},t)$ depends on the way of considered fields generation (for instance, due to external mean spatial gradients of these fields, energy forcing in a certain range of wave numbers, etc.). Assuming that fine turbulent mixing in viscous fluids or gases at $R_l >> 1$ proceeds in vortex tube-mixers, parameter $z(\vec{x},\tau)$ will define the ensemble of events referred only to vortex tubes, acquiring the form

$$z^{(\omega)}(\vec{x},\tau) = (c_p^{(\omega)}(\vec{x},\tau) - c_l^{(\omega)}(\vec{x},\tau) + c_l^0)/(c_p^0 + c_l^0) \tag{2}$$

where $c_p^{(\omega)}(\vec{x},\tau)$ and $c_l^{(\omega)}(\vec{x},\tau)$ are the concentrations of a scalar and high-vorticed fluid per unite of the volume of the medium; $c_p^0$ and $c_l^0$ are initial values of the concentrations of "fresh" scalar admixture and low-vorticed fluid, respectively. It is assumed that the burgons are absent at $t = t_0$. It is supposed also that the events related to vortex tubes are statistically correlated, i.e. space-time realizations of $c_p^{(\omega)}(\vec{x},t)$ and $c_l^{(\omega)}(\vec{x},t)$ may be described by continuous transfer equation

$$\frac{\partial c^{(\omega)}(\vec{x},\tau)}{\partial \tau} + w_k^{(\omega)}(\vec{x},\tau)\frac{\partial c^{(\omega)}(\vec{x},\tau)}{\partial x_k} = \frac{\partial q_k^{(\omega)}(\vec{x},\tau)}{\partial x_k}, \qquad (3)$$

where $q_k^{(\omega)}$ is the flux of the scalar related to the burgons in unite volume. It is supposed that the constitution equation for $q_k^{(\omega)}$ is of the form of Fourier law, i.e.

$$q_k^{(\omega)} = D_c(\vec{x},\tau)\frac{\partial c^{(\omega)}(\vec{x},\tau)}{\partial x_k},$$

where $D_c(\vec{x},\tau)$ is the coefficient of mechanical diffusion of scalar $c^{(\omega)}(\vec{x},\tau)$ in turbulent ambient generated by burgon's wakes, i.e.

$$D_c(\vec{x},\tau) \sim l_c \langle u_i^2 \rangle^{1/2}.$$

Here, $l_c(\vec{x},\tau)$ is statistically mean "free path" of scalar structures related to burgons and $\langle u_i^2 \rangle$ is doubled kinetic energy of ambient turbulence. The characteristic length $l_c$ is assumed to be proportional to Taylor's macroscale of a scalar field, i.e.

$$l_c(\vec{x},\tau) \sim L_c = 6\langle c^2 \rangle \langle u_i^2 \rangle^{1/2} / \varepsilon_c,$$

where $\langle c^2 \rangle$ is the variance of scalar fluctuations in ambient medium and $\varepsilon_c = \kappa\langle(\partial c/\partial x_k)^2\rangle$ is the rate of their "smearing" by molecular diffusion. The adopted model of mechanical diffusion of burgons into ambient low-vorticed (i.e. unstructured) fluid is based on the supposition that velocity field in a viscous fluid at large turbulence Reynolds numbers is due to both random migration of vortical structures [3] and interfering wakes behind the structures producing turbulent "noise" of weak vorticity (to our mind, the mechanism of the "noise" generation by burgon's movements is essentially similar to that of turbulence generation in the wake of moving solid body with the only considerable difference that in proposed model of turbulent medium the moving object is an *elementary vortical structure*). It is worthy noting that since the proper frequency of burgon's Karman street is individual for each of burgons, then the interference of the wakes produced by the collection of burgons can be considered as a physical motivation for the harmonic analysis of turbulent samples.

Velocity field of turbulent medium including the burgons is described by momentum differential equation of the form

$$\frac{\partial w_i^{(\omega)}}{\partial \tau} + w_k^{(\omega)}\frac{\partial w_i^{(\omega)}}{\partial x_k} + \frac{1}{\rho}\frac{\partial p^{(\omega)}}{\partial x_i} = \frac{\partial q_{ik}^{(\omega)}(\vec{x},\tau)}{\partial x_k} \qquad (4)$$

and continuity equation,

$$\frac{\partial w_i^{(\omega)}}{\partial x_i} = 0. \qquad (5)$$

where $q_{ik}^{(\omega)}$ is momentum flux related to burgons per unite volume. The constitution equation for $q_{ik}^{(\omega)}$ is assumed to be of the form of Newton's law, i.e.

$$q_{ik}^{(\omega)} = D_w(\vec{x},\tau)\frac{\partial w_i^{(\omega)}(\vec{x},\tau)}{\partial x_k},$$

where $D_w(\vec{x},\tau)$ is the coefficient of mechanical diffusion of burgons in turbulent ambient of law vorticity, i.e.

$$D_w(\vec{x},\tau) \sim l_w \langle u_i^2 \rangle^{1/2}.$$

Here, $l_w(\vec{x},\tau)$ is statistically mean "free path" of burgons which to be proportional to Taylor's macroscale of velocity field, i.e.

$$l_w(\vec{x},\tau) \sim L_u = 5\langle u_i^2 \rangle^{3/2} / \varepsilon_w,$$

where $\varepsilon_w = \nu \langle (\partial u_i / \partial x_k)^2 \rangle$ is the rate of the kinetic energy dissipation.

Emphasize that the forming of the field of scalar pulsation gradients by burgon-mixers is purely inviscid process proceeding at characteristic time interval $\Delta \tau \sim \tau_\omega$. During this time interval, *full mixing* of an admixture involved into the core of a vortex tube goes on, as well as the forming of the "double layer" of scalar gradient in the periphery of a burgon-mixer. The task of the theory proposed is the description of the evolution in space and time of a scalar substance in turbulence *influenced* by vortex structures.

## BASIC EQUATIONS

In the present work the parameter of scalar micromixing by burgons, $z^{(\omega)}(\vec{x},\tau)$, is chosen as the *parameter of influence*, which by its definition (2), is a space-time random function varying in the range

$$0 \le z^{(\omega)}(\vec{x},\tau) \le 1 \qquad (6)$$

and being governed by differential equation

$$\frac{\partial}{\partial \tau} z^{(\omega)}(\vec{x},\tau) + w_k^{(\omega)}(\vec{x},\tau) g_k^{(\omega)}(\vec{x},\tau) = \frac{1}{c_p^0 + c_l^0} \frac{\partial}{\partial x_k} (q_k^{p(\omega)}(\vec{x},\tau) - q_k^{l(\omega)}(\vec{x},\tau)), \qquad (7)$$

where $g_k^{(\omega)} = \partial z^{(\omega)} / \partial x_k$ with burgon-induced velocity field $w_i^{(\omega)}$ and burgon-mixed passive admixture concentrations being described by differential equations (3)-(5).

All hydro-thermodynamic parameters influenced by vortex structures may be represented in the form of functionals of random argument $z^{(\omega)}(\vec{x},\tau)$ as well as space $\vec{x}$ and time $\tau$, i.e.

$$w_i^{(\omega)}(\vec{x},\tau) = u_i \left[ z^{(\omega)}(\vec{x},\tau), \vec{x}, \tau \right], \qquad g_k^{(\omega)}(\vec{x},\tau) = s_k \left[ z^{(\omega)}(\vec{x},\tau), \vec{x}, \tau \right].$$

It is obvious that the differential operators in $\vec{x}$-space (or time) of thermodynamic functionals considered may be presented as

$$\left. \begin{array}{l} \dfrac{\partial}{\partial x_i} \varphi^{(\omega)} = \left( \dfrac{\partial}{\partial x_i} + s_i \dfrac{\partial}{\partial z^{(\omega)}} \right) \psi \left[ z^{(\omega)}(\vec{x},\tau), \vec{x}, \tau \right], \\[2mm] \dfrac{\partial^2}{\partial x_k^2} \varphi^{(\omega)} = \left( \dfrac{\partial^2}{\partial x_k^2} + s_k^2 \dfrac{\partial^2}{\partial (z^{(\omega)})^2} + 2 s_k \dfrac{\partial^2}{\partial x_k \partial z^{(\omega)}} + \dfrac{\partial s_k}{\partial x_k} \dfrac{\partial}{\partial z^{(\omega)}} \right) \psi \left[ z^{(\omega)}(\vec{x},\tau), \vec{x}, \tau \right]. \end{array} \right\} \qquad (8)$$

As far as the statistical description of scalar substance micromixing with structural elements of high-vorticed liquid is the subject of the present study, we shall ignore all factors which are able to "camouflage" the forming of scalar field microstructures. In particular, we shall not differ the coefficients of mechanical diffusion of velocity and scalar fields generated by burgons. Besides, the coefficients are supposed to be constant in space and time. In this case, differential equation (7) takes on the form of passive scalar transfer equation, i.e.

$$\frac{\partial z^{(\omega)}(\vec{x},\tau)}{\partial \tau} + u_k g_k^{(\omega)} = D\frac{\partial}{\partial x_k} g_k^{(\omega)}. \tag{9}$$

Note that the forming of spatial fine structure of a scalar, i.e. field of its spatial gradients, $g_i(\vec{x},\tau) = \partial c_p(\vec{x},\tau)/\partial x_i$, playing qualitatively the same role as the strain tensor $s_{ij}$ in the forming of fine velocity field structure, is a principal feature of the mixing of any transportable scalar by vortex tubes at $R_l \gg 1$.

Following from (9), differential equation for vector $g_k^{(\omega)}(\vec{x},\tau) = \partial z^{(\omega)}(\vec{x},\tau)/\partial x_k = s_k\left[z^{(\omega)}(\vec{x},\tau),\vec{x},\tau\right]$ may be presented, regarding relations (8), as

$$\frac{\partial s_i}{\partial \tau} + u_k \frac{\partial s_i}{\partial x_k} + s_k \frac{\partial u_k}{\partial x_i} + s_i s_k \frac{\partial u_k}{\partial z^{(\omega)}} = D\left(\frac{\partial^2 s_i}{\partial x_k^2} + s_k^2 \frac{\partial^2 s_i}{\partial (z^{(\omega)})^2} + 2 s_k \frac{\partial^2 s_i}{\partial x_k \partial z^{(\omega)}}\right). \tag{10}$$

Velocity functional $u_i\left[z^{(\omega)}(\vec{x},\tau),\vec{x},\tau\right]$ is described by the equation

$$\frac{\partial u_i}{\partial \tau} + u_k \frac{\partial u_i}{\partial x_k} + \frac{1}{\rho}\left(\frac{\partial \pi}{\partial x_i} + s_i \frac{\partial \pi}{\partial z^{(\omega)}}\right) = D\left(\frac{\partial^2 u_i}{\partial x_k^2} + s_k^2 \frac{\partial^2 u_i}{\partial (z^{(\omega)})^2} + 2 s_k \frac{\partial^2 u_i}{\partial x_k \partial z^{(\omega)}}\right). \tag{11}$$

To develop a statistical model, the form of equations (10) and (11) for the functionals $u_i$ and $s_i$ is unacceptable, since the equations contain derivatives with respect to the random argument $z^{(\omega)}(\vec{x},\tau)$. However, if we consider only the events associated with a certain level of mixture fraction parameter $z^{(\omega)}(\vec{x},\tau)$ arbitrarily taken from range (6), then corresponding functionals turn to the functions of higher dimension, i.e.

$$\left.\begin{array}{l}u_i\left[z^{(\omega)}(\vec{x},\tau),\vec{x},\tau\right]\Big|_{z^{(\omega)}=\eta} = U_i(\eta,\vec{x},\tau), \\ s_i\left[z^{(\omega)}(\vec{x},\tau),\vec{x},\tau\right]\Big|_{z^{(\omega)}=\eta} = g_i^{(\omega)}(\vec{x},\tau)\Big|_{z^{(\omega)}=\eta} = \frac{\partial z^{(\omega)}(\vec{x},\tau)}{\partial x_i}\Big|_{z^{(\omega)}=\eta} = S_i(\eta,\vec{x},\tau),\end{array}\right\} \tag{12}$$

and derivatives of functionals with respect to random argument $z^{(\omega)}(\vec{x},\tau)$ become derivatives with respect to variable $\eta$ of functions $U_i(\eta,\vec{x},\tau)$ and $S_i(\eta,\vec{x},\tau)$, i.e.

$$\left.\begin{array}{l}\dfrac{\partial u_i\left[z^{(\omega)}(\vec{x},\tau),\vec{x},\tau\right]}{\partial z(\vec{x},\tau)}\bigg|_{z^{(\omega)}=\eta} = \dfrac{\partial U_i(\eta,\vec{x},\tau)}{\partial \eta}, \\ \dfrac{\partial s_i\left[z^{(\omega)}(\vec{x},\tau),\vec{x},\tau\right]}{\partial z(\vec{x},\tau)}\bigg|_{z^{(\omega)}=\eta} = \dfrac{\partial S_i(\eta,\vec{x},\tau)}{\partial \eta}.\end{array}\right\} \tag{13}$$

The meaning of transformations (12) and (13) consists in *geometrodynamical* description of random fields: variable $\eta$ determines in 3-D space $\vec{x}$ the family of selected random surfaces. Any function $\Phi(\eta,\vec{x},\tau)$ defined on surface is random one too. In other words, the relations (12), (13) mean the replacement of the turbulence parameters description *in* 3-D space by that *on* the surfaces defined by progressive variable $\eta$ and appeared randomly 3-D space at given moment of time. Space-time realizations of the events defined by above way are governed by the differential equations derived from equations (10) and (11) with the use of relations (12) and (13). These conditioned equations have the following form:

$$\frac{\partial U_i(\eta,\vec{x},\tau)}{\partial \tau} + U_k(\eta,\vec{x},\tau)U_k^i(\eta,\vec{x},\tau) + \frac{1}{\rho}\left[\frac{\partial P(\eta,\vec{x},\tau)}{\partial x_i} + \frac{\partial P(\eta,\vec{x},\tau)}{\partial \eta}S_i(\eta,\vec{x},\tau)\right] =$$
$$D\left[\frac{\partial^2}{\partial x_k^2}U_i(\eta,\vec{x},\tau) + S_k^2(\eta,\vec{x},\tau)\frac{\partial^2 U_i(\eta,\vec{x},\tau)}{\partial \eta^2} + 2S_k(\eta,\vec{x},\tau)\frac{\partial}{\partial \eta}U_k^i(\eta,\vec{x},\tau)\right], \quad (14)$$

$$U_i^i(\eta,\vec{x},\tau) + \frac{\partial U_i(\eta,\vec{x},\tau)}{\partial \eta}S_i(\eta,\vec{x},\tau) = 0, \quad (15)$$

$$\frac{\partial S_i(\eta,\vec{x},\tau)}{\partial \tau} + U_k(\eta,\vec{x},\tau)S_k^i(\eta,\vec{x},\tau) +$$
$$S_k(\eta,\vec{x},\tau)U_i^k(\eta,\vec{x},\tau) + S_i(\eta,\vec{x},\tau)S_k(\eta,\vec{x},\tau)\frac{\partial}{\partial \eta}U_k(\eta,\vec{x},\tau) =$$
$$D\left[S_{kk}^i(\eta,\vec{x},\tau) + S_k^2(\eta,\vec{x},\tau)\frac{\partial^2 S_i(\eta,\vec{x},\tau)}{\partial \eta^2} + 2S_k(\eta,\vec{x},\tau)\frac{\partial}{\partial \eta}S_k^i(\eta,\vec{x},\tau)\right], \quad (16)$$

where

$$S_k^i(\eta,\vec{x},\tau) = \frac{\partial s_i\left[z^{(\omega)}(\vec{x},\tau),\vec{x},\tau\right]}{\partial x_k}\bigg|_{z^{(\omega)}=\eta} \quad ; \quad U_i^k(\eta,\vec{x},\tau) = \frac{\partial u_k\left[z^{(\omega)}(\vec{x},\tau),\vec{x},\tau\right]}{\partial x_i}\bigg|_{z^{(\omega)}=\eta}.$$

Differential equation (16) describes the dynamics of spatial gradients of mixture fraction parameter on isosurface $z^{(\omega)}(\vec{x},\tau)$ of level $\eta$ taken from the range (6). Velocity field related to these surfaces is governed by equations (14) and (15).

## SECOND-ORDER MODEL OF MIXTURE FRACTION GRADIENT $S_i(\eta,\vec{x},\tau)$ DEFINED ON ISOSCALAR "VORTEX" SURFACE $z^{(\omega)}(\vec{x},\tau) = \eta$

Consider statistics of the gradients of conditioned mixture fraction parameter having presented random function $S_i(\eta,\vec{x},\tau)$ as well as velocity vector $U_i(\eta,\vec{x},\tau)$ as the sum of the ensemble averaged values and the fluctuations, i.e.

$$S_i(\eta,\vec{x},\tau) = \langle S_i \rangle(\eta,\vec{x},\tau) + s_i(\eta,\vec{x},\tau), \quad U_i(\eta,\vec{x},\tau) = \langle U_i \rangle(\eta,\vec{x},\tau) + u_i(\eta,\vec{x},\tau),$$
$$S_i^k(\eta,\vec{x},\tau) = \langle S_i^k \rangle(\eta,\vec{x},\tau) + s_i^k(\eta,\vec{x},\tau), \quad U_i^k(\eta,\vec{x},\tau) = \langle U_i^k \rangle(\eta,\vec{x},\tau) + u_i^k(\eta,\vec{x},\tau).$$

Then, differential equation for mathematical expectation of $S_i(\eta,\vec{x},\tau)$ following from equation (16) may be presented as

$$\frac{\partial \langle S_i \rangle}{\partial \tau} + \langle U_k \rangle \langle S_k^i \rangle + \langle u_k s_k^i \rangle + \langle S_k \rangle \langle U_i^k \rangle + \langle s_k u_i^k \rangle + (\langle S_i \rangle \langle S_k \rangle + \langle s_i s_k \rangle) \frac{\partial}{\partial \eta} \langle U_k \rangle +$$

$$\langle S_k \rangle \left\langle s_i \frac{\partial u_k}{\partial \eta} \right\rangle + \langle S_i \rangle \left\langle s_k \frac{\partial u_k}{\partial \eta} \right\rangle + \left\langle s_i s_k \frac{\partial u_k}{\partial \eta} \right\rangle = D \left[ \langle S_{kk}^i \rangle + (\langle S_k \rangle^2 + \langle s_k^2 \rangle) \frac{\partial^2}{\partial \eta^2} \langle S_i \rangle + \right.$$

$$\left. \left\langle s_k^2 \frac{\partial^2}{\partial \eta^2} s_i \right\rangle + 2 \left( \langle S_k \rangle \left\langle s_k \frac{\partial^2}{\partial \eta^2} s_i \right\rangle + \langle S_k \rangle \frac{\partial}{\partial \eta} \langle S_k^i \rangle + \left\langle s_k \frac{\partial}{\partial \eta} s_k^i \right\rangle \right) \right], \quad (17)$$

where

$$\langle S_{kk}^i \rangle = \left\langle \frac{\partial^2}{\partial x_k^2} S_i \Big|_{z^{(\omega)} = \eta} \right\rangle.$$

Using equation (17), one can derive differential equation for the variance of vector $s_i$ in the form of

$$\frac{\partial}{\partial \tau} \langle s_i^2 \rangle + 2 \langle U_k \rangle \langle s_i s_k^i \rangle + 2 \langle S_k \rangle \langle s_i u_k^k \rangle + 2 \langle u_k s_i \rangle \langle S_k^i \rangle + 2 \langle s_i s_k \rangle \langle U_k^i \rangle +$$

$$2 \langle u_k s_i s_k^i \rangle + 2 \langle s_i s_k u_i^k \rangle + 2 \left( \langle S_k \rangle \langle s_i^2 \rangle + \langle S_i \rangle \langle s_i s_k \rangle + \langle s_i^2 s_k \rangle \right) \frac{\partial \langle U_k \rangle}{\partial \eta} +$$

$$2 \left( \langle S_i \rangle \langle S_k \rangle \left\langle s_i \frac{\partial u_k}{\partial \eta} \right\rangle + \langle S_i \rangle \left\langle s_i s_k \frac{\partial u_k}{\partial \eta} \right\rangle + \langle S_k \rangle \left\langle s_i^2 \frac{\partial u_k}{\partial \eta} \right\rangle \right) + 2 \left\langle s_i^2 s_k \frac{\partial u_k}{\partial \eta} \right\rangle =$$

$$D \left\{ 2 \langle s_i s_{kk}^i \rangle + 2 (\langle s_i s_k \rangle \langle S_k \rangle + \langle s_i s_k^2 \rangle) \frac{\partial^2}{\partial \eta^2} \langle S_i \rangle + \right.$$

$$\langle S_k \rangle^2 \left[ \frac{\partial^2}{\partial \eta^2} \langle s_i^2 \rangle - 2 \left\langle \left( \frac{\partial s_i}{\partial \eta} \right)^2 \right\rangle \right] + 2 \langle S_k \rangle \left[ \left\langle s_k \frac{\partial^2}{\partial \eta^2} s_i^2 \right\rangle - 2 \left\langle s_k \left( \frac{\partial s_i}{\partial \eta} \right)^2 \right\rangle \right] +$$

$$\left. 2 \left\langle s_k^2 s_i \frac{\partial^2}{\partial \eta^2} s_i \right\rangle + 4 \left[ \langle s_i s_k \rangle \frac{\partial}{\partial \eta} \langle S_k^i \rangle + \langle S_k \rangle \left\langle s_i \frac{\partial}{\partial \eta} s_k^i \right\rangle + \left\langle s_i s_k \frac{\partial}{\partial \eta} s_k^i \right\rangle \right] \right\}, \quad (18)$$

where

$$s_{kk}^i = S_{kk}^i - \langle S_{kk}^i \rangle.$$

Incompressibility equation (15) gives the equations for mean value, $\langle S_i \rangle$, and for conditioned pulsation, $s_i$, as follows

$$\langle U_k^k \rangle + \frac{\partial \langle U_k \rangle}{\partial \eta} \langle S_k \rangle + \left\langle \frac{\partial u_k}{\partial \eta} s_k \right\rangle = 0, \quad (19)$$

$$u_k^k + \frac{\partial u_k}{\partial \eta} s_k - \left\langle \frac{\partial u_k}{\partial \eta} s_k \right\rangle = 0. \quad (20)$$

It follows from equation (17) and (18) that differential description of the statistics of conditioned mixture fraction parameter gradient $s_i(\eta, \vec{x}, \tau)$ assumes the knowledge of the statistics of conditioned velocity field. Supposing that corresponding velocity field model is available (second-order model of conditioned turbulence will be published elsewhere), refer further to the statistics of the vector $s_i(\eta, \vec{x}, \tau)$.

Since the concept of turbulence influenced by vortex structures (burgons) refers to the quality inherent to all turbulent flows without exception, from spatially homogeneous and isotropic turbulence to nonhomogeneous one, it is worth considering basic model which would constitute the basis of conditioned turbulence correlation model of any complexity.

## The Statistics of Conditioned Gradients of the Mixture Fraction Parameter on the Surface $z^{(\omega)}(\bar{x},\tau) = \eta$ in Isotropic Turbulence

In isotropic turbulence, equations of incompressibility (19) and (20) reduce to the equation

$$u_k^k + \frac{\partial u_k}{\partial \eta} s_k = 0,$$

allowing to represent equation (18) for $\langle s_i^2 \rangle$ in the form of

$$\frac{\partial}{\partial \tau}\langle s_i^2 \rangle + 2\langle u_k s_i s_k^i \rangle + 2\langle s_i s_k u_i^k \rangle + 2\langle s_i^2 s_k \frac{\partial u_k}{\partial \eta} \rangle =$$

$$D\left[2\langle s_i s_{kk}^i \rangle + 2\langle s_k^2 s_i \frac{\partial^2}{\partial \eta^2} s_i \rangle + 4\langle s_i s_k \frac{\partial}{\partial \eta} s_k^i \rangle\right]. \quad (21)$$

As follows from the DNS data [9], the most probable direction of $\bar{s}(\eta,\bar{x},\tau)$ - vector coincides with that of contraction eigenvector $\bar{e}_1(\bar{x},\tau)$ normal to the surface $z^{(\omega)} = \eta$. Consequently, in coordinate system $\bar{x}$ attached to the surface, vector $\bar{s}(\eta,\bar{x},\tau)$ has one component only, i.e. $\bar{s}(\eta,\bar{x},\tau) = [s_n(\eta,\bar{x},\tau),0,0]$.

Then, the equation (21) can be given in the form of

$$\frac{\partial}{\partial \tau}\langle s_n^2 \rangle + 2\langle u_k s_n s_k^n \rangle + 2\langle s_n^2 u_n^n \rangle + 2\langle s_n^3 \frac{\partial u_n}{\partial \eta} \rangle =$$

$$D\left[2\langle s_n s_{kk}^n \rangle + 2\langle s_n^3 \frac{\partial^2}{\partial \eta^2} s_n \rangle + 4\langle s_n^2 \frac{\partial}{\partial \eta} s_n^n \rangle\right].$$

Using incompressibility equation given above, one can represent conditioned normal gradient of velocity vector in the form of

$$u_n^n(\eta,\bar{x},\tau) = \frac{\partial u_n\left[z^{(\omega)}(\bar{x},\tau),\bar{x},\tau\right]}{\partial x_n}\bigg|_{z^{(\omega)}=\eta} = u_k^k(\eta,\bar{x},\tau) - \left(\frac{\partial u_{\tau 1}}{\partial x_{\tau 1}} + \frac{\partial u_{\tau 2}}{\partial x_{\tau 2}}\right)\bigg|_{z^{(\omega)}=\eta} =$$

$$-\left[\frac{\partial u_n(\eta,\bar{x},\tau)}{\partial \eta} s_n(\eta,\bar{x},\tau) + 2\frac{\partial u_\tau}{\partial x_\tau}\bigg|_{z^{(\omega)}=\eta}\right].$$

Taking into account above relationships, one can give the equation for variance $\langle s_n^2 \rangle$ as follows

$$\frac{\partial}{\partial \tau}\langle s_n^2 \rangle + \left\langle u_k \frac{\partial}{\partial x_k} s_n^2 \bigg|_{z^{(\omega)}=\eta} \right\rangle - 4\left\langle s_n^2 \frac{\partial u_\tau}{\partial x_\tau}\bigg|_{z^{(\omega)}=\eta} \right\rangle =$$

$$D\left[2\left\langle s_n^2(\eta,\bar{x},\tau) \frac{\partial^2}{\partial x_k^2} s_n\left[z^{(\omega)}(\bar{x},\tau),\bar{x},\tau\right]\bigg|_{z^{(\omega)}=\eta} \right\rangle + \right.$$

$$\left. 2\left\langle s_n^3(\eta,\bar{x},\tau) \frac{\partial^2}{\partial \eta^2} s_n(\eta,\bar{x},\tau) \right\rangle + 4\left\langle s_n^2(\eta,\bar{x},\tau) \frac{\partial}{\partial \eta}\left(\frac{\partial}{\partial x_n} s_n\left[z^{(\omega)}(\bar{x},\tau),\bar{x},\tau\right]\bigg|_{z^{(\omega)}=\eta}\right) \right\rangle\right]. \quad (22)$$

If from the whole ensemble of burgons-mixers those are chosen that move normally to the surface $z^{(\omega)}(\vec{x},\tau) = \eta$, then the velocity vector in attached coordinate system is one component too, i.e.

$$\vec{u}(\eta,\vec{x},\tau) = \left[ u_n(\eta,\vec{x},\tau),0,0 \right]. \tag{23}$$

In this case, the surface $z^{(\omega)} = \eta$ does not subjected by stretch, i.e.

$$STR = 4\left\langle s_n^2 \frac{\partial u_\tau}{\partial x_\tau}\bigg|_{z^{(\omega)}=\eta} \right\rangle = 0. \tag{24}$$

Note that relation (24) refers to *flat* surface composed by a collection of burgons moving as an undivisible material object in the direction of the contraction vector $\vec{e}_1(\vec{x},\tau)$.

Finally, if from above defined subensemble of burgons-mixers those are chosen that generate the surface $z^{(\omega)} = \eta$ inside of their layers of constant (in the direction of the normal to the surface) mixture fraction gradient, then the following relations on the surface hold

$$\left. \begin{array}{r} \dfrac{\partial s_n\left[z^{(\omega)}(\vec{x},\tau),\vec{x},\tau\right]}{\partial x_n}\bigg|_{z^{(\omega)}=\eta} = 0, \\[2mm] \dfrac{\partial^2 s_n\left[z^{(\omega)}(\vec{x},\tau),\vec{x},\tau\right]}{\partial x_n^2}\bigg|_{z^{(\omega)}=\eta} = 0. \end{array} \right\} \tag{25}$$

In the above relations the first means that the surface $z^{(\omega)} = \eta$ is material one moving with the same velocity as that of burgon (see, Gibson's relation given above). Regarding the existence of the "vortex" layer ("vortlet") where the gradient $s_n$ is constant with respect to the coordinate $x_n$, one can consider it as a hypothesis based on the mechanism of turbulent mixing outlined above that provides the complete mixing inside the burgon core and the forming of mixed scalar concentration "jump" in its periphery.

Under conditions (23) and (25), differential equation (22) is reduced to the following simplest form

$$\frac{\partial}{\partial \tau}\langle s_n^2 \rangle = 2D\left\langle s_n^3 \frac{\partial^2}{\partial \eta^2} s_n \right\rangle.$$

With the quasinormality hypothesis being applied to the right-hand side term, i.e.

$$2\left\langle s_n^3 \frac{\partial^2}{\partial \eta^2} s_n \right\rangle = 6\langle s_n^2\rangle\left\langle s_n \frac{\partial^2 s_n}{\partial \eta^2} \right\rangle = 3\langle s_n^2 \rangle\left(\left\langle \frac{\partial^2}{\partial \eta^2} s_n^2 \right\rangle - 2\left\langle \left(\frac{\partial s_n}{\partial \eta}\right)^2 \right\rangle\right).$$

the above equation can be presented in the form of

$$\frac{\partial}{\partial \tau}\langle s_n^2 \rangle(\eta,\tau) + 6D\left\langle \left(\frac{\partial s_n}{\partial \eta}\right)^2 \right\rangle\langle s_n^2 \rangle = 3D\langle s_n^2 \rangle\frac{\partial^2 \langle s_n^2 \rangle}{\partial \eta^2}. \tag{26}$$

This differential equation governs the time-dependent diffusion in one-dimension $\eta$-space of conditioned mixture fraction gradient normal to the surface $z^{(\omega)} = \eta$. Obviously, the equation is valid at $\tau \geq \tau_\omega$ and for $\eta$ in the range $0 \leq \eta \leq 1$. In this equation, the rate of dissipation of $s_n$ in $\eta$-space, i.e. the term $\varepsilon_s^{(\eta)} = D\langle(\partial s_n/\partial\eta)^2\rangle$, has to be described in general by the corresponding differential equation. In this paper, however, for the simplicity, we will give an approximation of the term. Keeping this aim in mind, we introduce Burger's variables in $\eta$-space, i.e.

$$\eta_{AB} = \frac{1}{2}(\eta_A + \eta_B), \quad \vartheta = \eta_B - \eta_A,$$

where $\eta_A$ and $\eta_B$ are coordinates of two adjacent points in $\eta$-space. The parameter $\varepsilon_s^{(\eta)}$ can be presented in terms of the above coordinates as follows

$$\varepsilon_s^{(\eta)} = \frac{1}{4} D \frac{\partial^2}{\partial\eta^2}\langle s_n^2\rangle + DK_s^{(\eta)}, \qquad (27)$$

where

$$K_s^{(\eta)} = \left(-\frac{\partial^2}{\partial\vartheta^2}\langle s_n s_n'\rangle\right)_{\vartheta=0}$$

is the curvature in $\eta$-space of the layer of constant (in $\bar{x}$-space) gradient of the mixture fraction parameter $s_n$; $\langle s_n s_n'\rangle$ is two-point correlation in $\eta$-space of the gradients $s_n$. Worth to noting that the parameter $\delta_s^2 = \langle s_n^2(\eta,\tau)\rangle^{-1}$ is the measure of the layer's thickness. Then, the ratio

$$\frac{K_s^{(\eta)}}{\langle s_n^2\rangle} = \left(\frac{\delta_s}{R_s^{(\eta)}}\right)^2 = F^{(s)} \qquad (28)$$

can be considered as that of the layer's thickness (in $\bar{x}$-space) to its curvature (in $\eta$-space) with both parameters referred to the surface $z^{(\omega)} = \eta$.

As known, the sharpness of the velocity gradients and that of a scalar increase with turbulence Reynolds number. Hence, at $R_l$ increasing, $\delta_s$ decreases. The curvature of the layer in $\eta$-space is likely to decrease too. At $R_l \to \infty$ some limit value of above ratio presumably exists, i.e. $F_s^{(s)} = const$, with the constant undetermined since the description of the curvature is aboard the paper.

In view of relationships (27) and (28) equation (26) can be represented as follows

$$\frac{\partial}{\partial t}S(\eta,t) + F^{(s)}S^2(\eta,t) = \frac{1}{2}S\frac{\partial^2}{\partial\eta^2}S, \qquad (29)$$

where $t = 3(\tau - \tau_\omega)D\langle s_n^2\rangle(1/2,0)$; $S = \langle s_n^2\rangle/\langle s_n^2\rangle(1/2,0)$; $\langle s_n^2\rangle(1/2,0)$ is the value of $\langle s_n^2\rangle$ in the initial moment of time at $\eta = 1/2$.

The solution of equation (29) can be found by the separation-variable method seeking the variable in the form

$$s(\eta,\tau) = X(\eta)T(t). \qquad (30)$$

Then equation (29) is reduced to the pair of ordinary differential equations as follows

$$\frac{1}{T^2(t)}\frac{dT(t)}{dt} + \lambda = 0,$$

$$\frac{d^2 X(\eta)}{d\eta^2} - 2F_s^{(s)} X(\eta) + 2\lambda = 0,$$

where $\lambda$ is some constant.

Having regard to homogeneous boundary conditions at $\eta = 0$ and $\eta = 1$, i.e. $X(\eta = 0) = 0$ and $X(\eta = 1) = 0$, solution (30) can be represented in the form of

$$S(\eta,\tau) = \frac{1}{1+t_*}\frac{\left(1 - ch\eta\sqrt{2F^{(s)}} + \dfrac{ch\sqrt{2F^{(s)}} - 1}{sh\sqrt{2F^{(s)}}} sh\eta\sqrt{2F^{(s)}}\right)}{\left(1 - ch\dfrac{\sqrt{2F^{(s)}}}{2} + \dfrac{ch\sqrt{2F^{(s)}} - 1}{sh\sqrt{2F^{(s)}}} sh\dfrac{\sqrt{2F^{(s)}}}{2}\right)}, \qquad (31)$$

where

$$t_* = tF^{(s)} \Big/ \left(1 - ch\frac{\sqrt{2F^{(s)}}}{2} + \frac{ch\sqrt{2F^{(s)}} - 1}{sh\sqrt{2F^{(s)}}} sh\frac{\sqrt{2F^{(s)}}}{2}\right)$$

is nondimensioned time.

Solution (31) is valid at large values of the turbulence Reynolds number related to the conditioned parameters of velocity field, i.e.

$$R_\lambda\big|_{z^{(\omega)}=\eta} = \frac{\left\langle u_i^2\big|_{z^{(\omega)}=\eta}\right\rangle^{1/2} \left(\lambda_u^2\big|_{z^{(\omega)}=\eta}\right)^{1/2}}{\nu} \gg 1,$$

where

$$\lambda_u^2\big|_{z^{(\omega)}=\eta} = 5\frac{\left\langle u_i^2\big|_{z^{(\omega)}=\eta}\right\rangle}{\left\langle \omega_u^2\big|_{z^{(\omega)}=\eta}\right\rangle}$$

is squared Taylor's microscale of length and

$$\left\langle \omega_u^2\big|_{z^{(\omega)}=\eta}\right\rangle = \left(-\frac{\partial^2}{\partial \xi_s^2}\left\langle u_i u_i'\big|_{z^{(\omega)}=\eta}\right\rangle\right)_{\xi=0}$$

is the variance of the vorticity estimated on the surface $z^{(\omega)} = \eta$.

As for the turbulence Peclet number related to the layer of a scalar gradient,

$$P_\lambda\big|_{z^{(\omega)}=\eta} = \frac{\left\langle u_i^2\big|_{z^{(\omega)}=\eta}\right\rangle^{1/2} \delta_s}{\nu} \sim \Pr R_\lambda\big|_{z^{(\omega)}=\eta} \left(\frac{\left\langle \omega_u^2\big|_{z^{(\omega)}=\eta}\right\rangle}{\left\langle s_\eta^2\right\rangle\left\langle u_i^2\big|_{z^{(\omega)}=\eta}\right\rangle}\right)^{1/2} = \Pr R_\lambda\big|_{z^{(\omega)}=\eta}\left(\frac{\delta_s}{\lambda_u\big|_{z^{(\omega)}=\eta}}\right),$$

it can be small even at $R_\lambda\big|_{z^{(\omega)}=\eta} \gg 1$ since the ratio $\left(\delta_s/\lambda_u\big|_{z^{(\omega)}=\eta}\right)$ is rather small. In this case one can neglect the effect of the surface stretch as compared with the molecular diffusion of the layer irrespective of the direction of the burgon's velocity vector to the surface's normal. In our calculation of the evolution in time of $\langle s_n^2 \rangle(\eta)$-profile the asymptotic value of $F_s^{(s)}$ equal 2 was adopted, i.e. the same value as in the model of conventional scalar isotropic field decaying in strong isotropic turbulence [15]. The results of the calculation are presented in Fig. 3. The study of the dependence of conditioned scalar dissipation on time and the parameter $\eta$ in homogeneous turbulence is given, e.g., in the work of Mell et al. [16]. The data presented in Fig. 3 correspond qualitatively to those of the mentioned work. The quantitative comparison is difficult since the data of direct numerical simulation in the work are normalized on unconditioned parameters of decaying turbulence.

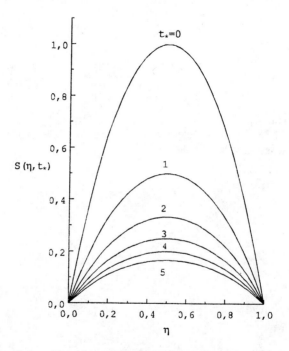

Fig.3  Time history of the profile of the variance of conditioned normal mixture fraction gradient.

## DISCUSSION AND CONCLUSIONS

In the lecture, the fundamentals of turbulent mixing theory based on the supposition that main mechanism of turbulent mixing at high turbulence Reynolds numbers is attributed to persistent vortex structures is outlined. As known at present [1, 2], such structures are vortex tubes, or Burger's vortices, produced in the turbulence irrespective of its form (i.e. isotropic, homogeneous anisotropic and nonhomogeneous). Hence, all the thermodynamics parameters considered in any statistical theory have to carry the information on the burgon's presence in turbulence at $R_\lambda \gg 1$. In other words, the functions have to be dependent on a parameter accounting for an inherent property of the burgons. By virtue of the fact that an individual burgon, during its forming, is able to involve a scalar admixture contained in nonpremixed medium [3,5], the mixture fraction parameter is chosen as a *parameter of influence*. In this case, the thermodynamic parameters of the turbulence are the functionals depending on the mixture fraction parameter $z^{(\omega)}$ [6], which, in turn, is random function of space and time. In the paper, the differential equations for the functionals are derived. To use practically the equations the procedure for the filtering of the events on the surface $z^{(\omega)} = \eta$ is developed.

Formally, this procedure results in the random functions of higher dimension depending both on $\bar{x}, \tau$ and progressive variable $\eta$ characterizing the level of an isoscalar "vortex" surface, instead of thermodynamic functionals. Physically, the filtering by means of progressive variable corresponds to selecting of the scalar substance formations produced by burgons (we have named these formations as "vortlets"). In other words, the transition of functionals conditioned by the level $z^{(\omega)} = \eta$ to the functions of higher dimension means the replacement of turbulence parameters description *in* 3-D space by that on the surfaces defined by progressive variable $\eta$.

In the lecture, the techniques for the deducing of the differential equations governing the conditioned statistical moments is developed. A number of simplifying suppositions based on DNS-data are introduced to elucidate the diffusion of statistical parameters of $s_i$-vector in $\eta$-space. In particular, it was supposed that the layer of the scalar gradient $g_i = \left(\partial c / \partial x_i\right)\big|_{z^{(\omega)} = \eta}$ "accompanying" the rate of dissipation layer is generated in the peripheric part of a burgon-mixer. The layer of the scalar gradient ("vortlet") is a scalar field structure persisting during burgon's life time, up to its annihilation due to the collision with other burgons or its dissipation by viscosity. Authors believe that the existence of such a layer as the persistent scalar structure, if this were proved by DNS, could shed the light on the "vortical" nature of turbulent mixing in viscous fluids and gases.

In the lecture, one-dimensional differential equation governing the diffusion in $\eta$-space of the conditioned variance $\langle s_i^2 \rangle$ is derived, under particular restrictions imposed on the kinematics of burgons-mixers. This equation can be considered as that related to the surface with its elements moving in the direction normal to them. The authors consider the model proposed as a zero order approach to the general "discrete-wave" theory of turbulence.

Quite a natural question comes to mind: how does the proposed concept of "vortex" turbulent mixing relate to the statistical turbulence theory in general and statistical models (correlational, spectral or PDF) in particular? We believe that creation of the local gradients of a scalar by burgons-mixers is the fundamental mechanism of turbulent mixing in accordance with the modern knowledge about the role of vortex tubes in the forming of the fine structure in the velocity field, i.e. the vorticity. Therefore, the treatment of the turbulence as a random process ignoring the presence of regular structures can be adequate only in fluids with "smeared" (due to viscosity) burgons, for example, during the final stage of homogeneous turbulence decay or in the far region of round wakes or jets where turbulence Reynolds numbers are rather small. In other words, absolutely stochastic turbulence can be regarded as a relict of the developed turbulence containing persistent vortex structures.

As to developed turbulence, i.e. turbulence at large values of $R_\lambda$, the presence of vortex structures for turbulence description is scarcely to be ignored. Indeed, as evidenced by direct numerical simulations of isotropic turbulence, the presence of vortical structures with spatial-time coherence is inherent in the turbulence at $R_\lambda \gg 1$. Occupying 3-5% of the whole volume

[1], vortical structures contribute crucially to the total vorticity and probably to the fine structure of a scalar field [9]. The presence of vortical structures does not allow realizations of thermodynamic functions to be considered purely random. The statistical description of developed turbulence has therefore to take into account its "discrete-wave" nature. In this case, the field discreteness is associated with the presence of burgons as moving material objects with high vorticity and its wave character is associated with turbulent "background" formed by burgon's wake interference (we use the term "wave" in the sense of possibility to expand continuous realization of the parameters in the spectrum of wave modes).

Development of general "discrete-wave" theory of turbulence is likely to be problematic at present. Nevertheless, some bypassing approaches to the problem of adequate turbulence modeling are seen. Probably, one of them is that based on the method of Proper Orthogonal Decomposition [17]. In the present paper, the alternative bypassing approach is proposed based on conditionality of thermodynamic functions by a random variable associated with the inherent property of strong vortical structures - their ability to mix a scalar contaminant. In this case, the identification of the thermodynamic functionals by a certain level of the random argument resulting mathematically in the functions of higher dimension means physically the replacing the turbulence description in three-dimensional $\bar{x}$-space by the description of the random surfaces conditioned by a *progressive* variable $\eta$ and randomly spaced in the Cartesian coordinate system $\bar{x}$.

This work was supported in part by the European Commission in the framework of PECO-project, grant Nr: ERBCIPDCT940512.

*The authors are grateful to Prof. Norbert Peters for the useful discussion.*

## REFERENCES

[1] SHE, Z.S., JACKSON, E., ORSZAG, S.A.: "Structure and dynamics of homogeneous turbulence: models and simulations". Proc. R. Soc. London, A. 434 ( 1991) pp. 101-121.

[2] VINCENT, A., MENEGUZZI, M.: "The spatial structure and statistical properties of homogeneous turbulence", J. Fluid Mech., 225 (1991) pp. 1-20.

[3] VINCENT, A., MENEGUZZI, M.: "The dynamics of vorticity tubes in homogeneous turbulence", J. Fluid Mech., 258 (1994) pp. 245-254.

[4] BRACHET, M.E., MENEGUZZI, M., VINCENT, A., POLITANO, H., SULEM, P.L.: "Numerical evidence of smooth self-similar dynamics and possibility of subsequent collaps for three-dimensional ideal flows", Phys. Fluids, A. 4 N12 (1992) pp. 2845-2854.

[5] PASSOT, T., POLITANO, H., SULEM, P.L., ANGILELLA, J.R., MENEGUZZI, M.: "Instability of strained vortex layers and vortex tube formation in homogeneous turbulence", J. Fluid Mech., 282 (1995) pp. 313-338.

[6] KERR, R.: "Higher-order derivative correlations and the alignment of small-scale structures in isotropic turbulence", J. Fluid Mech., 153 (1985). pp. 31-54.

[7] ASHURST, W.N.T., KERSTEIN, A.R.. KERR, R.M., GIBSON, C.H.: "Alignment of vorticity and scalar gradient with strain rate in simulated Navier-Stokes turbulence", Phys. Fluids, 30 N8 (1987) pp. 2343-2353.

[8] RUETSCH, G.R., MAXEY, M.R.: "Small-scale features of vorticity and passive scalar fields in homogeneous isotropic turbulence", Phys.Fluids, A. 3 N6 (1991) pp. 1587-1597.

[9] RUETSCH, G.R., MAXEY, M.R.: "The evolution of small-scale structures in homogeneous isotropic turbulence", Phys.Fluids, A. 4 N12 (1992) pp. 2747-2760.

[10] REID, W.H.: "On the stretching of material lines and surfaces in isotropic turbulence with zero fourth cumulants", Proc. Cambr. Phil. Soc., 51 (1955), pp. 350-362.

[11] VIEILLEFOSSE, P.: "Local interaction between vorticity and shear in a perfect incompressible fluid", Journal de Physique, 43 (1982), pp. 837-842.

[12] OTTINO, J.M.: "Description of mixing with diffusion and reaction in terms of the concept of material surfaces", J. Fluid Mech., 114 (1982) pp. 83-103.

[13] GIBSON, C.H.: "Fine structure of scalar fields mixed by turbulence. 1. Zero-gradient points and minimal gradient surfaces", Phys. Fluids, 11 N11 (1968) pp. 2305-2327.

[14] PETERS, N.: "Laminar diffusion flamelet models in non-premixed turbulent combustion" Prog. Energy Combust. Sci., 10 (1984) pp. 319-339.

[15] KOLOVANDIN, B.A.: "Modelling the dynamics of turbulent transport processes", Advances in Heat Transfer, 21 (1991) pp. 185-237.

[16] MELL, W.E., NILSON, V., KOSALY, G., RILEY, J.J.: "Investigation of closure models for nonpremixed turbulent reacting flows", Phys. Fluids, 6 N3 (1994) pp. 1331-1356.

[17] LUMLEY, J.L.: "The structure of inhomogeneous turbulent flows", in Atm. Turb. And Radio Wave Prop. (A.M. Yaglom, V.I. Tatarsky, eds.), Nauka, Moscow, (1967) pp. 166-178, (in russian).

# FLOW CONTROL AND HEAT TRANSFER ENHANCEMENT WITH VORTICES

K. Suzuki, K. Inaoka
Department of Mechanical Engineering
Kyoto University, Kyoto 606-01, Japan

M. Kobayashi, H. Maekawa and K. Matsubara
Department of Mechanical Engineering
Niigata University, Igarashi 2-8050, Niigata 950-21, Japan

## SUMMARY

Three types of longitudinal vortices are discussed in this article. Effectiveness of a vortex generating jet and a triangle vortex generator attached to a LEBU manipulator are first demonstrated. Heat transfer enhancement can be achieved over a long streamwise distance with the generated vortices. Entrainment of cold fluid near to the heated wall from the main stream is the basic cause of the achieved heat transfer enhancement. In the second part, it is shown that Taylor-Goertler vortices can be generated even when a fully developed turbulent flow is introduced into a curved channel. However, they cannot easily be identified in statistic studies of the turbulent flows since they sway spanwise rather slowly keeping basically their cross-sectional structure. The cross-sectional structure can basically be predicted with $k-\varepsilon$ type turbulence model. Momentum and heat transfer due to the induced secondary flow is important as well as their turbulent counterpart.

## INTRODUCTION

Longitudinal vortices to be generated near the wall entrain the cold high momentum fluid near to the heated wall from the main stream so that, by their generation near the wall, flow separation from the wall can effectively be suppressed and the wall heat transfer can significantly be enhanced. Therefore, a lot of research works have been done on the longitudinal vortices both in the fluid mechanics and heat transfer fields [1-7]. Recent review of Fiebig [8] is a good source of the works on longitudinal vortices. The present paper treats three topics related to the longitudinal vortices. In the first, discussion will be developed for the longitudinal vortices artificially induced downstream the vortex generator attached to a large eddy break-up manipulator (hereafter abbreviated as a LEBU plate), which is inserted into a flat plate turbulent boundary layer. How and how effectively the wall heat transfer can be enhanced will be discussed in the next. Secondly, attention will be turned to the longitudinal vortices artificially attempted but without mounting any vortex generator, namely to the longitudinal vortices to be generated with a vortex generating jet. In the last discussed will be the most typical type of naturally induced longitudinal vortices; i.e. the Taylor-Goertler vortices appearing in a curved channel. Especially, attention will be paid to the case where a fully developed turbulent flow is introduced into a curved channel.

As discussed previously [9,10], transverse vortices give also interesting means for controlling of near wall flow and enhancement of wall heat transfer. However, they are not discussed here. The present article is devoted to the discussion only on the effectiveness of the longitudinal vortices on the flow control and heat transfer enhancement.

## A TURBULENT BOUNDARY LAYER DISTURBED BY AN INSERTION OF A LEBU PLATE ATTACHED WITH A TRIANGULAR WINGLET-TYPE VORTEX GENERATOR

Discussion will first be given to the longitudinal vortices artificially induced downstream a triangular winglet-type vortex generator attached to a Large Eddy Break-Up manipulator (LEBU plate) which is inserted into a flat plate turbulent boundary layer. The complex body composed of the LEBU plate and the vortex generator is schematically illustrated in Figure 1. The reason why studies have been made with this complex body will first be stated in the following.

Marumo et al. [11-14] and Suzuki et al. [15,16] studied the characteristics of a flat plate turbulent boundary layer disturbed by a cylinder mounted near the flat plate. Flat plate skin friction was found to be reduced downstream the inserted cylinder while heat transfer from the flat plate was enhanced in the same streamwise region [15], and this phenomenon was called the dissimilarity between the momentum transfer and heat transfer. This is an interesting feature of the disturbed boundary layer because it implies the possibility of separate manipulation of turbulent momentum transfer and heat transfer from each other. It is also interesting from a practical view point because it suggests a possibility that heat exchangers of higher heat transfer performance can be developed with smaller pressure loss penalty. Using a cylinder as a body to insert, however, total momentum loss cannot be reduced since its drag is not small. Therefore, another body of different shape having smaller drag is necessary to be introduced. The LEBU plate is best in this sense because its form drag is small and it can additionally reduce the wall skin friction. It was found, however, that heat transfer enhancement could not be achieved with the LEBU plate [17]. As an effort to level up the lowered heat transfer coefficient, a study was initiated with the LEBU plate attached with the vortex generator and the geometric arrangement illustrated in Figure 1 was found best, i.e. simple but most effective in heat transfer enhancement, among several ways of different geometric arrangement of the LEBU plate and the vortex generator [18].

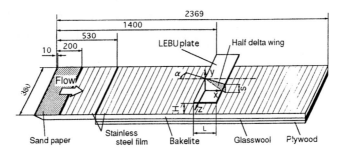

Figure 1  Schematic view of the geometric arrangement and coordinate systems.

Figure 2 shows an example of the measured spatial distribution of the wall heat transfer coefficient, $h$. In this figure, $h$ is normalized with the counterpart of a normal flat plate boundary layer obtained at the same position at the same Reynolds number, $h_0$. It is seen that the complex body illustrated in Figure 1 can work effectively in enhancing the flat plate heat transfer. The base level of $h/h_0$ is found to be lower than unity by an amount of the heat transfer suppression produced by the LEBU plate. However, its value is conspicuously larger than unity, i.e. heat transfer is effectively enhanced, in a certain region downstream the insertion position of the complex body. The region has the spanwise width of only three to four times the approaching boundary layer thickness, $\delta$. Therefore, in practical application, multiple vortex generators should be mounted spanwise at an appropriate interval. However, heat transfer enhancement can be achieved over long streamwise distance. In the first region up to $x/\delta = 8.1$ where the longitudinal vortex develops, the value of $h/h_0$ increases downstream and, after that position, it keeps almost constant value down to the last measured station. Although any supporting figure is not shown here because of the space limitation, higher effectiveness in heat transfer enhancement was found to be achieved with the vortex generator of larger height $s$, and at larger angle of attack, $\alpha$, so far as the vortex generator size does not exceed the boundary layer thickness and when $\alpha$ is smaller than thirty degree.

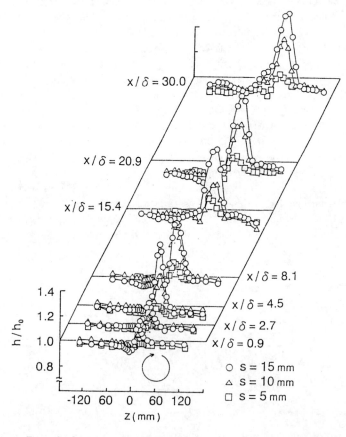

Figure 2   Spatial distribution of heat transfer coefficient on the flat plate.

Figure 3 presents the velocity map of the secondary flow and the temperature contours measured in a cross-section [19]. Along the abscissa of the velocity map, the region where heat transfer enhancement is achieved is shown by the thick line and the peak position of heat transfer coefficient is shown by a circle drawn on it. It is clearly observed that, in the region where downwash flow is produced by the generated vortex, lower temperature fluid is entrained near to the wall from the main stream and therefore wall heat transfer is enhanced. Turbulence intensification incurred by the vortex generator mounted on the LEBU plate at a certain angle of attack also contributes to the wall heat transfer enhancement. However, it is effective only in a small part of the upwash region so that it is not a major contributor [19]. Therefore, it is concluded that heat transfer enhancement can be achieved disregarding the type of geometric arrangement of the vortex generator as far as the longitudinal vortex is effectively generated. Inaoka examined the cases where similar complex bodies of combining LEBU plate and vortex generators are introduced into a turbulent duct flow and found that they are also effective in heat transfer enhancement [20]. However, net reduction of pressure drop penalty could not be achieved and its possibility has still to be studied in future.

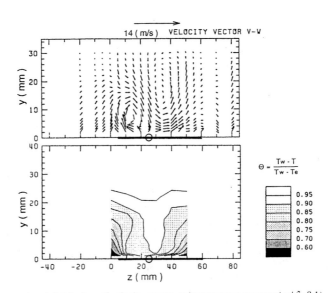

Figure 3  Cross-sectional distribution of velocity vectors and temperature contours ($x/\delta=8.1$).

## HEAT TRANSFER ENHANCEMENT WITH A VORTEX GENERATING JET

Impinging jet heat transfer used in many practical applications like the inner surface cooling of gas turbine vane is effective in the jet stagnation region. However, heat transfer coefficient is much smaller outside the stagnation region and it becomes smaller even in the stagnation region when jets are injected into a cross flow just as in the case of gas turbine vane inner surface cooling. Therefore, heat transfer enhancement is important in such cases. The complex body discussed in the previous section gives a means to enhance the inner flow heat transfer as has already been discussed, but its mounting is not easy in a small space like the one between the insert and the inner surface of the gas turbine vanes, which is normally less than one millimeter. Here is discussed the possibility of using a vortex generating jet. Johnston and Nishi studied a possibility of generating a longitudinal vortex in a cross flow by controlling the direction of jet [21]. They tried to suppress the flow separation by the generated vortex, therefore, they tried to generate the vortex near the surface from which the jet is issued. In our studies [9, 22], generation of the vortex is attempted near the target plate so as to enhance the target plate heat transfer.

Figure 4 shows the schematic diagram of the test section used in the study and Figure 5 presents the spatial distribution of heat transfer coefficient for the impinging jet with cross flow. The measured heat transfer coefficient is presented in a form normalized with its counterpart of duct flow having no jet impingement. The right side data are for the normal jet, impinging perpendicularly to the target plate, and the left data for the jet injected in a cross-section with an inclination of 45 degree from normal direction. VR in the figure denotes the ratio between the jet velocity and the cross flow velocity. At the positions near the jet nozzle, peak heat transfer coefficient is lower with the inclined jet than with the normal jet. This is inevitable because the impinging distance is larger for the inclined jet than with the normal jet. However, enhancement can be obtained in wider region with the inclined jet. Furthermore, peak heat transfer coefficient becomes larger with the inclined jet than with the normal jet at more downstream positions. Therefore, number density of jet nozzles can be reduced and, with the reduced number of jet nozzles, jet velocity becomes higher if the cooling air total mass flow rate is kept constant. This additionally results in the leveling up of peak heat transfer coefficient to some extent. Therefore, use of the vortex generating jet is promising in two ways as a means to achieve highly effective inner surface cooling of gas turbine vanes.

Detailed flow and turbulence measurements with Laser Doppler Velocimeter are under development. Some preliminary flow field measurement has been made with a Particle Tracking Velocimetry. Figure 6 shows again the distribution of heat transfer coefficient in comparison with the velocity vector map of secondary flow obtained at the cross-sections near the positions where heat transfer measurement was made. It is seen that heat transfer is enhanced not only in the jet impingement region but also in the region where the secondary flow accompanied with one of the two generated longitudinal vortices washes the target plate. This is different from the case of the longitudinal vortex produced by the vortex generator. This difference results from the larger secondary flow velocity in the present case.

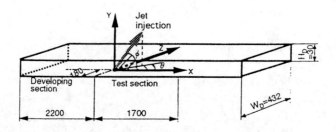

Figure 4  Schematic diagram of the test section.

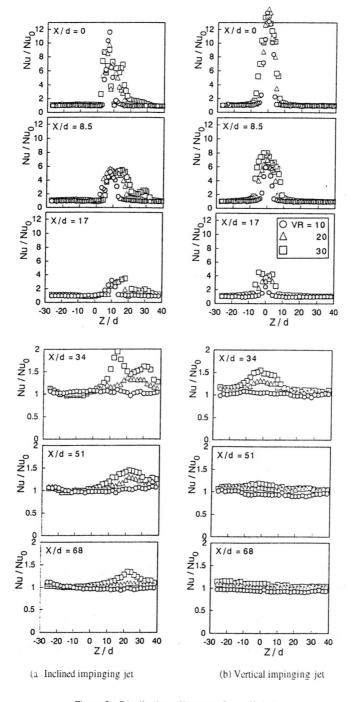

(a) Inclined impinging jet  (b) Vertical impinging jet

Figure 5 Distribution of heat transfer coefficient.

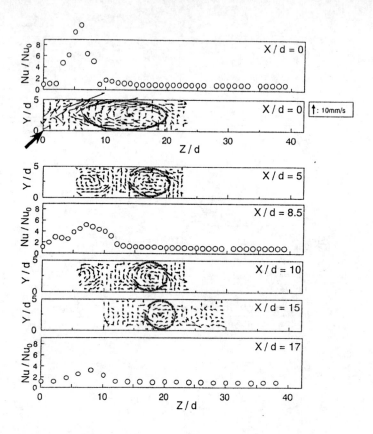

Figure 6  Distribution of heat transfer coefficient and cross-sectional distribution of velocity vectors (VR=10).

## LONGITUDINAL VORTICES IN A CURVED CHANNEL

When a low-turbulence flow is introduced into a curved channel, it is well established that longitudinal vortices are incurred and the vortices are called Dean vortices or Taylor-Goertler vortices [23]. However, it is also known that, when a fully turbulent flow is introduced into a curved channel, the flow can be regarded statistically two-dimensional in the case where the channel has an aspect ratio larger than seven [24]. Thus, there has been a controversial question on if longitudinal vortices can be generated in such cases [25]. The last part of the present article is devoted to the discussion on this point based on the work of Kobayashi [6].

Kobayashi et al. did thorough studies on such a flow in a channel having the aspect ratio of fifteen and the inner wall to outer wall radius ratio of 0.92. In their experiments, a fully developed turbulent flow was established in a straight channel preceding the curved channel which was served as the test section. In Figure 7 presented are the streamwise changes of the static pressure, the wall shear stresses and the wall heat fluxes measured on the channel walls [26, 27]. Pressure field develops most quickly and $\phi=15°$ is enough for the wall pressure gradient to reach its asymptotic value relevant for a fully developed flow in a curved channel. Wall shear stress and wall heat flux are found to reach their asymptotic values at $\phi=75°$ which is shorter compared to the longer distance necessary for turbulence quantities to reach their asymptotic values. Radial distributions of all the turbulence quantities reach their asymptotic ones at the position of $\phi=135°$. Unless the measured position is specified in the figure, the data discussed below are the ones measured at a position of $\phi=177°$ where the flow was regarded to be in a fully developed state.

Figure 8 shows for instances the spanwise distributions of radial component of the Reynolds normal stresses measured at several different radial positions. They do not show any spatial non-uniformity or more specifically any periodical spanwise waviness as found in a turbulent boundary layer on a concave wall [28, 29]. Therefore, the present flow to be discussed in the following was statistically two-dimensional so that they do not directly suggest

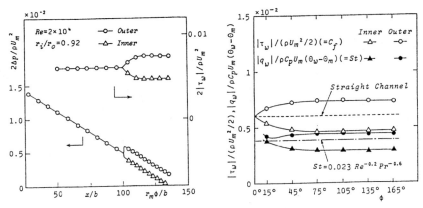

Figure 7  Streamwise changes of the static pressure, shear stress and heat flux on the channel wall.

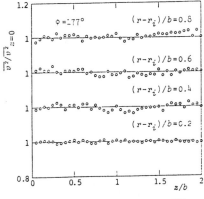

Figure 8  Spanwise distributions of $\overline{v^2}$.

the existence of the longitudinal vortices. Figure 9 compares the measured radial distribution of the time mean streamwise velocity multiplied by $r$, the radial coordinate, with other available results for different inner wall to outer wall radius ratio [30-35]. Among the plotted profiles, the experiments of Ellis and Joubert [33] and Hunt and Joubert [35] were done by leading a low turbulence flow into the curved test section. The studied flow should not have been statistically two-dimensional in these two experiments. Ellis and Joubert were actually aware of the spanwise non-uniformity of the measured time-mean streamwise velocity. This also seems to be the case of the data reported in [30]. Radial peak position of the plotted quantity moves inward with an increase of the inner wall to outer wall radius ratio except for the three profiles of [30, 33, 35]. In all other cases, both of inner and outer region velocity data are found to merge in one curve if they are replotted in a manner proposed by Wattendorf [32] as shown in Figure 10. Flows having statistical two-dimensionality like the ones to be discussed here have to be clearly distinguished from the ones having spanwise non-uniformity as the ones reported in [30, 33, 35].

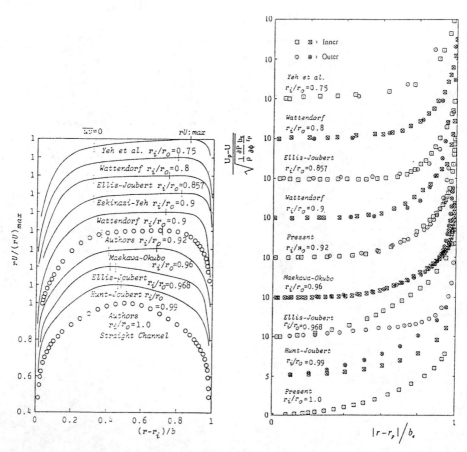

Figure 9   Time mean angular velocity for different inner wall to outer wall radius ratio.

Figure 10   Wattendorf's similarity.

Figure 11 shows the radial distribution of the measured value of $\overline{uv}$. The dash and dot line represents the position where the production rate of $\overline{uv}$ is equal to zero while solid lines show the distribution calculated from the following momentum equation (1).

$$\overline{uv} = -\frac{r_i^2}{r^2}\frac{\tau_i}{\rho} + vr\frac{\partial}{\partial r}\left(\frac{U}{r}\right) - \frac{1}{r^2}\frac{\partial}{\partial \phi}\int_{r_i}^{r} rU^2 dr - UV - \frac{r^2 - r_i^2}{2r^2}\frac{\partial P}{\partial \phi} \qquad (1)$$

where $r$ is the radial coordinate, $r_i$ the inner wall radius, $\tau_i$ the inner wall shear stress and $U$ and $V$ the time-mean streamwise and radial velocity components, respectively. $V$ was calculated from the following continuity equation.

$$V = -\frac{1}{r}\frac{\partial}{\partial \phi}\int_{r_i}^{r} U dr. \qquad (2)$$

The calculated results agree well with the measured data, indicating the accuracy and consistency of the measured data.

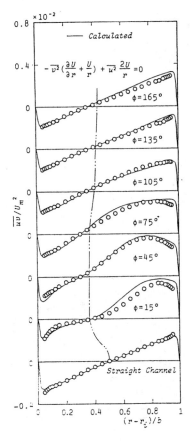

Figure 11    Radial distribution of the measured value of $\overline{uv}$.

In addition to the detailed measurement of the second and third order velocity, velocity-temperature correlations, one dimensional spectra, spatial correlation functions and length scales were measured [36]. Figure 12 presents some of such results, i.e. the radial distributions of the integral scales defined as follows:

$$L_{ij} = U \int_0^{\tau_0} R_{ij}(\tau) / R_{ij}(0) d\tau \tag{3}$$

where $R_{ij}(\tau)$ is the time correlation function of $u_i(t)$ and $u_j(t+\tau)$, the fluctuating components of velocity, and $\tau_0$ the time difference where $R_{ij}(\tau)$ takes zero value. The presented results indicate the appearance of flow structure having the length scale approximately equal to the channel width. Figure 13 shows the effective power spectra $\kappa E_{ij}(\kappa)$ obtained at different radial positions at a streamwise position in a fully developed flow region. $\kappa$ is the wave number and $E_{ij}(\kappa)$ is defined as follows:

$$E_{ij}(\kappa) = \frac{U}{2\pi} \int_{-\infty}^{\infty} u_i(t) u_j(t) \exp(-i\kappa U t) dt . \tag{4}$$

The results obtained at a position of $(r - r_i)/b = 0.1$ are very similar to the counterparts of straight channel having zero curvature. The results at other radial positions have a peak at a low wave number. This indicates the existence of almost periodically changing unsteady flow. Based on this, Kobayashi et al. presumed that the longitudinal vortices are generated in the channel even when high turbulence flows are introduced into curved channels and that the

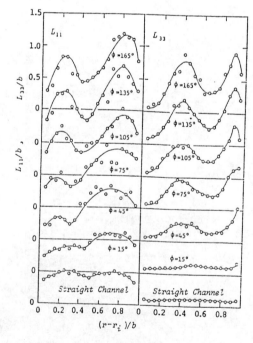

Figure 12  Radial distributions of the integral scales ($L_{11}$, $L_{33}$).

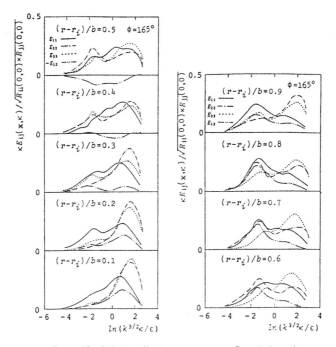

Figure 13  Relative effective power spectra (Curved channel).

vortices sway almost periodically in spanwise direction roughly keeping their basic cross-sectional structure characterized by the length scale $L$ to be obtained from Figure 12. Here will be discussed some other results which support their presumption.

In the experiments reported in reference [36], Kobayashi et al. mounted 0.6mm thickness wires just upstream the inlet to the curved channel at a spanwise interval equal to 0.9 times the channel width, keeping each wire in a position normal to the flow direction and parallel to the radial direction. The spanwise interval roughly corresponds to the length scale $L$ observed in Figure 12. With this small trick, they succeeded in fixing the spanwise position of Taylor-Goertler vortices and could identify the cross-sectional structure of the vortices in their statistical measurement. The measurement was made at the Reynolds number of $2 \times 10^4$ with the test section of the same geometry as the one used in the above experiments. In addition to these experiments, they also did numerical computation to predict the cross-sectional structure of the longitudinal vortices with a simple $k-\varepsilon$ type turbulence model [38]. In the computation, the following three-component decomposition of velocity $U_i$ were introduced:

$$U_i = U_i^* + \tilde{u}_i + u_i' \qquad (5)$$

where $U_i^*$ is the steady, two-dimensional part of the velocity components, $\tilde{u}_i$ the steady velocity component associated with the incurred longitudinal vortex and $u_i'$ the stochastic, irregular fluctuating components of velocity related to turbulence fields. Introducing the velocity components $(U, V, W)$ for the cylindrical coordinate system $(\phi, r, z)$. $V^* = W^* = 0$ should hold from the base flow two-dimensionality, and $\tilde{v}$ and $\tilde{w}$ can be related to a stream function $\psi$ and the streamwise vorticity $\Omega$ aligned with the angular direction $\phi$ as follows:

$$\tilde{v} = -\frac{1}{r}\frac{\partial \psi}{\partial z}, \tag{6}$$

$$\tilde{w} = \frac{1}{r}\frac{\partial \psi}{\partial r}, \tag{7}$$

$$\frac{\partial}{\partial r}\left(\frac{1}{r}\frac{\partial \psi}{\partial r}\right) + \frac{\partial}{\partial z}\left(\frac{1}{r}\frac{\partial \psi}{\partial z}\right) = -\Omega. \tag{8}$$

Then, the time-averaged equations of motion are expressed as follows.

$$\left(\frac{\tilde{v}}{r}\right)\frac{\partial}{\partial r}\{r(U^*+\tilde{u})\} + \tilde{w}\frac{\partial \tilde{u}}{\partial z} = \left(\frac{1}{r^2}\right)\frac{\partial}{\partial r}\left[r^2\left\{\frac{\mu}{\rho}r\frac{\partial}{\partial r}\left(\frac{U^*+\tilde{u}}{r}\right) - \overline{u'v'}\right\}\right]$$
$$+ \frac{\partial}{\partial z}\left(\frac{\mu}{\rho}\frac{\partial \tilde{u}}{\partial z} - \overline{u'w'}\right) - \frac{1}{\rho r}\frac{\partial P}{\partial \phi}, \tag{9}$$

$$\tilde{v}\frac{\partial \Omega}{\partial r} + \tilde{w}\frac{\partial \Omega}{\partial z} = \frac{\mu}{\rho}\left\{\frac{\partial}{\partial r}\left(\frac{1}{r}\right)\frac{\partial(r\Omega)}{\partial r} + \frac{\partial^2 \Omega}{\partial z^2}\right\} + 2\frac{(U^*+\tilde{u})}{r}\frac{\partial \tilde{u}}{\partial z}$$
$$+ \frac{1}{r}\frac{\partial \overline{u'^2}}{\partial z} + \frac{\partial^2 \overline{w'^2}}{\partial r \partial z} - \frac{\partial^2 \overline{v'w'}}{\partial z^2} + \frac{\partial}{\partial r}\left\{\frac{1}{r}\frac{\partial(r\overline{v'w'})}{\partial r}\right\} - \frac{\partial}{\partial z}\left\{\frac{1}{r}\frac{\partial(r\overline{v'^2})}{\partial r}\right\}. \tag{10}$$

The Reynolds stresses are related to velocity gradient with the isotropic eddy viscosity concept in the following manner.

$$-\overline{u'v'} = \frac{\mu_t}{\rho}r\frac{\partial}{\partial r}\left(\frac{U^*+\tilde{u}}{r}\right), \tag{11}$$

$$-\overline{u'w'} = \frac{\mu_t}{\rho}\frac{\partial \tilde{u}}{\partial z}, \tag{12}$$

$$-\overline{v'w'} = \frac{\mu_t}{\rho}\left(\frac{\partial \tilde{v}}{\partial z} + \frac{\partial \tilde{w}}{\partial r}\right), \tag{13}$$

$$\overline{v'^2} = -2\frac{\mu_t}{\rho}\frac{\tilde{v}}{r} + \frac{2}{3}k, \tag{14}$$

$$\overline{v'^2} = -2\frac{\mu_t}{\rho}\frac{\partial \tilde{v}}{\partial r} + \frac{2}{3}k, \tag{15}$$

$$\overline{w'^2} = -2\frac{\mu_t}{\rho}\frac{\partial \tilde{w}}{\partial z} + \frac{2}{3}k, \tag{16}$$

where $\mu_t$ is related to the turbulence kinetic energy $k$ and its viscous dissipation rate $\varepsilon$ by the following expression as usual:

$$\mu_t = C_\mu f_\mu \frac{\rho k^2}{\varepsilon} \tag{17}$$

and $k$ and $\varepsilon$ are determined by simultaneously solving $k-\varepsilon$ type turbulence model similar to the one proposed by Nagano and Tagawa [38].

In the following, the cross-sectional structure of the longitudinal vortex identified in the experiments of fixing its spanwise position are discussed in comparison with its numerical counterpart [37]. Figures 14, 15, 16 and 17 show some examples of such comparison. In Figure 14 is plotted the iso-contours of the stream function $\psi$ in a cross-section, in Figure 15 are presented the radial distributions of $\tilde{v}$ and $\tilde{w}$, the time-mean radial and streamwise velocity components of the secondary flow measured along the lines slicing the fixed vortex at several different spanwise positions, and in Figure 16 are shown the radial distributions of the time-mean streamwise velocity ($U^*+\tilde{u}$) obtained again at several different vortex-slicing positions. In Figure 17 is compared the distribution of the calculated value of $U^*$ with other three similar profiles. Two of them are the experimental ones and one of them is the ensemble average of the measured radial profiles of the time-mean streamwise velocity obtained at different vortex-slicing positions. If the presumption noted above is correct and all the profiles measured at different vortex-slicing positions are visited at equal rate in the swaying motion of the vortices, the resulting ensemble average should be equal to the time-mean profile obtained in a curved channel without mounting any thin wires mentioned above. The latter should be the direct experimental counterparts of $U^*$. The last one plotted in the figure is a numerical one calculated with the same procedure as the one described above but with the two-dimensionality assumption for the time-mean velocity field, i.e. assuming that $\tilde{v} = \tilde{w} = 0$. The last one does not assume the existence of longitudinal vortices at all.

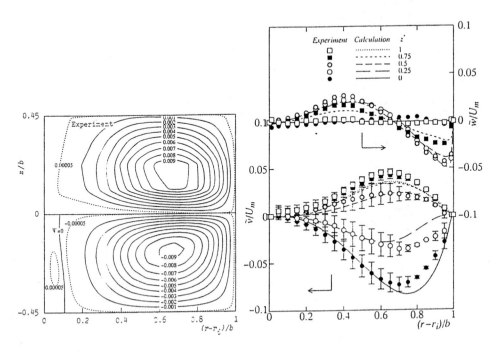

Figure 14  Iso-contours of the stream function.   Figure 15  Secondary flow velocity.

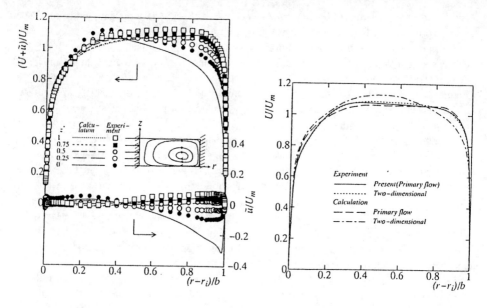

Figure 16  Streamwise time-mean angular velocity.    Figure 17  Primary-flow velocity.

    The results plotted in the four figures indicate that the cross-sectional structure of the Taylor-Goertler vortices can roughly be predicted with the used $k - \varepsilon$ type turbulence model. More important is that the numerical computation presuming the existence of the vortices supplies better results for the time-mean two-dimensional velocity field, which is naturally achieved in a normal channel preceded by a long inlet straight channel. This suggests the validity of such presumption that, when a fully developed turbulent flow is introduced into a curved channel, Taylor-Goertler vortex is generated in the channel but sways spanwise slowly and almost periodically keeping its basic cross-sectional structure. In Figure 18, another comparison is made among three apparent shear stresses: namely first the Reynolds shear stress $\overline{u'v'}$ related to the stochastic velocity fluctuation, secondly the apparent shear stress $\overline{\tilde{u}\tilde{v}}$ incurred by the secondary flow in a cross-section and thirdly the sum of such two shear stresses. The figure reveals that both of the stochastic velocity fluctuation and the structural secondary flow in a cross section equally contribute to the radial momentum transfer. The latter should be quite different in nature from the first one: namely the turbulent momentum transfer. Therefore, it is certainly difficult to predict the total momentum transfer only with a turbulence model approach without assuming the existence of longitudinal vortices. This should also be true for the total heat transfer in a curved channel. Therefore, it is highly recommended to develop a calculation procedure in which turbulence modeling approach and prediction of cross-sectional flow structure are combined with each other. The computation discussed in the above is one of them but it needs an information about the size of the vortices. In this point, another elaborate approach has to be developed for general use.

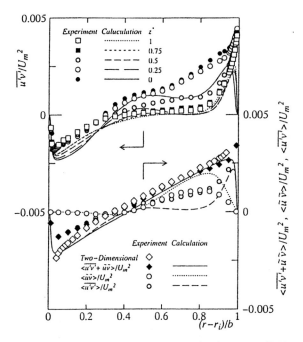

Figure 18  Turbulent shear stress and axial average ($\overline{u'v'}$, $\langle \overline{u'v'} \rangle$, $\langle \overline{\tilde{u}\tilde{v}} \rangle$, $\langle \overline{u'v'} + \tilde{u}\tilde{v} \rangle$).

## CONCLUDING REMARKS

Complex body of triangle winglet type vortex generators attached to a LEBU plate works well in generating longitudinal vortices, therefore, it is effective to enhance the wall heat transfer when it is inserted in a turbulent boundary layer. Heat transfer enhancement is achieved over a large distance in streamwise direction. More effective heat transfer enhancement can be achieved with a larger size of the vortex generator and at its larger angle of attack. Heat transfer enhancement is achieved in the downwash flow region where entrainment of cooler fluid from the main stream near to the heated surface is produced. Intensification of turbulence produced by the body contributes minorly to the heat transfer enhancement.

Impinging jet heat transfer is effective around the stagnation region but its effectiveness becomes lower in cases with a cross flow. If the jet direction is properly adjusted, longitudinal vortices can be generated and heat transfer can be enhanced except at positions close to the nozzle. At the same time, the spanwise width of high heat transfer region is widen. This enhancement is attained both at the downwash flow region and where the target plate is swept by the spanwise flow associated with the induced secondary flow.

When a fully developed flow is introduced into a curved channel, a statistically two-dimensional flow is achieved in the channel. Taylor-Goertler vortices are generated in this case but cannot be identified easily in the statistical measurements of turbulence quantities. It is because the vortices sway slowly and almost periodically in spanwise direction keeping their basic cross-sectional structure. With a small trick to mount thin wires at a spanwise interval in a cross section upstream the inlet to the curved channel, the spanwise position of the vortices can be fixed. Both of momentum and heat transfer produced by the secondary flow accompanied by the generated Taylor-Goertler vortices and of turbulent momentum and heat transfer equally contribute to the radial total momentum and heat transfer in the channel. Thus, a computation assuming the two-dimensionality of the basic flow cannot capture the flow characteristics correctly. The cross-sectional structure of the vortices can be predicted rather well with a simple $k - \varepsilon$ type turbulence model.

## REFERENCES

[1] NAKAMURA, I., OSAKA, H., KUSHIDA, T., OHKUBO, N., "A study on the interaction between longitudinal vortices and a turbulent boundary layer", Trans. JSME - B, 53-492 (1987), pp.2340-2347.
[2] LIGRANI, P.M., CHOI, S., SCHALLET, A.R., SKOGERBOE, P., "Effects of Dean vortex pairs on surface heat transfer in curved channel flow", Int. J. Heat Mass Transfer, 39-1 (1996), pp.27-37.
[3] PAULEY, W. R., EATON, J. K., "The effect of embedded longitudinal vortex arrays on turbulent boundary layer heat transfer", J. Heat Transfer, 116-4 (1994), pp.871-879.
[4] EIBECK, P. A., EATON, J. K., "The effects of longitudinal vortices embedded in a turbulent boundary layer on momentum and thermal transport", Proc. 8th Int. Heat Transfer Conf., 3 (1986), pp.1115-1120.
[5] WROBLEWSKI, D.E., EIBECK, P. A., "Measurements of turbulent heat transport in a boundary layer with an embedded streamwise vortex", Int. J. Heat Mass Transfer, 34 (1991), pp.1617-1631.
[6] KOBAYASHI, M., "An Experimental study on the turbulent transport mechanism in the two-dimensional curved channel", Doctor Thesis at Kyoto University, 1995.
[7] WROBLEWSKI, D.E., "An experimental investigation of turbulent heat transport in a boundary layer with an embedded streamwise vortex", Doctor Thesis at University of California, Berkeley, 1990.
[8] FIEBIG, M., "Vortex generators for compact heat exchangers", J. Enhanced Heat Transfer, 1 - 2, 2 (1995), pp.43 -61.
[9] SUZUKI, K., "Advances in turbulent heat transfer - control and enhancement -", Proc. 3rd ASME-JSME Thermal Joint Conference, 1 (1995), pp.1-9.
[10] SUZUKI, K., "Flow modification and heat transfer enhancement with vortices", Proc. 9th Int. Symp. on Transport Phenomena, 1 (1996), pp.72-83.
[11] MARUMO, E., SUZUKI, K., SATO, T., "A turbulent boundary layer disturbed by a cylinder", J. Fluid Mech., 87-1 (1978), pp.121-141.
[12] MARUMO, E., SUZUKI, K., SASAKI, T., SATO, T., "A turbulent boundary layer disturbed by a cylinder", Trans. JSME - B, 46-407 (1980), pp.1211-1219.
[13] MARUMO, E., SUZUKI, K., SASAKI, T., KINUTA, H., SATO, T., "A turbulent boundary layer disturbed by a cylinder located near the wall (2nd report, Production and dissipation rates of turbulent kinetic energy)", Trans. JSME 46-407 (1980), pp.1220-1228.
[14] MARUMO, E., SUZUKI, K., SATO, T., "Turbulent heat transfer in a flat plate boundary layer disturbed by a cylinder", Int. J. Heat and Fluid Flow, 6-4 (1985), pp.241-248.
[15] SUZUKI, H., SUZUKI, K., SATO, T., "Dissimilarity between heat and momentum transfer in a turbulent boundary layer disturbed by a cylinder", Int. J. Heat Mass Transfer, 31 (1988), pp.259-265.
[16] SUZUKI, H., SUZUKI, K., KIKKAWA, Y., KIGAWA, H., KAWAGUCHI, Y., "Dissimilarity between heat and momentum transfer in a turbulent boundary layer disturbed by a cylinder", Turbulent Shear Flows 7, Springer-Verlag (ed. Durst. F. et al.), (1991), pp.119-135.
[17] INAOKA, K., SUZUKI, K., SUZUKI, H., KIGAWA, H., "Heat transfer in a turbulent boundary layer disturbed by means of a manipulator for breaking up large eddies", Heat Transfer Japanese Research, 21 (1993), pp.705-720.
[18] INAOKA, K., SUZUKI, K., SUZUKI, H., HAGIWARA, H., SUZUKI, K., "Augmentation of turbulent heat transfer with a vortex generator attached to a LEBU plate", Trans. JSME - B, 58-551 (1992), pp.2241-2247.
[19] INAOKA, K., SUZUKI, K., "Structure of the turbulent boundary layer and heat transfer downstream of a vortex generator attached to a LEBU plate", Turbulent Shear Flows 9, Springer-Verlag (ed. Durst. F. et al.), (1994),

pp.365-382.
[20] INAOKA, K., "A study on turbulent flow control and heat transfer enhancement with vortices", submitted as Doctor Thesis at Kyoto University, 1996.
[21] JOHNSTON, J. P., NISHI, M., "Vortex generator jets - Means for flow separation control", AIAA Journal, 28 (1990), pp.989-994.
[22] NAKABE, K., INAOKA, K., Ai, T., SUZUKI, K., "An experimental study on flow and heat transfer characteristics of longitudinal vortices induced by an inclined impinging jet in a cross flow", Proc. of 3rd KSME /JSME Joint Thermal Engineering Conf., 3 (1996), pp.59-64.
[23] TANI, I., "Production of longitudinal vortices in the boundary layer along a concave wall", J. Geophy. Res., 67-8 (1962), pp.3075-3080.
[24] DEAN, R. B., "Reynolds number dependence of skin friction and other bulk flow variables in two-dimensional rectangular duct flow", Trans. ASME, J. Fluid Engineering, 100 (1978), pp.215-223.
[25] PATEL, V. C., KSME 50th Anniversary Meeting Invited Lecture, 1995.
[26] KOBAYASHI, M., MAEKAWA, H., TAKANO, T., KOBAYASHI, M., "Experimental study of turbulent heat transfer in a two-dimensional curved channel", JSME Int., 37-3 (1994), pp.545-553.
[27] KOBAYASHI, M., MAEKAWA, H., TAKANO, T., HAYAKAWA, H., "An experimental study on a turbulent flow in a two-dimensional curved channel (Time-mean velocity and multiple velocity correlations in the entrance section)", Trans. JSME - B, 57-544 (1991), pp.4064-4071.
[28] BARLOW, S. R., JOHNSTON, J. P., "Structure of a turbulent boundary layer on a concave surface", J. Fluid Mech., 191 (1988), pp.137-176.
[29] HOFFMANN, P. H., MUCK, K. C., BRADSHAW, P., "The effect of concave surface curvature on turbulent boundary layers", J. Fluid Mech., 161 (1985), pp.371-403.
[30] ESKINAZI, S., YEH, H., "An investigation on fully developed turbulent flow in a curved channel", J. Aeronaut. Sci., 23 (1956), pp.23-34 & p.75.
[31] YEH, H., WILLIAM, G.R., HWACHII, L., "Further investigations on fully developed turbulent flows in a curved channel", ONR Contract NONR - 248 (1956), John Hopkins Univ.
[32] WATTENDORF, F. L., "A study of the effect of curvature on fully developed turbulent flow", Proc. Roy. Soc., 148 (1953), pp.565-598.
[33] ELLIS, L. B., JOUBERT, P. N., "Turbulent shear flow in a curved duct", J. Fluid Mech., 62-1 (1974), pp.65-84.
[34] MAEKAWA, H., OHOKUBO, J., "Turbulent flow in a curved channel (Experiment)", Proc. JSME Niigata Regional Meeting, (1972), pp.53-56.
[35] HUNT, I. A., JOUBERT, P. N., "Effects of small streamline curvature on turbulent duct flow", J. Fluid Mech., 91-4 (1979), pp.633-659.
[36] KOBAYASHI, M., MAEKAWA., H., SHIMIZU, Y., UCHIYAMA, K., "Experimental study on turbulent flow in two-dimensional curved channel", Trans. JSME - B, 57-545 (1992), pp.119-126. or JSME Int., 37-1 (1994), pp.38-46.
[37] KOBAYASHI, M., MAEKAWA, H., "Turbulent flow accompanied by Taylor-Goertler vortices in a two-dimensional curved channel", Flow Meas. Instrum., 6-2 (1995), pp.93-100.
[38] NAGANO, Y., TAGAWA, M., "An improved $k - \varepsilon$ model for boundary layer flows", J. Fluid Engng, ASME, 112 (1990), pp.33-39.

# LIST OF PUBLICATIONS AND DISSERTATIONS FROM THE FORSCHERGRUPPE

A large number of papers have already been published and some are in the process of being published by the members of the research group. Many of these papers are appeared in refereed journals and others in proceedings of conferences. Besides these paper a number of dissertations resulted from the projects of the research group. Below we present the list of publications in journals and conference proceedings and the list of dissertations.

## Papers in Journals

**Lau, S., Schulz, V., Vasanta Ram, V.**, A Computer Operated Traversing Gear for Threedimensional Flow Surveys in Channels, Experiments in Fluids, Vol. 14, No. 6, 475-476, 1993.

**Laschefski, H., Potthast, F., Biswas, C., Mitra, N. K.**, Numerical Investigation of Flow Structure and Mixed Convection Heat Transfer of Impinging Radial and Axial Jets, Numerical Heat Transfer, Part A, Vol. 26, No. 2, 123-140, 1994.

**Laschefski, H., Braess, D., Haneke, H. and Mitra, N. K.**, Numerical Investigations of Radial Jet Reattachement Flows, Int. J. for Numerical Methods in Fluids, Vol. 18, 629-649, 1994.

**Fiebig, M.**, Vortex Generators for Compact Heat Exchangers, J. Enhanced Heat Transfer, Vol. 2, No. 1-2, 43-61, 1995.

**Braun, H., Fiebig, M. and Mitra, N. K.**, Large Eddy Simulation in Ribbed Channels, CFD Journal, Vol. 4, 79-88, 1995.

**Fiebig, M., Grosse-Gorgemann, A. and Weber, D.**, Experimental and Numerical Investigation of Self-Sustained Oscillations in Channel with Periodic Structures, Experimental Thermal and Fluid Science, Vol. 11, 226-223, 1995.

**Laschefski, H., Cziesla, T. and Mitra, N. K.**, Influence of Exit Angle on Radial Jet Reattachement and Heat Transfer, J. Thermophysics and Heat Transfer (AIAA), Vol. 9, No. 1, 169-174, 1995.

**Fiebig, M.**, Embedded Vortices in Internal Flow: Heat Transfer and Pressure Loss Enhancement, Int. J. Heat and Fluid Flow, Vol. 16, 376-388, 1995.

**Lau, S.**, Experimental Study of the Turbulent Flow in a Channel with Periodically Arranged Longitudinal Vortex Generators, Experimental Thermal and Fluid Sciences, Vol. 11, No. 3, 255-261, 1995.

**Vieth, D., Kiel, R.**, Experimental and Theoretical Investigations of the Near-Wall Region in a Turbulent Separated and Reattached Flow, Experimental Thermal and Fluid Science, Vol. 11, Elsevier Science Inc., New York, 243-254, 1995.

Fiebig, M., Behle, M., Schulz, K. and Leiner, W., Color-Based Image Processing to Measure Local Temperature Distributions by Wide-Band Liquid Crystal Thermography, Applied Scientific Research 56, 113-143, Kluwer Academic Publishers, 1996.

Laschefski, H., Cziesla, T., Biswas, G. and Mitra, N. K., Numerical Investigation of Heat Transfer by Rows of Rectangular Impinging Jets, Numerical Heat Transfer, Part A, Vol. 30, No. 1, 87-101, July 1996.

Lorenz, S., Nachtigall, C. and Leiner, W., Permanent Threedimensonal Patterns in Turbulent Flows with Essentially Twodimensional Wall Configurations, Int. J. Heat Mass Transfer, Vol. 39, 373 - 382, 1996.

Braess, D. and Sarazin, R., An efficient smoother for the Stokes problem, Applied Numerical Mathematics 23, 3-19, 1997.

Owsenek, B. L., Cziesla, T., Mitra, N. K. and Biswas, G., Numerical Investigation of Heat Transfer in Impinging Axial and Radial Jets with Superimposed Swirl, Int. J. Heat and Mass Transfer, Vol. 40, No. 1, 141-147, 1997.

Cziesla, T., Tandogan, E., Mitra, N. K., Large Eddy Simulation of Heat Transfer from Impinging Slot Jets, Numerical Heat Transfer, A, Vol. 32, No. 1, 1-18, 1997.

Laschefski, H., Cziesla, T., Mitra, N. K., Evolution of Flow Structure in Impinging Three-Dimensional Axial and Radial Jets, Int. J. for Numerical Methods in Fluids, Vol. 25, 1083-1103, 1997.

Lau, S., Meiritz, K., Vasanta Ram, V., Measurement of Momentum and Heat Transport in the Turbulent Channel Flow with Embedded Longitudinal Vortices, Accepted for Publication in International Journal of Heat and Fluid Flow, 1997.

## Papers in Proceedings

Lau, S., Vasanta Ram,V., Longitudinal Vortices Imbedded in the Turbulent Channel Flow, Proceedings of the 13th Symposium on Turbulence, Missouri/Rolla, USA, September 21-23, 1992.

Fiebig, M., Grosse-Gorgemann, A. und Mitra, N. K., Selbsterregte Schwingungen bei laminarer Strömung in genuteten Kanälen, ZAMM 73, T493-T496, 1993.

Potthast, F., Laschefski, H., Mitra, N. K., Fiebig, M. and Biswas, G., Influence of Free Convection on Flow Structure and Heat Transfer of Impinging Radial and Axial Jets, in Enhanced Cooling Techniques for Electronics Applications, HTD-Vol. 263, (ASME), Editors: S.V. Garimella, M. Greiner, M. M. Yovanovich, V. W. Antonette, 19-32, 1993.

Riemann, K.-A., Fiebig, M., Leiner, W. and Mielenz, O., Turbulent Local Heat Transfer in Channels with Rectangular Vortex Generators, in Vortices and Heat Transfer, Proceedings of Eurotherm-Seminar 31, Editors: M. Fiebig, N.K. Mitra, Bochum, 1993.

Lorenz, S., Neumann, H. and Leiner, W., Distribution of the Heat Transfer Coefficient in a Channel with Periodic Transverse Grooves, in Vortices and Heat Transfer, Proceedings of Eurotherm-Seminar 31, Editors: M. Fiebig, N.K. Mitra, Bochum 1993.

Lorenz, S., Braun, H. and Bai, L., Wall Pressure Distribution in a Channel with Transverse Grooves, Proceedings of Eurotherm-Seminar 31, Editors: M. Fiebig, N.K. Mitra, Bochum 1993.

Lau, S., Experimental Study of the Turbulent Flow in a Channel with periodically arranged Longitudinal Vortex Generators, Proceedings of Vortices and Heat Transfer, Eurotherm 31, Bochum, Germany, May 24-26, 1993.

Braun, H., Fiebig, M. and Mitra, N. K., Large-Eddy Simulation of Separated Flow in a Ribbed Duct, AGARD-CP-551, 74$^{th}$ Fluid Dynamics Symposium, Chania, Crete, Greece, 9.1-9.9, 1994.

Fiebig, M., Grosse-Gorgemann, A., Hahne, W., Leiner, W., Lorenz, S., Mitra, N. K. and Weber, D., Local Heat Transfer and Flow Structure in Grooved Channels: Measurement and Computations, Proceedings 10th Int. Heat Transfer Conf., Brighton, UK, Vol. 4, 237-242, Aug. 1994.

Mitra, N. K., Fiebig, M., Winkelsträter, M., Soest, C., Laschefski, H., Numerical Investigations of Flow and Heat Transfer of Laminar 2D and 3D Impinging Jets, Proceedings 10th Int.Heat Transfer Conf. Brighton, UK, Vol. 3, 119-124, Aug. 1994.

Owsenek, B. L., Cziesla, T., Biswas, G. and Mitra, N. K., Influence of Swirl on Heat Transfer of Impinging Axial and Radial Jets, Proc. 2. ISHMT-ASME Tagung in Suratkhal, Indien, Dez. 28-30, 1995.

Fiebig, M., Grosse-Gorgemann, A. and Müller, U., Swirl and Flow Destabilisation for Heat Transfer Enhancement: A Numerical Investigation of Laminar Finned Channel Flow, Proceedings Eurotherm Seminar 46, Heat Transfer in Single Phase Flows 4, University of Pisa, Italy, 1-7, 1995.

Fiebig, M., Grosse-Gorgemann, A. and Hahne, H.-W., Heat Transfer Enhancement by Flow Destabilization in Ribbed Channels, Proceedings 2nd Baltic Heat Transfer Conference, Jurmala, Riga, Latvia, Aug. 1995, Computational Mechanics Publications, Southampton, 365-374, 1995.

Vasanta Ram, V., Longitudinal Vortices Embedded in Turbulent Couette Flow, in Proceedings of IUTAM Symposium on 'Asymptotic Methods for Turbulent Shear Flows at High Reynolds Numbers', Bochum, Germany, June 28-30, 1995.

Fiebig, M., Hahne, W. and Grosse-Gorgemann, A., Self-Sustained Oscillations, Vortices and Heat Transfer in Channels with Wing-Type Vortex Generators, 3rd Minsk International Heat and Mass Transfer Forum, Minsk, Belarus, 20.-24. Mai, 1996.

Fiebig, M., Vortices: Tools to Influence Heat Transfer - Recent Developments, 2nd European Thermal-Siences and 14th UIT Heat Transfer Conference, Vol. 1, 41-56, Rome, Italy, 29.-31. Mai, 1996.

Fiebig, M., Vortex Generators for Heat Transfer Enhancement, 9th Int. Symposium on Transport Penomena (ISTP-9), Singapore, 25.-28. Juni, 1996.

Fiebig, M., Self Sustained Oscillations and Transition in a Periodically Ribbed Channel, 19th Int. Congress of Theoretical and Applied Mechanics, Kyoto, Japan, 25.-28. August, 1996.

Fiebig, M., Hahne, W. and Weber, D., Heat Transfer and Drag Augmentation of Multiple Rows of Winglet Vortex Generators in Transitional Channel Flow: A Comparison of Numerical and Experimental Methods, Notes on Numerical Fluid Mechanics (NNFM), Vol. 53, Vieweg, Braunschweig 1996.

Fiebig, M., Hahne, W. and Weber, D., Computation of Three Dimensional Complex Flows, Proceedings of the IMACS-COST Conference on Computational Fluid Dynamics, Lausanne, September 13-15, 1995, Notes on Numerical Fluid Mechanics (NNFM), Vol. 53, Braunschweig 1996.

Cziesla, T., Braun, H., Biswas, G. and Mitra, N. K., Large Eddy Simulation in a Channel with Exit Boundary Conditions, NASA-ICASE, Rep. 96-18, 1996.

Cziesla, T. and Mitra, N. K., Large Eddy Simulation with Dynamic Subgrid Stress Model of a Rectangular Impinging Slot Jet, erscheint in Proc. 15 Int. Conf. on Numerical Methods in Fluid Dynamics, Monterey, USA, Springer, 24.-26. Juni 1996, Editor: P. Rutler, Springer, 382-287.

Cziesla, T. and Mitra, N. K., Large Eddy Simulation of Rectangular Jets, erscheint in Proc. 2nd ERCOFTAC on Direct and Large Eddy Simulation, Grenoble, 16.-19. Sep., 1996, Editor J.P. Chollet et al, Kluwer, 267-278.

Leiner, W., Imaging Techniques to Measure Local Heat and Mass Transfer, in Optical Methods and Data Processing in Heat and Fluid Flow, IMechE Conference Transaction 1996-3, 1 - 14, London, 1996.

Schulz, K., Behle, M. and Leiner, W., High Local Resolution Measurement of Heat Transfer Coefficients by Liquid Crystal Thermography, in Optical Methods and Data Processing in Heat and Fluid Flow, IMechE Conference Transaction 1996-3, 57 -70, London 1996.

Lau, S., Meiritz, K., Vasanta Ram, V., Measurement of Momentum and Heat Transport in the Turbulent Channel Flow with Embedded Longitudinal Vortices, Proceedings of Turbulent Heat Transfer Conference in the Series 'Engineering Foundation Conferences', San Diego, USA, March 10-15, 1996.

Lau, S., Bangert, N., Kaniewski, M., Vasanta Ram, V., Eingebettete Längswirbel hinter einer Wingletreihe, Presented at Gamm-Conference, Prag, May 27-31, 1996.

Cziesla, T., Mitra, N. K., Application of Dynamic Subgrid Stress Model on Rectangular Impinging Slot Jet Flows, AIAA Paper 97-1997, presented at 28 AIAA Fluid Dynamics Conf., Snowmass, Colorado, June 29 - July 2, 1997.

Fiebig, M., Vortices and Heat Transfer, ZAMM: Z. angew. Math. Mech. 77, 1, 3-18, 1997.

**Vasanta Ram, V., Kaniewski, M., Lau, S.**, Visualization of Interation Patterns of Longitudinal Vortices Embedded in a Channel Flow, Accepted for Presentation at SCART 97, Second International Conference on Flow Interaction, Berlin, Germany, July 20-25, 1997.

## Dissertations

**Riemann, K.-A.**, Experimentelle Untersuchungen zu Wärmeübergang, Strömungsverlust und selbsterregten Schwingungen in Kanälen mit periodisch angeordneten Rechteckwingletreihen, 1992.

**Laschefski, H.**, Numerische Untersuchung der dreidimensionalen Strömungs-struktur und des Wärmeübergangs bei ungeführten und geführten, laminaren Freistrahlen mit Prallplatte, Fortschritt-Berichte VDI, Reihe 19: Wärmetechnik/Kältetechnik, Nr.72, VDI-Verlag, Düsseldorf, 1993.

**Grosse-Gorgemann, A.**, Numerische Untersuchung der laminaren Strömungrund des Wärmeübergangs in Kanälen mit rippenförmigen Einbauten, Fortschritt-Berichte VDI, Reihe 19: Wärmetechnik/Kältetechnik, Nr. 87, VDI-Verlag, Düsseldorf, 1995.

**Lorenz, S.**, Lokaler Wärmeübergang und Strömungsstruktur bei turbulenter Strömung in einseitig querberippten Kanälen, Fortschritt-Berichte VDI, Reihe 7: Strömungstechnik, Nr. 285, VDI-Verlag, Düsseldorf, 1996.

**Weber, D.**, Experimente zu selbsterregt instationären Spaltströmungen mit Wirbelerzeugern und Wärmeübertragung, Cuvillier-Verlag, Göttingen, 1996.

**Kiel, R.**, Experimentelle Untersuchungen einer Strömung mit beheiztem lokalen Ablösewirbel an einer geraden Wand, Fortschritt-Berichte VDI, Reihe 7: Strömungsmechanik, Nr. 281, VDI-Verlag, Düsseldorf, 1995.

**Schäfer, P.**, Untersuchung von Mehrfachlösungen bei laminaren Strömungen, 1995.

**Vieth, D.**, Berechnung der Impuls- und Wärmeübertragung in ebenen turbulenten Strömungen mit Ablösung bei hohen reynolds-Zahlen, Fortschritt-Bericht VDI, Reihe 7: Strömungstechnik, Nr. 311, VDI-Verlag, Düsseldorf, 1996.

**Hemforth, F.**, Die Behandlung der instationären Navier-Stokes-Gleichungen mit P1/P1-Elementen: Diskretisierung und Löser, 1996.

**Sarazin, R.**, Eine Klasse von effizienten Glättern vom Jacobi-Typ für das Stokes-Problem, 1996.

**Braun, H.**, Grobstruktursimulation turbulenter Geschwindigkeits- und Temperaturfelder in Spaltströmungen mit Wirbelerzeugern, Cuvillier-Verlag, Göttingen, 1996.

**Neumann, H.**, Experimentelle Untersuchungen von hydrodynamischen und thermischen Anlaufströmungen mit periodisch angeordneten Wirbelerzeugern, Cuvillier-Verlag, Göttingen, 1997.

**Cziesla, T.**, Grobstruktursimulation der Strömungs- und Temperaturfelder von Prallstrahlen aus Schlitzdüsen, 1997.

## Addresses of the Editors of the Series "Notes on Numerical Fluid Mechanics"

Prof. Dr. Ernst Heinrich Hirschel (General Editor)
Herzog-Heinrich-Weg 6
D-85604 Zorneding
Federal Republic of Germany

Prof. Dr. Kozo Fujii
High-Speed Aerodynamics Div.
The ISAS
Yoshinodai 3-1-1, Sagamihara
Kanagawa 229
Japan

Prof. Dr. Bram van Leer
Department of Aerospace Engineering
The University of Michigan
3025 FXB Building
1320 Beal Avenue
Ann Arbor, Michigan 48109-2118
USA

Prof. Dr. Michael A. Leschziner
UMIST-Department of Mechanical Engineering
P.O. Box 88
Manchester M60 1QD

Prof. Dr. Maurizio Pandolfi
Dipartimento di Ingegneria Aeronautica e Spaziale
Politecnico di Torino
Corso Duca Degli Abruzzi, 24
I-10129 Torino
Italy

Prof. Dr. Arthur Rizzi
Royal Institute of Technology
Dept. of Aeronautics
Aerodynamics Division
S-10044 Stockholm
Sweden

Dr. Bernard Roux
Institut de Recherche sur les Phénomènes Hors d'Equilibre
(IRPHE)
Technopole de Chateau-Gombert
F-13451 Marseille Cedex 20
France

## Brief Instruction for Authors

Manuscripts should have well over 100 pages. As they will be reproduced photomechanically they should be produced with utmost care according to the guidelines, which will be supplied on request.
In print, the size will be reduced linearly to approximately 75 per cent. Figures and diagrams should be lettered accordingly so as to produce letters not smaller than 2 mm in print. The same is valid for handwritten formulae. Manuscripts (in English) or proposals should be sent to the general editor, Prof. Dr. E. H. Hirschel, Herzog-Heinrich-Weg 6, D-85604 Zorneding.